Computer Program Synthesis Methodologies

NATO ADVANCED STUDY INSTITUTES SERIES

Proceedings of the Advanced Study Institute Programme, which aims at the dissemination of advanced knowledge and the formation of contacts among scientists from different countries

The series is published by an international board of publishers in conjunction with NATO Scientific Affairs Division

A	Life Sciences	Plenum Publishing Corporation
B	Physics	London and New York
C	Mathematical and Physical Sciences	D. Reidel Publishing Company Dordrecht, Boston and London
D	Behavioural and Social Sciences	Martinus Nijhoff Publishers
E	Engineering and Materials Sciences	The Hague, London and Boston
F	Computer and Systems Sciences	Springer Verlag Heidelberg
G	Ecological Sciences	

Series C – Mathematical and Physical Sciences

Volume 95 – Computer Program Synthesis Methodologies

Computer Program Synthesis Methodologies

Proceedings of the NATO Advanced Study Institute held at Bonas, France, September 28 - October 10, 1981

edited by

ALAN W. BIERMANN

Duke University, Department of Computer Science, Durham, NC, U.S.A.

and

GÉRARD GUIHO

Laboratoire de Recherche en Informatique, Université de Paris-Sud, Orsay, France

D. Reidel Publishing Company

Dordrecht : Holland / Boston : U.S.A. / London : England

Published in cooperation with NATO Scientific Affairs Division

Library of Congress Cataloging in Publication Data

CIP

NATO Advanced Study Institute (1981 : Bonas, France)
Computer program synthesis methodologies.

(NATO advanced study institutes series. Series C, Mathematical and physical sciences ; v. 95)
"Published in cooperation with NATO Scientific Affairs Division."
Includes indexes.
1. Electronic digital computers–Programming–Congresses.
I. Biermann, Alan W., 1939– II. Guiho, Gerard, 1945–
III. Title. IV. Series.
QA76.6.N378 1981 001.64′2 82-16704
ISBN-13: 978-94-009-7021-2 e-ISBN-13: 978-94-009-7019-9
DOI: 10.1007/978-94-009-7019-9

Published by D. Reidel Publishing Company
P.O. Box 17, 3300 AA Dordrecht, Holland

Sold and distributed in the U.S.A. and Canada
by Kluwer Boston Inc.,
190 Old Derby Street, Hingham, MA 02043, U.S.A.

In all other countries, sold and distributed
by Kluwer Academic Publishers Group,
P.O. Box 322, 3300 AH Dordrecht, Holland

D. Reidel Publishing Company is a member of the Kluwer Group

TABLE OF CONTENTS

PREFACE

Ever since the early days of computing, programmers have been seeking ways to automate portions of the programming task. The first major step was the invention of assemblers which allowed mneumonic operation codes,handled certain addressing tasks and offered other functions. Next came a vast array of compiled and interpreted languages which implemented many data structures, flow-of-control features, and notational conveniences. In an attempt to continue this trend today, a number researchers are developing program synthesis methodologies which aim at making further automation possible. Most of these methodologies fall into one of the categories :

a) extensions to traditional programming methods,
b) synthesis of programs from examples,
c) synthesis from formal input-output specifications,
d) program construction from natural language dialogue.

These program synthesis methodologies are the subject of this book.

The first three chapters examine the program synthesis issue largely without regard to the automation problem. Systematic means are described for constructing programs with emphasis on what is the nature of the synthesis task and how can it be approached. M. BROY describes a transformational technique for deriving programs from their specifications. M. SINTZOFF describes central issues in the design of concurrent programs, and R. BACKHOUSE gives a methodology for decision making in program design that accounts both for correctness and efficiency.

The later chapters of the book sequentially address the four approaches to synthesis given above. Extensions to traditional programming methods often involve the construction of very high level languages to deal with larger objects (data structures) than traditional languages with appropriately more

A. W. Biermann and G. Guiho (eds.), Computer Program Synthesis Methodologies, vii–ix.

powerful operations on them. An early step in this direction was the development of APL, and more recent examples have been SETL which enables a user to code in terms of mathematical entities such as sets and BDL which allows a user, presumably a businessman, to specify a computation in terms of a series of tabular forms and a series of processing paths through which data flows. The design and implementation of such languages are examined in chapters by P. GOLDBERG.

Another extension to traditional methods is made possible by systems designed to automatically handle low level flow-of-control decisions. All the above higher level languages do this implicitly with their built in operators. PROLOG is a language which does this with a theorem proving mechanism employing primarily unification and backtracking. The programmer specifies the problem to be solved with a set of formal logic statements including a theorem to be proved. The theorem proving system finds a way to combine the axioms to prove the theorem, and in the process, it completes the desired calculation. H. GALLAIRE has contributed a chapter describing PROLOG giving many examples of its usage.

There are many situations in machine programming where it is more natural to give examples of what is wanted than to give programs. As an illustration, the operator of an automatic machine tool would probably prefer to manually direct the device through an example machining sequence and have it self program to do that sequence than to code the sequence using a symbolic language. The machine operator can be sure in the example sequence that all processes occur as desired and will probably produce a more reliable control program than if he or she had written the program directly. As a second illustration, a person formating a piece of text would probably much rather organize the page using a manual graphics device of some kind and then have an automatic editor create the command sequence to actually space out and format the text items correctly. A fascinating domain for these kinds of studies has been in the synthesis of LISP programs from examples of their input-output behaviors. J.P. JOUANNAUD and Y. KODRATOFF have included a chapter describing some of this work.

Another way to construct programs automatically is to deduce them from their input-output specifications. The specifications are given with a formal language but they do not give the method for doing the computation. The automatic system attempts to transform the specifications in a sequence of steps to the target program. The result program is guaranteed to implement the specifications and to terminate on legitimate inputs. Z. MANNA and R. WALDINGER give a deductive method for this type of synthesis where a theorem is proved and the target program is a by-product of the theorem proving process. The following chapter by J. DARLINGTON describes a transformational approach.

One of the ways in which nonprogrammers will be able to interact with computers of the future will be via typed or spoken natural language. While this may seem to be an impractical alternative in the near term, some progress has been made. The final chapter by A. BIERMANN discusses a system for natural language programming that has been developed to the point that college students can use it to solve problems typical of a first course in programming.

ACKNOWLEDGEMENTS

This book is the proceedings of the 1981 NATO Advanced Study Institute on Automatic Program Construction which was held in October 1981. This meeting has been made possible through the financial support and the framework provided by the NATO Scientific Affairs Division.

As Directors of the NATO Institute, we would like to thank all the participants for their valuable participation in the discussions and panel presentations which finally made a great success of this Institute. We are grateful to our host and hostess at the Château Professor and Mrs. J.C. SIMON, animateur of the Centre Culturel de Bonas, for providing a delightful environment for the Institute. We would also like to thank the chapter authors for their careful preparation of the chapters and the secretarial staff at the Laboratoire de Recherche en Informatique d'Orsay for their help in assembling this volume.

PROGRAM CONSTRUCTION BY TRANSFORMATIONS : A FAMILY TREE OF SORTING PROGRAMS.

Manfred BROY

Technische Universität München
Institut für Informatik
Postfach 20 24 20
D-8000 MÜNCHEN 2

The construction of a program is considered as a process consisting of a sequence of formal steps. Each step is characterized by a particular transformation rule. In general, the development starts with a formal specification and ends with an efficient program. In the first part of this paper this methodology and its formal foundations are briefly outlined. In the second part a number of sorting programs is developed along these lines leading to a decision tree of the program family of different sorting programs.

1. PROGRAM CONSTRUCTION BY TRANSFORMATIONS

In order to manage the enormous complexity of the activity of writing a program it seems indispensable to organize the development process in a formal way leading through a series of more and more detailed versions (cf. (4)). The switch from one version to the other is called a program transformation, an instruction how to perform this switch is called a transformation rule.

The reliability and understandability of the development and thus our confidence in the resulting piece of software obviously can be increased if both the steps and the single versions are represented in formal notation, such that all development stages can always be properly verified. A rigorous observation of this proceding leads to considering programming as a formal activity (cf. (17)). Using formal notations moreover allows conveniently to support the development process by an interactive development system.

This work was carried out within the Sonderforschungsbereich 49 - "Programmiertechnik" - Munich.

A. W. Biermann and G. Guiho (eds.), Computer Program Synthesis Methodologies, 1–49.

A first approach to a system for program development by transformations has been undertaken by R.M. Burstall and J. Darlington (see e.g. (28), (23)). In the meanwhile related efforts have been made elsewhere (33), (51), (54), (1), (2), (39), for a comprehensive survey see (46).

Of course, the basic motivations of the different groups working in the fiels of program transformations are rather different. First there are groups which are only interested in small local program transformations as part of an optimizing compiler. They only want to liberate the programmer from doing these tricky and burdensone local optimizations by hand. Second there are projects which consider program transformations as a tool in program design and development, i.e. as being part of the discipline of programming. Third there are people which try to automatize the process of program construction as much as possible leading to program synthesis.

Of course, there are no sharp boundaries between these three approaches. In their extreme cases the first approach is not very interesting for program development and the third one may not be very realistic on account of the undecidability results of formal logics.

In this paper we follow the second approach, trying to show the power of program transformations as a design tool in the process of program construction.

1.1. Significance of notation

Program development by transformation, of course, is independent from particular notations or programming languages. Both the rules and the program versions may be described informally in natural language. However, in man-machine communication or if a mathematically proper proof is aimed at, both the programs to be transformed and the resp. rules have to be represented by a formal language. In principle, any programming language could be used, but not all languages are equally well suited, at least for particular problem areas or particular levels of abstraction. To gain optimal freedom, flexibility and formal soundness by allowing smooth transitions between different levels of abstraction a special "Wide Spectrum Language" called CIP-L has been designed within the project CIP (cf (8), (10)). This language comprises constructs for problem specification, applicative and procedural programming. Hence the programmer can very flexibly choose the appropriate style for his specific problem.

1.2. Transformation rules

Basically, a <u>transformation</u> is the generation of a new piece of program from a given one ; it is said to be <u>correct</u>, if both programs are semantically equivalent. A transformation rule thus is mapping between sets of programs. In general, such a mapping is a partial one, as it is only defined for particular kinds of programs.

In principle, there are two possibilities for representing transformation rules. Both are strictly syntax-oriented as far as their application is concerned :

(i) The transformation rule can be described in the form of an algorithm, which takes a given program as input and produces an equivalent one as output - provided that the input program is in the domain of the rule (compilers behave in this way) ;

(ii) The transformation rule can be given as an ordered pair of program schemes, the "input template" S and the "output template" T (cf. (50), (31)), which correspond to the premise and conclusion of a post production rule.

We shall use case (ii) in the following, because of its better readability. The application of such a transformation rule to a concrete program resembles a substitution step of a Post system. We denote such rules in the form

$$\begin{array}{cc} S & \\ \Updownarrow & B \\ T & \end{array}$$

where S and T are the input and output templates mentioned above and B is a condition, called the applicability condition, mandatory for the applicability of the transformation ; the part B may be missing. If only one of the two arrows is present, the scheme is applicable in the indicated direction only, otherwise the scheme is symmetric with respect to S and T. The syntactic form of the input template together with the applicability condition specifies the exact domain of the transformation rule. (For a formal treatment of the equivalence of programs and program schemes see (21), (22), (48). For a classification see (9)).

Program schemes follow a particular (abstract) syntax, and the correctness of transformation rules depends heavily on a particular definition of the semantics ; therefore, we refer to the "wide spectrum language" CIP-L as a formal basis. Within CIP-L we distinguish several "styles" of programming, in particular the applicative level - including descriptive constructs - and the procedural level - comprising imperative constructs.

Note that transformation rules can be applied not only to programs but also to program schemes thus producing a new transformation rule consisting of the original program scheme as input template and of the generated program schemes as output template. The applicability conditions for the new rule are induced by those of

the applied rule. Thus, new transformation rules may be derived within the system.
One of the most basic rules is FOLD/UNFOLD. i.e. the replacement of an expression D by an identifier, defined by D or vice versa. Specialized forms of folding and unfolding are e.g.

UNFOLD / FOLD for object

$$\ulcorner (m_1\ x_1,\ldots,m_n\ x_n) = (D_1,\ldots,D_n)\ ;\ E \lrcorner$$

$$\Updownarrow \qquad \text{determinate } (D_i) \wedge \text{defined } (D_i)$$

$$E^{D_1,\ldots,D_n}_{x_1,\ldots,x_n}$$

where $E^{D_1,\ldots,D_n}_{x_1,\ldots,x_n}$ denotes the expression that is generated by simultaneously substituing all free occurrences of $x_1,\ldots,x_n$ in E by $D_1,\ldots,D_n$. (Or course, this substitution operation has to take care of name clashes between the free identifiers of the D_i and bound identifiers within E).

UNFOLD/FOLD for functions

$$f(A_1,\ldots,A_n)$$

$$\Updownarrow \quad (\underline{\text{funct}}\ f \equiv (\underline{m}_1\ x_1,\ldots,\underline{m}_n\ x_n)\ \underline{r}\ :\ E) \wedge \neg\, \text{occurs}(x_i, A_j)$$

$$\ulcorner (\underline{m}_1\ x_1,\ldots,\underline{m}_n\ x_n) \equiv (A_1,\ldots,A_n)\ ;\ E \lrcorner$$

Similar rules exist for procedures, and also rules like "unrolling of a loop" are in principle UNFOLD-operations.
Note, that these two rules simply may be composed to a single one (which often also is called UNFOLD/FOLD), since that output template of the rule for functions matches the input template of UNFOLD/FOLD for objects.

1.3. Formal specifications

Generally we use here two possibilities for formal specifications. Algebraic specification by abstract data types (cf (35)) and explicit specifications by predicate logic and description (cf. (20)) The first one is used to specify computation structures (by their algebraic properties, cf. the appendix) the second one is used to specify programs (their "functional behaviour"), i.e. it is used to specify the input-output relation (also called the "extensional" behaviour). Both approachs allow the programmer to abstract extensively from the particular algorithmic representation of a computation structure or program resp. To maintain flexibility the specification language has to be integrated into an algorithmic language such that specification constructs and algorithmic construct may be optimally mixed (cf. (10)).

1.4. Transformation of specifications

Having elaborated an initial specification one should not immediately start to look for an algorithmic solution, since a great deal of improvement often can be achieved by transforming the specification. For different reasons, for instance, it might be useful to generalize a given problem by adding further parameters. This proceeding is called embedding (also "generalization" ; cf. (7), (24), (54), (20), (41)).

It will suffice here to present one example e.g. that of embedding on the descriptive level :

```
funct f ≡ (m x) r : some r y : B(x,y)

                                    ∀ m x, r y :
      ⇕
                                       B(x,y) ⇔ C(x,y,E)
funct f ≡ (m x) r :
     [g(x,E) where
      funct g ≡ (m x, n z) r :
            some r y : C(x,y,z)]
```

Here g represents a generalization of the problem f. However even transformations of specifications may be of interest which do not introduce new auxiliary functions. The design of an algorithm can be expressed by a series of design decisions transforming the "prealgorithmic" problem specification into an algorithm. Frequently the particular form of the problem specification will influence the programmer's decisions and lead to a particular class of algorithms, whereas other classes of algorithms which also solve the given problem may be not so close to the particular formulation of the problem specification and will therefore be missed. Thus the programmer should spend great care to get a clear, general (formal) problem specification as a basis for the program development.

1.5. From specifications to algorithms

The different steps of programming are of different complexity. Some of them can be done very schematically, others require a greater deal of inspiration and intuition, such as the proper design of algorithms.
Although it seems possible to state some basic algorithm principles, such as induction, enumeration, exhaustion, etc., on which a lot of the known algorithms can be based on, intuition is neces-

sary for the design of algorithms. But, however, aren't there systematic ways to derive algorithms solving a given problem ?

And in which ways can these algorithms be found ? What are the relations between equivalent algorithms, i.e. algorithms, which solve the same problem ? The answer is : algorithms are not to be discovered, they are to be developed, derived, deduced in systematic, small steps from an initial problem specification. And different design decisions may lead to different algorithms solving the same original problem.
Of course one may give formal transformation rules for the transformation of specification to algorithms.

Consider for instance

$$\underline{\text{funct}}\ f \equiv (\underline{m}\ x)\ \underline{r} : \underline{\text{some}}\ \underline{r}\ y : P(x,y)$$

$$\Downarrow \qquad \text{AC}$$

$$\underline{\text{funct}}\ f \equiv (\underline{m}\ x)\ \underline{r} : \underline{\text{if}}\ C(x)\ \underline{\text{then}}\ T(x)\ \underline{\text{else}}\ G(f(H_1(x)), f(H_2(x)))\underline{\text{fi}}$$

where there exists a termination function $\underline{\text{funct}}\ (\underline{m})\ \underline{\text{nat}}$ down and the applicability condition AC is of the form

$$\forall\ \underline{m}\ x : C(x) \Rightarrow P(x,T(x)),$$

$$\forall\ \underline{m}\ x\ \underline{r}\ y_1, y_2 : \neg C(x) \wedge P(H_1(x),y_1) \wedge P(H_2(x),y_2) \Rightarrow P(x,G(y_1,y_2))$$

$$\forall\ \underline{m}\ x : (\text{down}(x) = o \Rightarrow C(x)) \wedge$$

$$(\text{down}(x) > o \Rightarrow \text{down}(H_1(x)) < \text{down}(x) \wedge$$

$$\text{down}(H_2(x)) < \text{down}(x) \quad)$$

For applications of this scheme see section 5 and section 5.1.

Of course, here the particular difficulties are hidden in the "discovery" of the operations C, T, H_1, H_2 and also of the termination function.

The scheme may also be transformed into a simplified scheme in some cases. Assume we always have $C(H_1(x))$ then the lower scheme can be transformed by a simple UNFOLD into

$$\underline{\text{funct}}\ f \equiv (\underline{m}\ x)\ \underline{r} : \underline{\text{if}}\ C(x)\ \underline{\text{then}}\ T(x)\ \underline{\text{else}}\ G(T(H_1(x)), f(H_2(x)))\underline{\text{fi}}$$

i.e. we have switched to *linear recursion*.

Of course, the transition from a formal specification to a program (i.e. an algorithm) represents one of the important steps in a development process. It corresponds to the application of a sometimes quite complex mathematical theorem (cf. the applicability conditions of the rule above). For examples of deriving fixed point equations from specifications by the "adjunction of a tautology" see (34) (see also (27). For the application of structural induction in this process see (41).

1.6. Transformation of applicative programs

Once derived an applicative (algorithmic) version there are two basic motivations for further transformation : to increase the efficiency and to bring the program into a particular form, such that a transition to procedural programs is possible.
A very powerful rule is that of *transformational induction* (cf. (14), (15)) which may be seen as a special case of computational induction (cf. (40)) :

$\lceil$ __funct__ $f_1 \equiv (\underline{m_1}\ x)\ \underline{r_1}\ :\ R_1(f_1,x)\ ;\ Q(f_1,E)\rfloor$

$\Updownarrow$ AC

$\lceil$ __funct__ $f_2 \equiv (\underline{m_2}\ y)\ \underline{r_2}\ :\ R_2(f_1,y)\ ;\ f_2(E)\rfloor$

where AC :

$Q(\Omega,\bullet) = \Omega$, where Ω denotes the everywhere undefined function

$Q(R_1(f,\overset{n}{\bullet}),y) = R_2(Q(f,\bullet),y)$ for all y and f.

Here $Q(f,\bullet)$ is used as shorthand for the function abstraction

$(\underline{m_2}\ y)\ \underline{r_2}\ :\ Q(f,y)$.

Transformational induction can in particular be used to prove the correctness of certain embedding techniques (cf. (9)) :

If one wants to have tail recursion, one may try to restructure the resp. term appropriately using algebraic properties of the primitive operations (like associativity, right-commutativity, etc., cf . (25)). An example is the rule

```
funct f ≡ (m x) r :
      if B then f(E1) ° E2
           else E3 fi
      ⇕                          ∀ r a,b,c : (a ° b) ° c = a ° (b ° c)
                                         ∀ r a : a ° e = a
funct f ≡ (m x) r :
   ⌈g(x,e) where
    funct g ≡ (m x,r z) r :
         if B then g(E1,E2 ° z)
              else E3 ° z fi ⌋
```

Another possibility is to use special properties of the "decrementing function" E_1 : If there exists an inverse function E_1, then the argument can be computed in the reverse order :

```
funct f ≡ (m x) r :
      if x ≠ c then f(E1(x)) o E2(x)
               else E3(x)        fi
                                          ∀ m x : Ẽ1(E1(x)) = x
          ⇕
                                      <<c is constant>>
funct f ≡ (m x) r :
    ⌈g(c,E3(c)) where
     funct g ≡ (m y,r z) r :
        if y ≠ x then m yo ≡ Ẽ1(y) ;
                      g(yo,z o E2(yo))
                 else z              fi ⌋
```

This latter transformation can be used for the formal introduction of stacking mechanisms, (cf. (49), (57)).

Note that "transformational induction" can also be used to transform fixed point equations of functions into fixed point equations for procedures (cf. section 1.7) or fixed point equations for non-deterministic sequential programs into fixed point equations for concurrent programs (cf. (15)).

1.7. From applicative programs to procedural ones

The transition from applicative programs to procedural ones is motivated mainly by the fact, that this programs usually run more efficiently on von Neumann architectures. The transition means a switch to a strictly sequential innermost-evaluation (while destroying possibilities of implicit parallelism cf. (14)) and may in particular be done for *tail-recursive* functions (cf. (6)), i.e. functions in "outermost" recursion :

```
funct f ≡ (m x) r : if B(x) then f(H(x)) else G(x) fi
     ⇕
funct f ≡ (m x) r :
 ⌈var m v := x;  while B(v) do v := H(v) od ; G(v)⌋
```

Again this rule may be considered as a form of transformational induction. Another possibility to prove the correctness of the transformation rule is to add resp. assertions to the procedural version (cf. (16)). This leads to a "joint development" of program and proof by transformation. For the transformation of linear non-tail recursion to procedural programs see (53).
For a sufficiently complete set of transformation rules cf. (48), and (10).

1.8. Transformation of procedural programs

In a procedural version, a final polishing can be done by a lot of little optimizing transformations such as elementing superfluous assignments (for instance x := x), introducing the sharing of variables (cf. (48)) or constant propagation from loops (cf. also finite differencing in (43)). However, generally the overall structure of the program remains fixed. Nevertheless we may switch to particular, assembler-like notation like goto's etc. deriving a machine-oriented version of our program. For a collection of a number of such rules see (51).
Note that also on this level "transformational induction" can be used :

Assume f and p are defined by

<u>funct</u> f ≡ (<u>m</u> x) <u>m</u> : R_1(f,x)

<u>proc</u> p ≡ (<u>var</u> <u>m</u> v) : R_2(p,v)

then

y := f(y)

⇕ AC

p(y)

where AC :

y := $R_1(\tilde{f}$,y)

⇕

$R_2(\tilde{p}$,y)

and <u>proc</u> $\tilde{p}$ ≡ (<u>var</u> <u>m</u> v) : v := $\tilde{f}$(v).

2. FORMAL BACKGROUND OF PROGRAM TRANSFORMATIONS

Studying the validity (i.e. the correctness) of program transformation rules gives valuable insights into the nature of programs and algorithms as well as into the semantic structure and concepts of programming languages. Given a complete semantic specification of a language a transformation rule represents a theorem. However, it is also possible to give only a semantic specification for parts of the language (the "semantic kernel") and add "axiomatic" transformation rules as axioms, such that the remaining parts of the language can be reduced to this kernel. This way the semantic specification is extended to the whole language ("transformational semantics", cf. (48), (15)). However, the semantic specification of a programming language may even be given completely by axiomatic transformation rules ("conditional equations"). More technically speaking a programming language can be described by an abstract type (cf. (19)) leading to an "algebraic" view of programming languages.

Every semantic specification of a language induces a semantic relation between programs (cf. (21), (22)). These relations build the formal background of program development by transformation, where one starts from a given program P and tries to come up with a final program P' such that $P \sim P'$ holds for some interesting relation "~" (mostly an equivalence).

Relation for program development therefore have to meet the following requirements :

(I) Every relation between programs must be <u>reflexive</u> : $\forall P : P \sim P$.

During the development one generally proceeded through a series of intermediate versions $P = P_0, P_1, \ldots, P_n = P'$. In this case, it should be possible to conclude $P \sim P'$ from the fact that $P_i \sim P_{i+1}$ for all i. Therefore, we have to require :

(II) Every relation between programs must be <u>transitive</u> : $\forall P,Q,R : P \sim Q \wedge Q \sim R \Rightarrow P \sim R$

Thus ~ has to be a quasi-ordering. Usually we are not considering complete programs but only parts of programs (e.g. a single assignment, the body of a loop etc.). Therefore the "local" validity of a relation ~ has to ensure also the "global" validity of the relation :

(III) Any complete program P[Q] containing the program part Q must fullfil the <u>substitution-property</u> : $\forall Q, Q' : Q \sim Q' \Rightarrow P[Q] \sim P[Q']$

Therefore, equivalence relations actually have to be <u>congruence relations</u> with respect to all language constructs and for <u>partial orderings</u> all languages constructs have to be monotonic.

<u>Remark</u> : In connection with fixed points of functionals often the contuinity of the language constructs with respect to a partial ordering is required :

A program (scheme) P[x] is ~-continuous iff for all ~-chains $(E_i)_{i \in N}$ with $E_i \sim E_{i+1}$ $lub(P[E_i])$ exists if $lub(E_i)$ exists, and $lub(P[E_i]) = P[lub(E_i)]$.

Continuity plays an essential role for instance in fixed point semantics or for induction proofs. (End of remark).

There are quite a number of interesting monotonic semantic relations for program development (cf. (21), (22)). Besides the most important congruence relation "functionally equivalent", the implementation ordering (also called descendent relation cf. (42),

(22)) for nondeterministic programs is of particular interest. A nondeterministic program is a program with a set of possible behaviors instead of exactly one possible behavior as for deterministic programs. A program P_2 is called implementation of program P_1 and we write

$$\begin{array}{c} P_1 \\ \Downarrow \\ P_2 \end{array}$$

if every possible behavior of P_2 is functionaly equivalent to a possible behavior pf P_1. Simple examples for transformation rules which lead to implementations are e.g. for the nondeterministic comprehensive choice

$$\begin{array}{ll} \underline{\text{some}}\ \underline{m}\ x : p(x) & \\ & \forall\ \underline{m}\ x : (q(x) \Rightarrow p(x)) \\ \Downarrow & (\exists\ \underline{m}\ x : p(x)) \Rightarrow (\exists\ \underline{m}\ y : q(y)) \\ \underline{\text{some}}\ \underline{m}\ x : q(x) & \end{array}$$

The expression some m x : p(x) represents a nondeterministic expression for which every object x of mode m such that p(x) holds is a possible result. Note, that the monotonicity of the implementation relation allows to choose implementations locally and to be sure, that the refined program is an implementation of the given one.
Using nondeterminism in program development one gains a great deal of flexibility : nondeterminism can be used to keep particular decisions open in a fully formalized way. Thus it allows to work with delayed design decisions.

3. THE EXAMPLE PROBLEM : SORTING

In the following sections for the sorting problem several well-known algorithms are created stepwise, starting from some formal specification. Thus their relationship to each other becomes apparent, their fundamental algorithmic ideas are extracted and their mathematical background is illustrated. Furthermore the correctness of the derived algorithms with respect to the problem is guaranteed, since only correctness-preserving steps are used.
From the specification of the sorting problem at first every algorithm or algorithm scheme is derived in a recursive version. Then, in most cases, improved versions are obtained by transformations to demonstrate the relationship between the recursive versions and the well-known iterative formulations (cf. (38), (56)). The used transformations are only described informally, since we are more interested here in the process of creating an algorithm than in the process of developing an algorithm into an efficient, iterative program.

The basis for the formulation of the sorting algorithms is the computation structure ARRAY, which consists of the mode array and some operations which are used ot work with objects of this mode. For this computation structure certain algebraic laws are valid. These laws characterize the computation structure and are a sound basis in program development both for correctness proofs and for transformation rules.
The computation structure ARRAY is described in the appendix. The structure is related to a flexible array (cf. (6)) and specified by an abstract data type (cf. (35)). For a more detailed treatment of abstract data types see (55). For examples for their use in program development see (45).
At first a formal specification of the sorting problem is given.

3.1 Formal specification of the sorting-problem

The sorting problem "sort a given array" may be described informally

> "generate a sorted array of the same length and with the same elements as the given array".

This description may be written in the following way :

```
funct issorted ≡ (array A) bool :
      ∀ nat[2..upb(A)] k : A[k-1] ≤ A[k]

funct hassameelements ≡ (array A, array B) bool :
      if upb(A) ≠ upb(B) then false
                         else ∃ nat[1..upb(A)] row nat p ,
                                nat[1..upb(A)] row nat q :
                              perm(p) ∧ perm(q) ∧
                              (∀ nat[1..upb(A)] k :
                               A[p[k]] = B[q[k]]) fi ;

funct perm ≡ (nat row nat p) bool :
      ∀ nat[1..upb(p)] k : ∃ nat[1..upb(p)] i : p[i] = k ;

funct sort ≡ (array A) array :
      that nat[1..upb(A)] row m X :
            hassamelements (A,X) ∧ issorted (X) ;
```

This seems to be a quite natural way to specifiy the sorting problem in terms of the language CIP-L. In Hoare's and Floyd's sense of program correctness (cf. (36), (32)) the function sort specifies only (the precondition and) the postcondition for a program, which solves the sorting problem.

Another way to specify sorting algorithms is to establish the natural correspondence between arrays and multisets, i.e. the multiset of elements of an array. Then hassameelements is reduced to the equality of the corresponding multisets.
From the problem description a number of well-known algorithms will be derived in the following sections. Each design decision will be described explicitly to demonstrate how the different algorithms arise.

3.2. Some properties of the predicate hassameelements

The predicate hassameelements has some algebraic properties, which may be used for the development of algorithms :

(1) hassameelements is an equivalence relation

(2) ∀ array A :

(∀(nat[1..upb(A)] i, nat[1..upb(A)] j) :

hassameelements(A, exch(A,i,j))

i.e. hassameelements is invariant under the permutation of the argumentarrays

(3) ∀(array A, array B, array C, array D) ::

hassameelements(A,B) ∧ hassameelements(C,D) ⇒

hassameelements(conc(A,C), conc(B,D))

These laws can be proved by the fact, that the set of all permutations of a set of n elements is an algebraic group under the operation composition and the set of all permutations of a set of m elements ($m \leq n$) can be considered as a subgroup.
Note, that it would have been feasible to include the predicate hassameelements in the computation structure ARRAY, only specifying its properties as axioms of the abstract data type.

4. GRAPH OF DERIVED PROGRAMS

Starting the development process with some formal problem description, each design decision brings the program versions closer to a complete algorithm. Since in most cases several design decisions are possible and since the same final program may arise by different sequences of decisions, the connection between the different versions may be represented in a HASSE diagram. Each terminal node denotes a complete algorithm, which may be represented by a particular program.
A further program development is apporpriate and necessary to get a machine-oriented, efficient versions. This process of program development may be represented by a further HASSE diagram for each algorithms. In this diagram all versions, arising during the transformation process, are complete algorithms.

Furthermore it is often appropriate to mix the prealgorithmic design decisions and the optimizing transformations, since a design decision may be delayed or an algorithm possibility may become obvious during the process of efficiency improvement. Nevertheless in the following diagram only design decisions responsible for the creation of the algorithms are incorporated showing the connection between the different algorithms derived in the following sections.
If we drope some additional arcs weobtain a tree which might be called the *decision tree* of the family of sorting programs. Note that each path in this tree represents the most significant design decisions which lead to the resp. program. An example for a development of a program family in the sense of a class of related program which use the same algorithmic principles can be found in (18).

5. DECOMPOSITION OF A : Divide and Conquer

An obvious possibility to get a sorting algorithm is to use the following inductive notion :

- if the length of an array is equal to 1, then the array is sorted (the empty array is also sorted) ;
- if the length of an array is greater than 1, then the array may be divided into two arrays, whose lengths are both smaller than the length of the original array. Then both arrays may be sorted and finally merged.

```
funct mergesort ≡ (array A) array :

        if upb(A) ≤ 1 then A else
        (array A1,A2) ≡ some array A1,A2: A1 ≠ emptyarray ≠ A2
                                ∧ hassameelements(A,conc(A1,A2)) ;
                           merge(mergesort(A1),mergesort(A2)) fi


funct merge ≡ (array A,array B : issorted(A) ∧ issorted(B)) array :
      that nat[1..(upb(A)+upb(B))] row m x : issorted(X) ∧
                                  hassameelements(X,conc(A,B)) ;
```

Here a classical algorithmic principle is used, the induction principle. The induction is applied to the length of the array. The reduction of the length in each step guarantees the termination of the algorithm.

The correctness of mergesort can be proved by substitution of the parameters A and B in the function merge by A1 and A2. Induction permits the use of the predicate issorted ; furthermore hassameelements is invariant under the permutation of the elements of its arguments. Thus by the rules of section 3.2 we get the predicate issorted.

Lemma : mergesort (A) = sort (A) □

The procedure mergesort does not describe one algorithm but rather a class of algorithms. Since the choice of A1 and A2 is nondeterministic and merge is in descriptive form, a deterministic choice and the construction of an algorithm for merge requires design decisions, which finally determine an algorithm. As is shown in the following section the choice influences the implementation and transformation of merge and vice versa.

Another a little more specific possibility for mergesort is to choose one of the subarrays from the interior of A :

```
funct mergesort1 ≡ (array A) array :
      if upb(A) ≤ 1 then A
                    else nat k ≡ some nat x : (1≤x)∧(x≤upb(A));
                         nat j ≡ some nat x : (1≤x)∧(x≤k);
                         merge(mergesort1(trim(A,j,k)),
                         mergesort1(conc(trim(A,1,j-1),
                                         trim(A,k+1,upb(A))))) fi
```

The function mergesort1 will not terminate, if upb(A) and 1 are always chosen for k and j respectively, i.e. mergesort1 describes a class of programs, which are only partially correct since algorithms are included which do not terminate. The partial correctness of mergesort1 can be proved in the same way as for mergesort. Note, that mergesort1 is contained in mergesort (i.e. describes a subclass of algorithms).
The following equivalence may be used in the last line of the above program :

∀(array A,array B) : sort(conc(A,B)) = merge(sort(A),sort(B)).

5.1. Decomposition of the array A into an array of length upb(A)-1 and a single element

The following algorithm originates from mergesort1 ; if k and j are chosen equal, then only one element is selected from array A and the remaining array is sorted in the same way

```
funct sortdeco ≡ (array A) array :
      if upb(A) ≤ 1 then A
                    else nat j ≡ some nat x : (1≤x)∧(x≤upb(A))
                         merge(trim(A,j,j),sortdeco(remove(A,j)))
                                                                fi
```

If we assume the existence of an algorithm to compute the function merge, then the algorithm sortdeco is designed up to the choice of the index j, which is executed nondeterministically. Let us consider two possibilities to determine the choice of j :

(a) the selection of j may be done depending on the values of A, e.g. j may be chosen, such that A[j] is the maximum or minimum. Doing so, merge can be substituted by insert on the rightmost (i.e. upb(A)) or leftmost (i.e. 1) position of the array A,

(b) j is chosen regardless of the values of the array A. Then merge can be transformed to a function insort inserting A[j] into the sorted array.

Thus two well-known algorithms arise :

- sorting by selection
- sorting by insertion (where of course several insertion algorithms are possible).

5.1.1. Sorting by selection. Now j is chosen, such that A[j] is maximal. Thus we get (the function indmax gives an index j, such that A[j] is maximal) :

```
funct selectsort ≡ (array A) array :
   if upb(A) ≤ 1 then A
                  else nat j ≡ indmax(A) ;
                       insert(selectsort(remove(A,j)),
                                           upb(A),A[j]) fi
```

Now the function indmax will be developed. At first indmax is specified :

```
funct indmax ≡ (array A) nat :
      some nat [1••upb(A)]k : ∀ nat [1••upb(A)]i : A[i] ≤ A[k]
```

A recursive version of indmax may look like :

```
funct indmax ≡ (array A) nat :
   if upb(A) = 1 then 1
                 else nat k ≡ some nat x : (1≤x)∧(x<upb(A));
                      (nat j1,nat j2) ≡ (indmax(trim(A,1,k)),
                                    k+indmax(trim(A,k+1,upb(A))));
                      if A[j1] ≥ A[j2] then j1
                      ⟦  A[j1] ≤ A[j2] then j2 fi          fi ;
```

Note that indmax corresponds to the same recursion scheme as the first version of mergesort at the beginning of section 5. If always nat k = upb(A)-1, we get :

```
funct indmax ≡ (array A) nat :
   if upb(A) = 1 then 1
                 else (nat j,nat i) ≡ (indmax(trim(A1,upb(A)-1))
                                                           upb(A)) ;
                      if A[j] ≥ A[i] then j
                      ⫿ A[j] ≤ A[i] then i  fi          fi ;
```

Through the recursion only the last element of A is removed in each step. We may embed the function. Thus A becomes a global array and nat-arguments are used to indicate the boundaries of the inspected subarray

```
funct indmax ≡ (array A) nat :
      funct find ≡ (nat i,nat j) nat :
            if i = 0 then j
                     else if A[j] ≥ A[i] then find(i-1,j)
                          ⫿ A[j] ≤ A[i] then find(i-1,i) fi fi ;
            find(upb(A)-1,upb(A)).                                ⌋
```

5.1.2. Sorting by selection with data structure. During the computation of indmax more information becomes available than finally is used for the result. This information can be stored by a stack for natural numbers (for the definition of stack see the appendix).

```
funct find ≡ (nat i,nat j,array A,stack s) stack :
   if i = 0 then push(s,j)
            else if A[j] ≥ A[i] then find(i-1,j,A,s)
                 ⫿ A[j] ≤ A[i] then find(i-1,i,A,push(s,j-i)) fi fi
```

Thus in stack s the index distances between the relative maxima are stored. The following equivalence is valid :

```
∀ nat n : ∀ nat[1..n] row m A :
   top(find(n-1,n,A,empty)) = indmax(A) ;
```

Now the function selectsort may be transformed to :

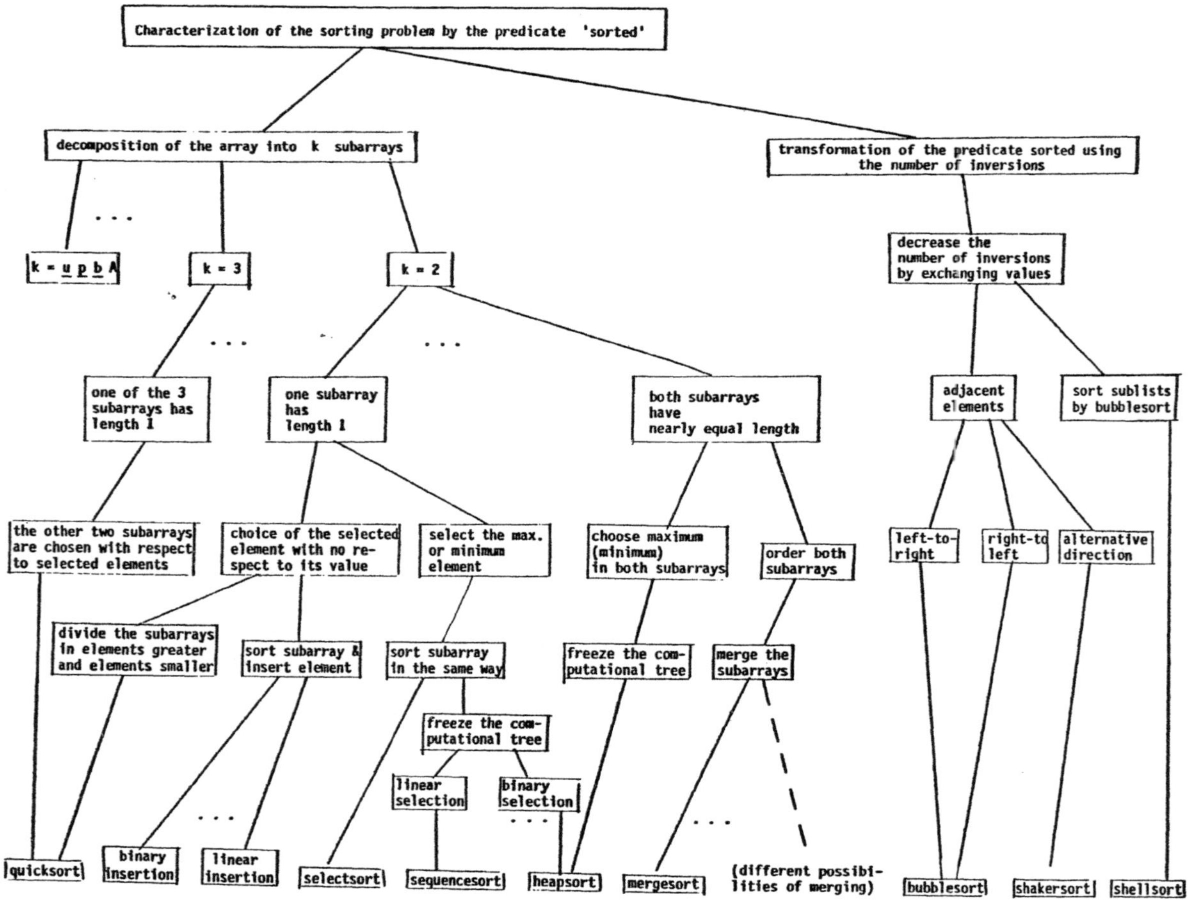

Fig. 1. Informal diagram of the design decisions that lead to particular algorithms

```
funct seqsort ≡(array A)array : selsort(A,find(upb(A)-1,
                                              upb(A),A,empty)),
   funct selsort ≡ (array A,stack s) array :
      if upb(A) ≤ 1
         then A
         else (nat j,array B) ≡ (top(s),remove(A,top(s))) ;
              stack h ≡
              if empty :: pop(s)
                 then find(j-1,j,B,empty)
                 else find(j-1,j+top(pop(s))-1,B,pop(pop(s))) fi;
              insert(selsort(B,h),upb(A),A[j])                  fi
```

In seqsort the maximum is only searched for in that part of the array, where the maximum was found during the previous searching. During the search the elements are composed to the last relative maximum in the previous searching pass.

5.1.3. Heapsort. Another way to choose an index of a maximum element in indmax of 5.1.1. is to break the array into two halves to search both halves and to compare the two indexes in the end :

```
funct indmax ≡ (array A) nat :
   if upb(A) = 1 then 1
                 else nat m ≡ (upb(A)+1) ÷ 2 ;
                      (nat m1,nat m2) ≡ (indmax(trim(A,1,m),
                                  m + indmax(trim(A,m+1,upb(A)));
                      if A[m1] ≥ A[m2] then m1
                      ▯ A[m1] ≤ A[m2] then m2 fi                 fi
```

or working on a global array, where j and k give the boundaries of the subarray.

```
funct indmax ≡ (array A) nat :
 ⌈ funct maxi ≡ (nat j,k) nat :
      if k = j then k
               else nat m ≡ (k+j) ÷ 2 ;
                    (nat m1,nat m2) ≡ (maxi(j,m),maxi(m+1,k));
                    if A[m1] ≥ A[m2] then m1
                    ▯ A[m1] ≤ A[m2] then m2 fi                fi
   maxi(1,upb(A))                                              ⌋
```

For this algorithm it is also possible to freeze the computational tree to conserve the information like in the previous section. There we worked with stack, because the recursion is of linear type, whereas now we use the mode tree :

```
mode  tree ≡ (tree left, m item, tree right) | empty ;
funct selecttree ≡ (array A) tree :
   if upb(A) = 1 then (empty,A[1],empty)
                 else nat m ≡ (upb(A)+1)÷ 2 ;
                 (tree x1, tree x2) ≡ (selecttree(trim(A,1,m)),
                                       selecttree(trim(A,m+1,upb(A))));
                 if item of x1 ≥ item of x2 then (x1,item of x1,x2)
                 ⟦  item of x1 ≤ item of x2 then (x1,item of x2,x2)
                                                                fi fi
funct formarray ≡ (tree t) array :
      if empty :: t then emptyarray
                    else conc(formarray(cut(t)),(item of t)) fi
funct cut ≡ (tree x) tree :
 ⌈ tree ℓ ≡ if left of x ≠ empty then
       if item of left of x = item of x then cut(left of x)
                                         else left of x   fi
                                 else left of x               fi;
   tree r ≡ if right of x ≠ empty then
       if item of right of x = item of x then cut(right of x)
                                          else right of x  fi
                                 else right of x              fi;
       if r = empty                then ℓ
       ⟦  ℓ = empty                then r
       elif item of ℓ ≥ item of r then (ℓ,item of ℓ,r)
       ⟦    item of ℓ ≤ item of r then (ℓ,item of r,r)   fi ⌋.
```

The algorithm has the form :

```
funct heapsort ≡ (array A) array :
      formarray(selecttree (A)) ;
```

Note : The data structure which is used to freeze the computational tree reflects the type of recursion of the function for which it is used.

Obviously this is only one possibility to use trees for sorting. Other methods (like e.g. replacement selection cf. (38)) can be obtained in similar way.

5.1.4. Quicksort. If in sortdeco j is chosen nondeterministically and afterwards the elements of the array A are divided into those greater than or equal to and others smaller than or equal to A[j] , then we get the following algorithm :

```
funct quicksort ≡ (array A) array :
      if upb(A) ≤ 1 then A
                    else nat j ≡ some nat i : 1≤i∧i≤upb(A)
                    (nat [1..upb(A)] row m X,nat m)=presort (A,j);
                    conc(insert(quicksort(trim(X,1,m-1)),m,X[m]),
                    quicksort(trim(X,m+1,upb(A))))              fi
funct presort  (array A,nat j)(array,nat) :
      if upb(A) = 1 then (A,1)
                    else
      if j ≠ 1      then (nat[1..upb(A)-1] row m X,nat m) ≡
                         presort(remove(A,1),j-1) ;
         if A[j]≤A[1] then (conc(X,(A[1])),m)
         ⫿ A[j]≥A[1] then (conc((A[1]),X),m+1) fi
      ⫿ j ≠ upb(A) then (nat[1..upb(A)-1] row m X,nat m) ≡
                         presort(remove(A,upb(A)),j) ;
         if A[j] ≤ A[upb(A)] then (conc(X,(A[upb(A)])),m)
         ⫿ A[j] ≥ A[upb(A)] then (conc((A[upb(A)]),X),m+1) fi fi fi
```

The function presort can also be formulated in a slightly different way, using the function exch :

```
funct presort ≡ (array A,nat j)(array,nat) :

⌈ funct presort1 ≡ (array A,nat min,nat max,nat j)(array,nat) :
    if min = max then (A,min)
  elif j = min then if A[min] ≤ A[max] then presort1(A,min,max-1,j)
                    ⟦  A[min] ≥ A[max] then presort1(exch(A,min,max),
                                                   min+1,max,max)  fi
  ⟦    j = max then if A[min] ≤ A[max] then presort1(A,min+1,max,j)
                    ⟦  A[min] ≥ A[max] then presort1(exch(A,min,max),
                                                   min,max-1,min)  fi
             else if A[j] ≤ A[min] then presort1(exch(A,min,max),
                                                 min,max-1,j)
                    ⟦  A[j] ≥ A[min] then presort1(A,min+1,max,j)
                    ⟦  A[j] ≤ A[max] then presort1(A,min,max-1,j)
                    ⟦  A[j] ≥ A[max] then presort1(exch(A,min,max),
                                                 min+1,max,j)      fi fi

      presort1(A,1,upb(A),j)
```

Two phases can be distinguished in the execution of presort : at first the conditions j = min of j = max may be false, but when one of them is true in one recursive call, one of them will hold in all succeeding calls, too. This used in the next version of presort where in the initial call the condition is made true :

```
funct presort ≡ (array A,nat j)(array,nat) :
⌈ funct jismax ≡(array A,nat min,nat j)(array,nat) :
        if min = j        then (A,j)
      elif A[min] ≤ A[j] then jismax(A,m+1,j)
        ⟦ A[min] ≥ A[j] then jismin(exch(A,min,j),min,j-1) fi,
  funct jismin  (array A,nat j,nat max)(array,nat) :
        if j = max        then (A,j)
      elif A[j] ≤ A[max] then jismin(A,j,max-1)
        ⟦ A[j] ≥ A[max] then jismax(exch(A,j,max),j+1,max) fi,
        jismax(exch(A,j,upb(A)),1,upb(A))                          ⌋
```

Of course other versions of quicksort are conceivable, e.g. if a parameter is used for the object A[j] instead of the parameter j (cf. (37)).

5.2. Sorting by insertion

If in the function sortdeco j is chosen regardless of the values of the array A, it is appropriate to choose j in a rather convenient way, for example to take upb(A).

```
funct insertsort ≡ (array A) array :
      if upb(A) ≤ 1 then A
                     else nat j ≡ upb(A) ;
                          insort(insertsort(remove(A,j)),A[j]) fi
```

The functional specification of insort can be derived from merge :

```
funct merge ≡ (array A,array B : issorted(A) ∧ issorted(B)) array :
   that nat[1..upb(A)+upb(B)] row m X : issorted(X)∧hassameelement
                                                      (X,conc(A,B))
```

If upb(B) = 1, then we may transform merge to insort, where y = B[1] :

```
funct insort ≡ (array A : issorted(A),m y) array :
   that nat [1..upb(A)+1] row m X :
        sorted(X) ∧ ∃ nat[1..upb(A)+1] j : X = insert(A,y,j) ;
```

For the creation of an algorithm which computes the function insort, the following considerations are used :

- if A is the empty array, then the result is simply that array, which has length 1 and has y as the only element.
- if A has at least length 1 then A is divided into two sub-arrays and y is insorted into one of the two arrays.

Thus function insort may be formulated in the following way :

```
funct insort ≡ (array A : issorted(A),m y) array :
      if upb(A) < 1 then insert(A,1,y)
                    else nat j ≡ some nat i : 1≤i∧i≤upb(A) ;
                    if A[j] ≤ y then conc(trim(A,1,j),
                                      insort(trim(A,j+1,upb(A)),y))
                    ⟦  A[j] ≥ y then conc(insort(trim(A,1,j-1),y)
                                      trim(A,j,upb(A)))   fi fi
```

Of course there are several possibilities to choose j in insort :

(i) we can take upb(A) or 1 for j and then simplify the second conditional clause, this means linear searching,

(ii) we can choose j in the middle between 1 and upb(A), i.e. nat j ≡ (1+upb(A)) ÷ 2 and so we get a binary search.

With (ii) we get (substituting the function trim by explicit trimmers) :

```
funct in2 ≡ (array A,nat min,nat max, m y) nat :
      if min > max then min
                   else nat j ≡ (min+max)÷2 ;
                        if A[j] ≤ y then in2(A,j+1,max,y)
                        ⟦  A[j] ≥ y then in2(A,min,j-1,y) fi fi
funct insort ≡ (array A:issorted(A),m y) array :
                        insert(A,in2(A,1,upb(A),y),y),
```

With (i) we obtain (j=max) :

```
funct in1 ≡ (array A,nat min,nat max,m y) nat :
      if min > max then min
                   else if A[max] ≤ y then in1(A,max+1,max)
                        □  A[max] ≥ y then in1(A,min,max-1) fi fi
```

which may be simplified to

```
funct in1 ≡ (array A,nat min,nat max,m y) nat :
      if min > max then min
    elif A[max] ≥ y then in1(A,min,max-1,y)
                    else max+1              fi .
```

5.3. Sorting by merging

For the function merge only a specification was given, but no algorithm. Now we want to develop a program for merge. At first we may use insort of section 5.2. :

```
funct merge ≡ (array A,array B : issorted(A) ∧ issorted(B)) array:
      if upb(B) < 1 then A
                    else merge(insort(A,B[1]),remove(B,1))  fi .
```

But there the fact, that B is already sorted, is not used. So a more symmetrical solution is preferable basing on the fact that A and B are sorted.

```
funct merge ≡ (array A,array B : issorted(A) ∧ issorted(B)) array :
      if upb(A) < 1              then B
      ⟦  upb(B) < 1              then A
    elif A[1] ≤ B[1]             then insert(merge(remove(A,1),B),1,A[1])
      ⟦  A[1] ≥ B[1]             then insert(merge(A,remove(B,1)),1,B[1])
      ⟦  A[upb(A)] ≤ B[upb(B)] then insert(merge(A,remove(B,upb(B))),
                                          upb(A)+upb(B),B[upb(B)])
      ⟦  A[upb(A)] ≥ B[upb(B)] then insert(merge(remove(A,upb(A)),B)
                                          upb(A)+upb(B),A[upb(A)]) fi
```

The last two lines may de deleted ; then the algorithm works by comparing only the first two elements.

With this merge procedure we get a complete mergesort, if i.e. the partition of the array is chosen, such that the arrays nearly have equal length.

```
funct mergesort ≡ (array A) array :
      if upb(A) ≤ 1 then 1
                    else nat k ≡ (upb(A)+1) ÷ 2 ;
                         merge(mergesort(trim(A,1,k)),
                         mergesort(trim(A,k+1,upb(A)))) fi.
```

Of course there is a similar variety of algorithms for merging as for sorting, but we shall consider no further algorithms for merging here. Nevertheless, you can also derive the class algorithms for merging in a systematic way.

6. TRANSFORMATION OF THE PREDICATE issorted

Another class of sorting algorithms may be derived by transforming the predicate issorted. To do this we use the function inversions which gives the number of inversions of an array :

```
funct inversions ≡ (array A) nat :
      |{(nat[1..upb(A)] i,nat[1..upb(A)]j) : (i<j)∧(A[i]>A[j])}|
```

It can easily be proved, that the following predicate is valid :

$$\text{issorted}(A) \Leftrightarrow \text{inversions}(A) = 0$$

The class of algorithms, that decrease the number of inversions stepwise to zero can be described by

```
funct inversort ≡ (array A) array :
      if inversions(A) = 0 then A
                           else inversort (some nat[1..upb(A)]row m X :
                                hassameelements(X,A) ∧
                                 inversions(X) < inversions(A)) fi
```

A simple method to choose X is to exchange two elements A[i], A[j] with i < j where A[i]>A[j]. Since hassameelements(exch(A,i,j),A) yields true, the correctness of this method is obvious. The problem now is to choose the indexes i,j.

A simple way is to take always adjacent elements, i.e. to choose j = i-1 and to consider all values i from upb(A) down to 2. This leads to a method which is not very efficient, since every exchange decreases the number of inversions only by 1.

Note, that if the number of inversions of an array A is greater than 0, then there are at least two adjacents elements A[j], A[j+1] such that A[j] > A[j+1].

6.1. Bubblesort

The above considerations lead to :

```
funct bubblesort ≡ (array A) array :
      if inversions(A) = 0 then A
                           else bubblesort(bubble(A,upb(A))) fi ,
funct bubble ≡ (array A,nat k) array :
      if k ≤ 1            then A
    elif A[k-1] > A[k]   then bubble(exch(A,k-1,k),k-1)
                          else bubble(A,k-1)           fi .
```

The number of inversions is only needed for testing for the termination of bubblesort. The function bubble also searches for inversions of adjacent elements. So we may combine the function bubble and the test.

```
funct bubblesort ≡ (array A) array :
  ⌈ (bool b,nat[1..upb(A)] row m X) ≡ bubbletest(A,upb(A)) ;
      if b  then A else bubblesort(X)                       fi ⌋

funct bubbletest ≡ (array A,nat k)(bool,array) :
      if k ≤ 1            then (true,A)
    elif A[k-1] > A[k]   then (false,bubble(exch(A,k-1,k),k-1))
                          else bubbletest(A,k-1)                fi.
```

6.2. Shakersort

Another possibility for the bubblesort is, instead of shifting the elements to the left (if they are smaller), rather to shift them to the right, if they are greater than their neighbours. In shakersort both the possibilities are mixed.

```
funct shakersort ≡ (array A) array :
⌈ (bool b1,nat[1..upb(A)] row m x1) ≡ bubbletestl(A,upb(A)) ;
     if b1 then A
           else (bool b2,nat[1..upb(A)] row m x2) ≡ bubbletestr(x1,1);
                if b2 then x1 else shakersort(x2) fi ;          fi⌋

funct bubbletestr ≡ (array A,nat k)(bool,array) :
      if k ≥ upb(A)      then (true,A)
    elif A[k] > A[k+1]   then (false,bubbler(exch(A,k,k+1),k+1))
                         else bubbletestr (A,k+1)                fi;

funct bubbler ≡ (array A,nat k) array :
      if k ≥ upb(A)      then A
    elif A[k] > A[k+1]   then bubbler(exch(A,k,k+1),k+1)
                         else bubbler(A,k+1)           fi ;

funct (array,nat) array bubbletestl ≡ bubbletest ;
                co bubbletest of section 6.1 co
```

6.3. Shellsort

If the function bubblesort is used to sort subsequences of the given array, whose elements have equal distance, we get

```
    funct subbubble ≡ (array A,nat k,nat h) array :
          if k+h > upb(A) then A
        elif A[k] > A[k+h] then subbubble(exch(A,k,k+h),k+h,h)
                           else subbubble(A,k+h,h)            fi.
```

Thus it is possible to sort all subarrays with distance h and then all subarrays with distance h' and so on.

If stack s contains a sequence of distance h we get

```
funct shellsort ≡ (array A,stack s) array :
      if ¬ empty :: s then shellsort(shell(A,1,last(s)),pop(s))
                      else A                                   fi
```

```
funct shell ≡ (array A,nat i,nat h) array :
      if i > h then A
              else (bool b,nat[1..upb(A)] row m X) ≡
                                  subbubbletest(A,1,h) ;
                   if b then shell(X,i+1,h)
                        else shell(X,i,h) fi          fi ,

funct subbubbletest ≡ (array A,nat k,nat h)(bool,array) :
     if k+h > upb(A)  then (true,A)
   elif A[k] > A[k+h] then (false,subbubble(exch(A,k,k+h),k+h,h))
                      else subbubbletest(A,k+h,h)               fi
```

If the stack s contains 1 then shellsort(A,s) will solve the sorting problem. For the advantages of shellsort see e.g. (38).

7. IMPROVEMENTS OF EFFICIENCY

The algorithms of section 5 and 6 are now further developed by program transformations. The particular transformations are given only informally. Finally iterative, more efficient programs are obtained, that may be compared to well-known iterative versions.

7.1.1. Sorting by selection

We are going to improve the program of section 5.1.1. :

```
   funct selectsort ≡ (array A) array :
         if upb(A) ≤ 1 then A
                       else nat j ≡ indmax(A) ;
                            insert(selectsort(remove(A,j)),upb(A),A[j])
                                                                      fi
```

A first transformation produces a version working with an explicit argument nat m which indicates that only trim(A,1,m) is yet unsorted.

```
   funct selsort ≡ (array X,nat m) nat :
         if m ≤ 1 then X
                  else nat j ≡ indmax(trim(X,1,m)) ;
                       selsort(insert(remove(X,j),m,X[j]),m-1)  fi
```

Now the recursion is of simple (repetitive) type and we obtain :

```
funct selectsort ≡ (array A) array :
⌈ (var nat[1..upb(A)] row m X,var nat m) := (A,upb(A)) ;
  while m > 1 do nat j ≡ indmax(trim(X,1,m)) ;
              (X,m) := (insert(remove(X,j),m,X[j]),m-1) od ;
                                                      X ⌋
```

The substitution of

```
insert(remove(X,j),m,X[j])
```

by

```
exch(X,j,m)
```

changes the algorithm, but allows the convenient transformation to an iterative version working with an array of variables, since the exchanging of two elements of an array is more efficiently executed on a machine with linear store than the execution of remove and insert. The tranformation influences only the order of the elements X[j] with $1 \leq j < m$ and does not destroy the correctness of the algorithms. Thus we get

```
funct selectsort ≡ (array A) array :
⌈ (var nat[1..upb(A)] row m X,var nat m) := (A,upb(A)) ;
  while m > 1 do nat j≡ indmax(trim(X,1,m)) ;
              (X,m) := (exch(X,j,m),m-1)            od ;
                                                  X ⌋ .
```

The use of an array of variables instead of a variable of mode array gives (cf appendix A.2) :

```
funct selectsort ≡ (array A) array :
⌈ (nat[1..upb(A)] row var m X,var nat m) := (A,upb(A)) ;
  while m > 1 do nat j ≡ indmax(trim(X,1,m)) ;
        if j ≠ m then (X[j],X[m],m) := (X[m],X[j],m-1)
                 else m := m-1
                                                fi od ;
                                                  X ⌋.
```

Note, that the test $j \neq m$ is now necessary to guarantee the alias taboo for the collective assignment (cf. (6)).

In section 5.1. the last version of indmax was of the form :

```
funct indmax ≡(array A) nat :
⌈ funct find ≡(nat i,nat j) nat :
        if i = 0 then j
                 else if A[j] ≥ A[i] then find(i-1,j)
                      ⟦ A[j] ≤ A[i] then find(i-1,i) fi fi ;

   find(upb(A)-1,upb(A))                                        ⌋
```

Since the recursion is of a simple (repetitive) type, it can be directly transformed to repetition :

```
funct indmax ≡ (array A) nat :
⌈ (var nat i,var nat j) := (upb(A)-1,upb(A)) ;
  while i > 0 do if A[j] ≥ A[i] then i := i-1
                 ⟦  A[j] ≤ A[i] then (i,j) := (i-1,i) fi od ;
                                                    j   ⌋.
```

Now the application of indmax can be substituted by its body in the function selectsort (this transformation is called unfolding) :

```
funct selectsort ≡ (array A) array :
⌈ (nat[1..upb(A)] row var m X,var nat m) := (A,upb(A)) ;
  while m > 1 do
        (var nat i,var nat j) := (m-1,m) ;
        while i > 0 do if X[j] ≥ X[i] then i := i-1
                       ⟦ X[j] ≤ X[i] then (i,j) :=(i-1,i) fi od;
        if j ≠ m then (X[m],X[j],m) := (X[j],X[m],m-1)
                 else m := m-1                         fi od ; X⌋
```

Further improvements of efficiency are possible by the use of a variable of mode m to store the values X[j]. This saves some computational time for the selection. Additional sequentialisation and elimination of the nondeterministic conditional clause gives :

```
funct selectsort ≡ (array A) array :
⌈ (nat[1..upb(A)] row var m X,var nat m) := (A,upb(A)) ;
  while m > 1 do (var nat i,var nat j) := (m-1,m) ;
        var m h := X[j] ;
        while i > 0 do if h ≥ X[i] then i := i-1
                                  else (i,j,h) :=(i-1,i,X[i]) fi od;
        if j ≠ m then X[j] := X[m] ;
                      X[m] := h    fi ;
        m := m-1
                                                              od ;
                                                              X ⌋
```

7.1.2. Sequencesort

A number of transformation steps like recursion removal, unfold etc. gives an iterative version of sequencesort :

```
funct seqsort ≡ (array A) array :
⌈ (var nat[1..upb(A)] row m X,var stack s) := (A,empty) ;
  (var nat i,var nat j,var nat z) := (upb(A)-1,upb(A),upb(A)) ;
  do while i > 0 do if X[j] ≥ X[i] then i := i-1
                    ⟦  X[j] ≤ x[i] then (i,j,s) :=
                                   (i-1,i,push(s,j-1)) fi
                                                       od ;
      if z = 1 then leave fi ;
      if empty :: s then (i,j,z) := (z-2,z-1,z-1)
                    else X := insert(remove(X,j),z,X[j]) ;
                         (s,i,j,z) := (pop(s),j-1,top(s) +
                                          j-1,z-1) fi od ;
                                                     X ⌋.
```

Since the transformations of sequencesort are of the same type as those applied for selectsort, they are not described here in detail. Note that for sequencesort the substitution of insert(remove(X,j),z,X[j]) by exch(X,j,z) is not possible, since the order of the elements, which is partly reflected in the stack s, is important.

7.1.3. Heapsort

Of course selecttree and cut of section 5.1.3 can be tranformed to iterative procedures. This is usually done using variables and references like in ALGOL 68. But since the recursion both of selecttree and of cut is of a rather complicated type and therefore the transformation process is voluminous, we renounce to present it here.

7.1.4. Quicksort

Now an iterative, more efficient version of quicksort is given. This version of quicksort is obtained from the version of section 5.1.4 by recursion removal for presort and quicksort and substituting the calls of presort, jismin and jismax by their bodies (all this functions are of simple, repetitive form, besides quicksort, where stacks are used for recursion removal).

```
funct quicksort ≡ (array A) array :
⌈ (nat[1..n] row var m X,var stack s1,var stack s2):=
  (A,push(empty,1), push(empty,upb(A))) ;
  while ¬ empty :: s1 do
  (var nat min,var nat max) := (top(s1),top(s2)) ;
 p : do
     do if min = max          then leave p
      elif X[min] ≤ X[max] then min := min+1
        ⟦  X[min] ≥ X[max] then (X[min],X[max],max) :=
                               (X[max],X[min],max-1) ;
                               leave                fi od ;
     do if min = max          then leave p
      elif X[min] ≤ X[max] then max := max-1
        ⟦  X[min] ≥ X[max] then (X[min],X[max],min) :=
                               (X[max],X[min],min+1) ;
                               leave                fi od
                                                       od ;
        if max - top(s1)>1 then s2 := push(push(pop(s2),max-1)),top(s2))
                           else s1 := pop(s1)                       fi;
        if top(s2) - min>1 then s1 := push(s1,min+1)
                           else s2 := pop(s2)                fi
                                                               od;
                                                               X⌋.
```

Note, that the notation p : do ... od and leave p permits to leave nested loops from the interior of one of the inner loops. This corresponds to do ... od; p and goto p. The particular recursive situation (mutual recursion of a simple repetitive type) is discussed in detail in (6) (in German : "schlichter Prozedurwechsel").

7.2. Insertsort

With in1 of section 5.2. we get (j and the remaining trimmer t can be melted into one variable) :

```
funct insort ≡ (array A : issorted(A),m y) array :
⌈ var nat t := upb(A) ;
  while t ≥ 1 do if A[t] > y then t := t-1
                              else leave fi od ;
                                   insert(A,t+1,y)    ⌋
```

Of course, the version in2 is more efficient and, using recursion removal, trimmers and an array variable, we get for insertsort

```
funct insertsort ≡ (array A) array :
⌈ (var nat[1..upb(A)] row m X,var nat t) := (A,2) ;
  while t ≤ upb(A) do (var nat min,var nat max) := (1,t-1) ;
       while min ≤ max do nat j ≡ (min+max) ÷ 2 ;
             if X[j] ≤ X[t] then min := j+1
             ⟦  X[j] ≥ X[t] then max := j-1              fi od ;
       (X,t) := (conc(insert(trim(X,1,t-1),min,X[t]),
                          trim(X,t+1,upb(A))),t+1)       od ;
                                                          X ⌋
```

On the machine oriented level where the use of rows of variables is appropriate the insertion is very costly. So for insertsort the use of a data structure like a binary insertion tree is a further improvement.

7.3. Mergesort

In section 5.3 the program had the final form :

```
funct mergesort ≡ (array A) array :
      if upb(A) ≤ 1 then A
                    else nat k ≡ (upb(A)+1) ÷ 2 ;
                         merge(mergesort(trim(A,1,k)),
                         mergesort(trim(A,k+1,upb(A))))    fi.
```

This algorithm may also be transformed into a recursive program working on a global variable of mode array.

```
funct mergesort ≡ (array A) array :
⌈ var nat[1..upb(A)] row m X := A ;
      proc msort ≡ (nat min,nat max) :
           if min < max then nat k ≡ (max+min) ÷ 2 ;
              msort(min,k) ; msort(k+1,max) ;
              merge(min,k,k+1),max)              fi ,

      proc merge ≡ (nat min1,nat max1,nat min2,nat max2) :
           if min1 ≤ max1 ∧ min2 ≤ max 2 then
              if X[min1] ≤ X[min2] then merge(min1+1,max1,min2,max2)
              ⟦  X[min1] ≥ X[min2] then X :=
                           insert(remove(X,min2),min1,X[min2]) ;
                           merge(min1+1,max1+1,min2+1,max2)    fifi;
      msort(1,upb(A))                                          X⌋
```

Using a stack, mergesort can be written without recursion :

```
funct mergesort ≡ (array A) array :
⌈ (var nat[1..upb(A)] row m X,stack s1,stack s2,stack c) :=
                A,push(empty,1),push(empty,upb(A)),push(empty,0)) ;
  while ¬ empty :: s1 do
    (var nat min,var nat max) := (top(s1),top(s2)) ;
    if top(c) = 0 then c := push(pop(c),1) ;
       if min < max then nat k ≡ (max+min) ÷ 2 ;
          (s1,s2,c) := (push(s1,min),push(s2,k),push(c,0)) ;
          (s1,s2,c) := (push(s1,k+1),push(s2,max),push(c,0)) fi
    ⟦ top(c) = 1 then
                (var nat min1,var nat max1,var nat min2,var nat max2) :=
                (min,(max+min) ÷ 2, (max+min) ÷ 2+1,max) ;
                while min1 ≤ max1  ∧ min2 ≤ max2 do
                   if X[min1] ≤ X[min2] then min1 := min1+1
                   ⟦  X[min1] ≥ X[min2] then
                      X := insert(remove(X,min2),min1,X[min2]) ;
                      (min1,min2,max1) := (min1+1,min2+1,max1+1) fi od;
                      (s1,s2,c) := (pop(s1),pop(s2),pop(c))      fi od;
                                                                   X⌋
```

Another class of mergesort programs can be derived, when already sorted subsequences of the given array are searched for and merged.

7.4. Bubblesort

Since the recursive definitions of bubble, bubblesort and bubbletest are all of simple (repetitive) form, the transformation to a repetition can easily be done :

```
funct bubblesort ≡ (array A) array :
⌈  (nat[1..upb(A)] row var m X,var bool b) := (A,false) ;
   while ¬ b do var nat k := upb(A) ;
      do if k = 1             then b := true ; leave fi ;
       elif X[k-1] > X[k] then (X[k-1],X[k],k) := (X[k],X[k-1],k-1);
                               while k > 1 do
                                if X[k-1] > X[k] then (X[k-1],X[k]) :=
                                                      (X[k],X[k-1])  fi ;
                                   k := k-1                           od ;
                                   leave
                               else k:= k-1                    fi od od ;
                                                                       X⌋
```

An analysis of this algorithm shows that some improvements are possible, since an increasing part of the array is not altered during the rest of the algorithm.

```
funct bubblesort≡(array A) array :
⌈ (nat[1...upb(A)] row var m X,var nat q) := (A,1)
  b : do var nat k := upb(A) ; nat r ≡ q ;
      do if k = r             then leave b
       elif X[k-1] > X[k] then (X[k-1],X[k-1],k,q) := (X[k],X[k-1],
                                                        k-1,q) ;
         while k > r do
            if X[k-1] > X[k] then (X[k-1],X[k],q) :=
                                  (X[k],X[k-1],k)   fi ;
               k := k-1                                     od ;
               leave
                         else k:= k-1                 fi od od ;
                                                              X ⌋
```

Finally the introduction of a variable var m h for X[k] may lead to some further improvements (cf. 7.1.1.).

7.5. Shakersort

For shakersort of section 6.3 the same improvements as for bubblesort are possible, finally leading to :

```
funct shakersort ≡ (array A) array :
⌈ (nat[1..upb(A)] row var m X,var nat q1,var nat q2 := (A,1,upb(A)) ;
b : do (var nat k,var nat r) := (q2,q1) ;
      do if k = r then leave b
       elif X[k-1] > X[k] then (X[k-1],X[k],k,q1) :=
                               (X[k],X[k-1],k-1,k)  fi ;
         while k > r do
            if X[k-1] > X[k] then (X[k-1],X[k],q1) :=
                                  (X[k],X[k-1],k)     fi ;
               k := k-1                                  od ;
         leave
                         else k := k-1                    fi ;
                                                          od ;

      (k,r) := (q1,q2) ;
      do if k = r then leave b
       elif X[k] > X[k+1] then (X[k],X[k+1],k,q2) :=
                               (X[k+1],X[k],k+1,k) ;
          while k > r do
             if X[k] > X[k+1] then (X[k],X[k+1],q2) :=
                                   (X[k+1],X[k],k)     fi ;
                     k := k+1                             od ;
             leave
               else k := k+1                        fi od od ;
                                                          X ⌋.
```

7.6. Shellsort

Recursion removal and unfold of subbubble in shellsort of section 6.3 gives with the additional improvements of bubblesort of section 7.4 :

```
funct shellsort ≡ (array A,stack s) array :
⌈ (nat[1..upb(A)] row var m X,var stack s) := (A,y) ;
  while ¬ empty :: s do
  (var nat h,var nat i) := (last(s),1) ;
  while i ≤ h do
  var nat q := upb(A)-h+1 ;

b : do var nat k := i ; nat r ≡ q ;
    do if k ≥ r            then leave b
     elif X[k] > X[k+h] then (X[k],X[k+h],k,q) := (X[k+h],X[k],k+h,q) ;
         while k < r do
            if X[k] > X[k+h] then (X[k],X[k+h],q) :=
                                  (X[k+h],X[k],k)    fi ;
            k := k+h                                           od ;
            leave
                      else h := k+h                      fi od od ;
                           i := i+1                            od ;
                           s := pop(s)                         od ;
                                                               X ⌋
```

8. FINAL REMARKS

In the previous sections some well-known sorting algorithms have been developed from a formal problem description. Some others have not been considered, i.e. different possibilities of merging and sorting by distribution (besides quicksort that can also be considered as sorting by distribution), since in general for distribution sorting some more structure has to be defined on mode m. Furthermore sorting algorithms that use already ordered subsequences of the given array have not been considered.

8.1. Some remarks on program synthesis

In the previous sections the development of the particular algorithms was done in small and I hope clear steps, guided and supervised by intuition at every moment.

Another approach is found by Darlington (29). Darlington uses the fact that sorting can be regarded as searching for the permutation under which the given array is ordered, i.e. looking for the inverse permutation.

A sorting algorithm then can be obtained from a simple exhaustive search through the set of all permutations. Every algorithm that enumerates all permutations with n elements (n = upb(A)) induces a search method. But algorithms created in this way are not very efficient, since the number of average comparisons is $\frac{n!}{2}$, if every permutation has the same probability.

So Darlington applies some correctness preserving transformations to optimize the searching. In this way, he obtains some well-known sorting algorithms (quicksort, selection sort, merge sort, insertion sort, exchange sort and bubble sort). Darlington could have found even other sorting algorithms by using other permutation generating algorithms. For example, had he used an algorithm that generates permutations by adjacent transpositions (cf. e.g. (11)), he would have found shaker sort.

For further remarks on (29), see (3). An approach which is to some extend closer than program construction in the previous section (which was mainly taken from (13)) can be found in (26).

It is a very interesting aspect to synthesize algorithms from other ones. But often this way of program synthesis seems to be unnatural and gives only very inefficient algorithms in the beginning. The main problem is then to transform the resulting algorithms to more efficient programs. This, however, requires a great deal of intuition.

8.2. Conclusion

The intention of this paper is to demonstrate how algorithms are developed in systematic ways supported by intuition. A further aspect is the study of algorithmic principles and the promising use of computation structures, i.e. modes, functions and their algebraic properties and laws.

The construction of a program should start with the formal specification of the problem using predicates and the analysis of the predicates properties with respect to the used computation structures. Then a sequence of design decisions may lead to a number of algorithms. In each step, a design decision leads to the substitution of prealgorithmic specifications (via predicates) by algorithms until a complete algorithm is constructed. Each design decision means the application of some mathematical theorem. The "half-algorithmic" versions, in which algorithmic formulations and specifications by predicates (prealgorithmic formulations) are mixed, may be considered as schemes describing classes of algorithms. Every design decision reduces the class of algorithms to a subset of the class described by the previous version.

In the final phase of a program development process, the original algorithms are optimized by program transformations with respect to a special machine or to special features of a language (if a language does not comprise the algorithmic constructs, in which the original algorithm is written). This part of the program development was described informally here, only giving the final iterative version, while more attention in this paper was paid to the creation of algorithms. Of course, sometimes it may be more appropriate to alter the algorithm during this optimization process to enable the programmer to establish further optimizations (c.f. 7.1.1). For an example of detailed program development with respect to efficiency improvements see e.g. (12), (44).

ACKNOWLEDGMENT

This study represents the way of program construction as developed by the CIP research group whose member the author is since several years. Therefore, the work of F.L. Bauer, W. Dosch, R. Gnatz, B. Krieg-Brückner, A. Laut, T. Matzner, B. Möller, H. Partsch, P. Pepper, K. Samelson, R. Steinbrüggen, M. Wirsing and H. Wössner are gratefully acknowledged. In particular, I like to thank H. Partsch for fruitful discussions during the preparation of this paper.

APPENDIX

A - THE USED COMPUTATION STRUCTURES

A.1. The computation structure nat

A basis for the formulation of algorithms with arrays is a linearly ordered set of indexes. For simplicity reasons we use intervals of natural numbers here (for a more general treatment see (45)).

The set of all natural numbers is characterized by the mode <u>nat</u>. Intervals of this set are denoted by the mode <u>nat</u>[m..n].

Furthermore, on <u>nat</u> and its submodes the usual operations are assumed to be available like addition, substraction, multiplication and integer division. Of course the computation structure of natural numbers may be described by an abstract data type (cf. for instance (6) for a detailed treatment).

A.2. The computation structure array

To define a computation structure we use algebraic methods. Since several functions are applied to manipulate arrays, we restrict the original definition to a smaller subset of these functions and define the others in terms of the defined ones. Let m be some given mode.

```
type ARRAY ≡ (mode m) array, select, insert, remove, empty array, upb :
     mode array,
     funct array empty array,
     funct (array) nat upb,
     funct (array A, nat n : 0 < n ≤ upb(A)) m select,
     funct (array A, nat n, m : 0 < n ≤ 1+upb(A)) array insert,
     funct (array A, nat n : 0 < n ≤ upb(A)) array remove,

law : upb(empty array) = 0,
law : 0 < n ≤ 1+upb(A) ⇒ upb(insert(A,n,x)) = upb(A)+1,

law : 0 < n ≤ 1+upb(A) ⇒
   select(insert(A,n,x),m) = if n = m then x
                             ⫿  n < m then select(A,m-1)
                             ⫿  n > m then select(A,m)  fi ,

law : 0 < n ≤ 1+upb(A) ⇒
   remove(insert(A,n,x),m) = if n = m then A
                             ⫿  n < m then insert(remove(A,m-1),n,x)
                             ⫿  n > m then insert(remove(A,m),n-1,x) fi,

law : insert(insert(A,m,x),n,y) = if n ≤ m then insert(insert(A,n,y),
                                                          m+1,x)
                                  ⫿  n > m then insert(insert(A,n-1,y),
                                                          m,x)  fi

                                              end of type.
```

Lemma : Every array A can be represented by a finite number (= upb(A)) of iterated applications of insert to the empty array.

Proof : Every term denoting an array object contains empty array. If the outer function application is different from insert, then one of the laws may be applied. □

As usual we denote select(A,n) by A[n].

We define some additional functions :

```
funct exch ≡ (array A, nat i, nat j : 0 < i ≤ upb(A) ∧
                                      0 < j ≤ upb(A)) array :
        insert(remove(insert(remove(A,i),i,A[j]),j),j,A[i]),

funct conc ≡ (array A, array B) array :
    if upb(A) = 0 then B
    ⟦  upb(A) > 0 then insert(conc(remove(A,1),B),1,A[1]) fi,

funct trim ≡ (array A, nat i, nat j : 1 ≤ i ≤ upb(A)+1 ∧
                                      0 ≤ j ≤ upb(A)) array :
    if i > j then emptyarray
    ⟦ i ≤ j then insert(trim(A,i+1,j),1,A[i]) fi.
```

Furthermore we define a (implicit) conversion between every array A with upb(A) = n to an object of mode nat[1..n] row m B, such that

$$\forall\ \underline{\text{nat}}[1..n]\ i : B[i] = A[i].$$

In the following we don't distinguish objects fo mode array and of mode nat[1..n] row m. Every array A is considered to be of mode nat[1..upb(A)] row m (nat[1..0] row m is the mode of the empty array).

Using the definitions above the following predicates hold (and can be proved) :

```
(1) ∀(array A,array B,nat[1..upb(A)] i,nat[1..upb(A)] j,m y) :
            upb(A) = upb(exch(A,i,j)) ∧
   j ≥ i ⇒ upb(trim(A,i,j)) = j-i+1 ∧
            upb(conc(A,B)) = upb(A)+upb(B) ∧
            upb(remove(A,i)) = upb(A)-1 ∧
            upb(insert(A,i,y)) = upb(A)+1 ;

(2) ∀(array A,nat[1..upb(A)] i,nat[1..upb(A)] j) :
            exch(exch(A,i,j),i,j) = A ∧
            exch(A,i,j) = exch(A,j,i) ∧
            exch(A,i,i) = A ;

(3) ∀(array A,array B,array C) :
            conc(A,conc(B,C)) = conc(conc(A,B),C);
```

(4) ∀(array A,nat[1..upb(A)] j) :

conc(trim(A,1,j-1), trim(A,j,upb(A))) = A ;

(5) ∀(array A,nat[1..upb(A)+1] j,m y) :

remove(insert(A,j,y),j) = A ;

(6) ∀(array A,nat[1..upb(A)] j) :

insert(remove(A,j),j,A[j]) = A

(7) ∀(array A,nat[1..upb(A)+1] j,m y) :

insert(A,j,y)[j] = y ;

These properties are only a small collection of algebraic laws. For the problem of complete characterization of a computation structure by axioms see (55). The computational structure ARRAY with the given functions is rather complex. Nevertheless algebraic properties of ARRAY can be analyzed and proved and they are a safe tool for the construction, formulation and verification of algorithms.

For ARRAY several theorems and lemmas can be proved as a basic for the development of many kinds of algorithms, not only for sorting problems. With these theorems and lemmas the construction of algorithms is as well supported as the construction of triangles etc. in geometry by the classical theorems.

The functions described above can be transformed onto an iterative level with array variables of arrays of variables. Let A be an array of mode nat[1..n] row m and V be a variable array of mode var nat[1..n] row m and W be an array (of variables) of mode nat[1..n] row var m, the following equivalence on the different levels of programming is sometimes used for exch (let i,j be some natural numbers of mode nat[1..n]) :

```
exch(A,i,j)
           ↕
─────────────────────────── { i≠j

⌈ nat[1..n] row var m W := A ;
      (W[i],W[j]) := (W[j],W[i];
                              W ⌋
```

The other functions cannot simply be transformed onto the iterative level. Note, that the use of the functions described above treats the array as one object and not, like usually done, as a row of objects. This helps to avoid some difficulties arising by selective updating (cf. (30), (56)). For example the collective assignment (A[i],A[j]) := (1,5) leads to a contradiction iff i = j.

Accordingly an object V of mode var nat[1..n] row m, i.e. a row variable, may only be changed as a whole. Thus the assignment to a component V[i] where i is some natural number between 1 and n,

V[i] := x

is not correct. Instead, we have to write

V := insert(remove(V,i)i,x) ;

An assignment to a component of an array can only be done for e.g. some W of mode nat[1..n] row var m :

W[i] := x.

Note, however, that this combination of insert and remove is not used very frequently in this paper.

The mode nat[1..n] row var m reflects the structure of a machine with a linear store and the operations that are more efficiently executed.

The consideration of an array as a row of variables is strongly influenced by the conception of a linear store. The functions conc, insert, remove and trim do not communicate directly with the rows of variables as they exist e.g. in ALGOL 60 etc.

Thus, matters very often get complicated by transformation from the "abstract" formulations with the computation structure array to "machine-oriented" formulations with rows of variables.

So it seems advantageous to avoid these complications by developing our algorithms on the abstract level - including all necessary analysis, and afterwards to switch to iterative, "machine-oriented" formulations. This helps to reduce the complexity of the programming task.

A.3. The computational structure STACK

For recursion removal sometimes the computation structure STACK is used with its typical operations :

```
mode empty ≡ atomic{empty},
mode stack ≡ (stack x,nat n) | empty ;
funct push ≡ (stack s,nat n) stack : (s,n) ;
funct pop  ≡ (stack s : ¬ empty :: s) stack : x of s ;
funct top  ≡ (stack s : ¬ empty :: s) nat : n of s ;
```

The computation structure STACK may be characterized by the following algebraic laws :

∀ stack s,nat n : pop(push(s,n)) = s,
top(push(s,n)) = n,

pop(empty) = error,
top(empty) = error,

(empty :: empty) = true,
(empty :: push(s,n)) = false,

REFERENCES

(1) Arsac, J. : La construction de programmes structurés. Paris Dunod 1977.

(2) Balzer, R.M. : Correct and efficient software implementation via semi-automatic transformations (Research Proposal, unpublished).

(3) Barstow, D. : Remarks on "A synthesis of several sorting algorithms" by John Darlington. Acta Informatica 13, 1980, pp. 225-227.

(4) Bauer F.L. : Programming as an evolutionary process. Technische Universität München, Institut für Informatik, Rep. n° 7617, 1976. Also : Proc. 2nd Int. Conf. on Software Engineering, Oct. 1976, San Francisco, CA, pp. 223-234.

(5) Bauer, F.L., Broy, M. (eds.) : Program construction. Lecture Notes of the International Summer School on Program Construction. Marktoberdorf 1978, Lecture Notes in Computer Science 69, Berlin-Heidelberg-New York : Springer 1979

(6) Bauer, F.L., Wössner, H. : Algorithmic language and program development. Berlin-Heidelberg-New York : Springer 1981 (to appear).

(7) Bauer, F.L., Partsch, H., Pepper, P., Wössner, H. : Techniques for program development. Technische Universität München, Institut für Informatik, Interner Bericht, Sept. 1976. Also in : Software Engineering Techniques. Infotech State of the Art Report 34, 1977, pp. 27-50.

(8) Bauer, F.L., Broy, M., Gnatz, R., Hesse, W., Krieg-Brückner B., Partsch, P., Pepper, P., Wössner, H. : Towards a wide spectrum language to support program specification and program development. SIGPLAN Notices 13:12, 1978, pp. 15-24.

(9) Bauer, F.L., Broy, M., Partsch, H., Pepper, P., Wössner, H. : Systematics of transformation rules. In : (5), pp. 273-283.

(10) Bauer, F.L., Broy, M., Dosch, W., Gnatz, R., Geiselbrechtiger, F. Hesse, W., Krieg-Brückner, B., Laut, A., Matzner, T., Möller, B., Partsch, H., Pepper, P., Samelson, K., Wirsing, M. Wössner, H. : Report on a wide spectrum language for program specification and development. Technische Universität München, Institut für Informatik, TUM-I8104, May 1981.

(11) Broy, M. : Program development for steinhaus type permutation generating programs. Technische Universität München, Institut für Informatik, TUM-INFO-07701, Januar 1977.

(12) Broy, M. : Program development : The ackermann function as an example. Technische Universität München, Institut für Informatik, TUM-INFO-7716, Juni 1977.

(13) Broy, M. : A case study in program development : sorting. Technische Universität München, Institut für Informatik, TUM-INFO-7831, November 1978.

(14) Broy, M. : Transformation parallel ablaufender Programme. Technische Universität München, Dissertation an der Fakultät für Mathematik, Februar 1980.

(15) Broy, M. : Transformational semantics for concurrent programs. IPL 11:2, October 1980, pp. 87-91.

(16) Broy, M., Krieg-Brückner, B. : Derivation of invariant assertions during program development by transformation. ACM TOPLAS 2:2, July 1980, pp. 321-337.

(17) Broy, M., Pepper, P. : Programming as a formal activity. IEEE Transactions on Software Engineering, SE-7:1, January 1981, pp. 14-22.

(18) Broy, M., Wirsing, M. : Program development : From enumeration to backtracking. IPL 10:4,5, July 1980, pp. 193-197.

(19) Broy, M., Wirsing, M; : Programming languages as abstract data types. In : M. Dauchet (ed.), 5ème Colloque de Lille "Les arbres en algèbre et en programmation": Lille 1980, pp. 160-177.

(20) Broy, M., Gnatz, R., Wirsing, M. : Problemspezifikation - eine Grundlage für Programmentwicklung. Workshop on Reliable Software, Bonn University, September 22-23, 1978. München : Hanser 1979, pp. 235-246.

(21) Broy, M., Pepper, P., Wirsing, M. : On relations between programs. In : B. Robinet (ed.) : Proc. 4th Symposium on Programming, Paris 1980, Lecture Notes in Computer Science 83, Berlin-Heidelberg -New York : Springer 1980, pp. 59-78.

(22) Broy, M., Partsch, H., Pepper, P., Wirsing, M. : Semantic relations in programming languages. In : S.H. Lavington (ed.) : Information Processing 80, Proc. of the IFIP Congress 80, Amsterdam-New York-Oxford : North Holland Publ. Comp. 1980, pp. 101-106.

(23) Burstall, R.M., Darlington, J. : Some transformations for developing recursive programs. Proc. of 1975 Int. Conf. on Reliable Software, Los Angeles 1975, pp. 465-472. Also (revised version), J. ACM 24, 1, pp. 44-67, 1977.

(24) Burstall, R., Feather, M. : Program development by transformations : an Overview. Proc. of Toulouse CREST Course on Programming, Toulouse, 1978.

(25) Cooper, D.C. : The equivalence of certain computations. Comp. J. 9, pp. 45-52, 1966.

(26) Clark, K.L., Darlington, J. : Algorithm classification through synthesis. The Computer Journal 23:1, 1979, pp. 61-65.

(27) Clark, K.L., Sickel, S. : Predicate logic : A calculus for deriving programs. IJCAI 5, 1977, pp. 419-420.

(28) Darlington, J., Burstall, R.M. : A system which automatically improves programs. Acta Informatica 6, pp. 41-60, 1976.

(29) Darlington, J. : A synthesis of several sorting algorithms D.A.I. Research Report n° 23, Dept. of Artificial Intelligence, University of Edinburgh, 1976, see also : Acta Informatica 11, 1978, pp. 1-30.

(30) Dijkstra, E.W. : A discipline of programming. Series in Automatic Computation, Prentice Hall, Englewood Cliffs, N.J. 1976.

(31) DeRemer, F.L. : Transformational grammars. In : F.L. Bauer, J. Eickel (eds.) : Compiler construction. An advanced course. Lecture Notes in Computer Science 21, Berlin, Heidelberg, New York : Springer 1974, pp. 121-145.

(32) Floyd, R.W. : Assigning meanings to programs. Proc. of Symposia in Applied Mathematics of the Amer. Math. Soc. 19, pp. 19-32, 1967.

(33) Gerhart, S.L. : Correctness-preserving program transformations. Conf. Rec. Second ACM Symp. on Principles of Programming Languages, Jan. 1975, pp. 54-66.

(34) Gnatz, R. : mod-Funktion und Maximumsbestimmung : Entwicklung zweier Programme. Technische Universität München, Institut für Informatik, TUM-INFO-7712, July 1977.

(35) Guttag, J. : The specification and application to programming of abstract data types. Ph.D. thesis, Univ. of Toronto

Depart. Comp. Sci., Rep. CSRG-59.

(36) Hoare, C.A.R. : An axiomatic basis for computer programming. Comm. ACM 12, pp. 576-583, 563, 1969.

(37) Hoare, C.A.R. : Quicksort. The Computer Journal 5:1, 1962, pp. 1015.

(38) Knuth, D.E. : The art of computer programming, vol. 3, Sorting and searching. Reading (Mass.) : Addison Wesley, 1973.

(39) Loveman, D.B. : Program improvement by source-to-source transformation. Journal ACM 24:1, pp. 121-145, 1977.

(40) Manna, Z. : Mathematical theory of computation. New York, McGraw Hill, 1974.

(41) Manna, Z., Waldinger, W. : A deductive approach to program synthesis. ACM TOPLAS 2:1, January 1980, pp. 90-121.

(42) McCarthy, J. : A basis for a mathematical theory of computation. In : P. Braffort, D. Hirschberg (eds.) : Computer programming and formal systems. Amsterdam : North Holland 1963, pp. 33-70.

(43) Paige, R., Koenig, S. : Finite differencing of computable expressions. Rutgers University, New Brunswick, N.J. 1980, LCSR-TR-8.

(44) Partsch, H., Pepper, P. : Program transformations on different levels of programming. Technische Universität München Institut für Informatik, TUM-INFO-7715, 1977.

(45) Partsch, H., Broy, M. : Examples for change of types and object structures. In : Bauer, Broy (5), pp. 421-463.

(46) Partsch; H., Steinbrüggen, R. : A comprehensive survey on program transformation systems. Technische Universität München, Institut für Informatik, TUM I8108, July 1981.

(47) Paterson, M.S., Hewitt, C.E. : Comparative schematology. In : Record of the Project MAC Conf. on Concurrent Systems and Parallel Computation, New York : ACM 1970, pp. 119-127.

(48) Pepper, P. : A study on transformational semantics. Dissertation, Munich 1979. Also in (5), pp. 322-405.

(49) Pepper, P., Partsch, H., Wössner, H., Bauer, F.L. : A transformational approach to programming. In : B. Robinet (ed.) Program transformations. Proc. of the 3rd International Symposium on Programming, March 28-30, 1978, Paris : Dunod 1978.

(50) Rosen, B.K. : Tree-manipulating systems and Church-Rosser theorems. Journal ACM 20:1, pp. 60-187, 1973.

(51) Standish, T.A., Harriman, D.C., Kibler, D.F., Neighbors, J.M. The Irvine program transformation catalogue. Dept. Inform. and Comp. Sc., University of California at Irvine, CA, Jan. 1976.

(52) Standish, T.A., Harriman, D.C., Kibler, D.F., Neighbors, J.M. Improving and refining programs by program manipulation. Proc. 1976 ACM Annual Conf., Oct. 1976, pp. 509-516.

(53) Strong, H.R. : Translating recursion equations into flow charts. In : Proc. 2nd Annual ACM Symposium on Theory of Computing, New York 1970, pp. 184-197. Also J. CSS 5, pp. 254-285, 1971.

(54) Wegbreit, B. : Goal-directed program transformation. IEEE Transactions on Software Engineering, vol. SE-2, n° 2, pp. 69-80, 1976.

(55) Wirsing, M., Pepper, P., Partsch, H., Dosch, W., Broy, M. : On hierarchies of abstract data types. Technische Universität München, Institut für Informatik, TUM-I8007, 1980.

(56) Wirth, N. : Algorithmen und Datenstrukturen. Teubners Studienbücher Informatik, B.G. Teubner Stuttgart, 1975.

(57) Wössner, H., Pepper, P., Partsch, H., Bauer, F.L. : Special transformation techniques. In : (5), pp. 290-321.

ISSUES IN THE METHODICAL DESIGN OF CONCURRENT PROGRAMS

Sintzoff Michel

Philips Research Laboratory Brussels

Abstract : An overview of typical issues in the systematic design of concurrent programs is presented. In a first group of problems, it is shown how to ensure desirable logical properties such as the absence of deadlock or starvation. A second set of issues is concerned with efficiency; the attention is focussed on the increase of the independence of program components. The proposed design techniques are developed in terms of concurrent programs expressed as nondeterministic iterations of guarded commands using shared variables. In a third part, a tentative method is presented for replacing the iterative derivation of predicates by easier techniques yielding weaker but safe versions of these predicates. The automation of design methods for concurrent programs is also evaluated.

1. INTRODUCTION

By presenting this overview of a number of typical issues in the methodical design of concurrent programs, our intention is to substantiate the following thesis : the eventual automation of program construction must depend on an understanding by human intelligence of the program design process. In other words, one should not attempt to automate concurrent programming whereas, or even because, the corresponding design techniques still are ill-known. We investigate two main classes of issues : how to ensure desirable logical properties of concurrent programs, and how to increase their potential efficiency. The logical properties we consider are the impossibility of deadlock, the certainty of termination, and the impossibility of starvation. As to efficiency-oriented characteristic, we focus our attention on

A. W. Biermann and G. Guiho (eds.), Computer Program Synthesis Methodologies, 51–78.

the decreasing of interactions between concurrent components, in order to foster a more economical use of distributed architectures. To facilitate the understanding of these issues by uninitiated readers, and to simplify the writing, we shall follow a rather elementary style of exposition. Thus, the model of concurrent programs we use is nothing more than that of nondeterministic iterations with shared variables : nondeterminism and sharing represent concurrency and communication, respectively. This model allows the application of well-established logical rules. Also, we do not consider here the issue of the formal and abstract specification of concurrency problems : the present author is still incompetent in this area. The design techniques will thus be expressed mainly in terms of transformation schemes between programs : we start with initial program versions, possibly supplemented with assertions, and we aim at deriving improved versions. Moreover, most of the examples will be simple illustrations, rather than significant case studies.

We think appropriate to stress a second thesis : a fundamental principle of obtaining constructive variants for proposed intellectual methods is the idea of approximation. For instance, to be automated, inductive principles must be realized by, e.g., finitistic means. This reminds of finite decimal expansions used to represent real numbers. The problem is to discover approximation schemes which are completely constructive and which satisfy desired correctness criteria. To exemplify this problem, we shall present a possible approximation scheme for eliminating dead-ends and also relate this technique to known methods for improving convergence in transition or rewriting systems.

The last part of the paper includes more general discussions about other issues in the area of concurrent programming : problem specification, very high level languages, design rules, proof rules.

Let us now sketch the way we express concurrent programs. Consider

$$
\begin{array}{rll}
P = \underline{\text{do}} & P1 : & B1 \rightarrow S1 \\
\| & P2 : & B2 \rightarrow S2 \\
\| & \ldots & \\
\| & P_N : & B_N \rightarrow S_N \\
\underline{\text{od}} & &
\end{array}
$$

The program P is an iteration of "guarded commands" $P1,\ldots,P_N$. Each P_i consists of a guard B_i, which is a logical expression, and of a command S_i, which expresses a basic action of the program. The iteration ends if and when all the guards become false ; during the iteration any command may be selected for evaluation provided its guard is then true. Since it is not required that the guards be mutually exclusive, nondeterministic selection is allowed : parallel execution corresponds then to the execution of two guarded commands whose guards are both true and which do not modify common variables. In order to express a finer grain of concurrency, it suffices to decompose a given program in terms of more elementary commands : this can be done systematically by using auxiliary variables as program counters, and will be shown later in examples. In case the guards never become all false, the program represents a non-terminating system.

The calculus wp of weakest preconditions [1] provides the logical framework in which to reason about these nondeterministic iterations of guarded commands. We recall the definition of the weakest precondition ensuring that program P does establish a post-assertion R :

$$wp(P,R) = \lim_{n\to\infty} H^n$$

$$H^0 \equiv R$$

$$H^{n+1} \equiv H^n \vee (\bigvee_i B_i) \wedge \bigwedge_i [B_i \supset wp(S_i, H^n)] \quad i=1...N$$

Thus H^{n+1} is H^n plus the set of states from which H^n can be reached by an application of any of the commands the guard of which is true. The term $wp(S_i, H^n)$ is the weakest precondition to establish H^n by S_i :

$$wp((S;T),Q) \equiv wp(S, wp(T,Q))$$

$$wp(x{:=}E, Q) \equiv Q^x_E \text{ , or } Q[E/x]$$

2. PREVENTING DEADLOCK

We begin here the presentation of methods [2] to ensure essential logical properties. The first one is the absence of deadlock. A deadlock occurs in a program which is intended not to terminate, when all the guards become false. Consider for instance the following program where $a < b$:

```
P : do P1 : x < b           → x:=x+1
    ‖  P2 : x ≥ a ∧ x < b → x:=x-1
    od
```

The state $x=b$ can be reached by iteratively applying P1 from, say, $x=a-1$. Then a deadlock occurs : both the guard of P1 and that of P2 are false for $x=b$.

We shall aim at strengthening the guards of the given program so as to eliminate any such risk of occurrence of deadlocks. The idea is quite simple. To prevent any deadlock, it suffices to ensure that the predicate "at least one guard is true" is an invariant of the program : thus, obviously, if the program can be started, it may never reach a state where all guards are false. The problem thus is to derive new guards C_i which satisfy the required property. For a program on the form

$$\begin{aligned} P : \underline{do}\ & P1 : B1 \rightarrow x:=f1(x) \\ \| \ & P2 : B2 \rightarrow x:=f2(x) \\ \underline{od} & \end{aligned}$$

the strengthened guards C1 and C2 are solutions of the equations

$$C1(x) \equiv B1(x) \wedge [C1(f1(x)) \vee C2(f1(x))]$$

$$C2(x) \equiv B2(x) \wedge [C1(f2(x)) \vee C2(f2(x))]$$

The equations express that each C_i strengthens B_i (viz $C_i \equiv B_i \wedge \ldots$) and that C1 $\vee$ C2 is an invariant of the new program

$$\begin{aligned} P' : \underline{do}\ & P1' : C1 \rightarrow x:=f1(x) \\ \| \ & P2' : C2 \rightarrow x:=f2(x) \\ \underline{od} & \end{aligned}$$

The union of the guards $C \equiv C1 \vee C2$ is invariant since the equations for C1, C2 ensure

$$C_i \supset (C1 \vee C2)\ [f_i(x)/x]$$

viz. $$C_i \supset wp(x:=f_i(x),\ C)$$

Thus at least one of the guards C_1, C_2 must hold after applying P_i' on a state where C_i holds.

It remains to give a technique for deriving solutions of the equations for C1,C2. A straightforward one is to compute them by successive approximations, starting with <u>true</u> as initial values :

$$C^0(x) \equiv \underline{true}$$

$$C_i^{n+1}(x) \equiv B_i(x) \wedge \bigvee_j C_j^n(f_i(x)) \qquad i,j=1\ldots N$$

$$C_i(x) \equiv \lim_{n\to\infty} C_i^n(x)$$

In easy cases, the solutions are obtained after a finite number of steps, say m, such that $C_i^{m+1} \equiv C_i^m$ for all i. In more serious applications, an induction of a general form $C_i(n) \equiv C_i^n$ is required ; then $C_i \equiv \forall n : C_i(n)$, which in turn may require non trivial simplifications.

The example given above is an easy case, as shown hereafter :

$$C_1^0 \equiv \underline{true},\ C_2^0 \equiv \underline{true}$$

$$C_1' \equiv x < b$$

$$C_2' \equiv x \geq a \wedge x < b$$

$$C_1^2 \equiv x < b \wedge [x+1 < b \vee x+1 \geq a \wedge x+1 < b]$$

$$\equiv x < b-1$$

$$C_2^2 \equiv x \geq a \wedge x < b \wedge [x-1 < b \vee x-1 \geq a \wedge x-1 < b]$$

$$\equiv x \geq a \wedge x < b$$

$$C_1^3 \equiv x < b \wedge [x+1 < b-1 \vee x+1 \geq a \wedge x+1 < b]$$

$$\equiv x < b-2 \vee x \geq a-1 \wedge x=b-2$$

$$\equiv x < b-2 \vee x=b-2 \qquad (a < b \text{ implies } b-2 \geq a-1)$$

$$\equiv x < b-1$$

$$C_2^3 \equiv x \geq a \wedge x < b \wedge [x-1 < b-1 \vee x-1 \geq a \wedge x-1 < b]$$

$$\equiv x \geq a \wedge x < b$$

We observe $C_1^3 \equiv C_1^2$ and $C_2^3 \equiv C_2^2$; hence the new, deadlock-free program is

$$\begin{aligned} P' : \underline{do}\ & P1': x < b-1 \rightarrow x:=x+1 \\ \| \ & P2': x \geq a \wedge x < b \rightarrow x:=x-1 \\ \underline{od} & \end{aligned}$$

We now illustrate the application of the method on a classical example, namely crossed competition for reusable resources. Two components S and T repeatedly compete for a combination of two resources a and b. Component S consists of three parts, S1 for taking a, S2 for taking b, and S3 for releasing both a and b after some time. T is similarly decomposed in T1, T2, T3, but takes the resources in reverse order, viz. first b then a ; this is why deadlock may occur.

```
P : do S1 : s=1 ∧ a=1 → s:=2 ; a:=0
    ∥  S2 : s=2 ∧ b=1 → s:=3 ; b:=0
    ∥  S3 : s=3       → s:=1 ; a:=1 ; b:=1
    ∥  T1 : t=1 ∧ b=1 → t:=2 ; b:=0
    ∥  T2 : t=2 ∧ a=1 → t:=3 ; a:=0
    ∥  T3 : t=3       → t:=1 ; a:=1 ; b:=1
    od
```

The variables a and b express the availability of the two resources. The variable s serves to specify the required sequencing of the parts S1, S2, S3, and similarly for t w.r.t. T. Assume we start from $s=1 \wedge t=1 \wedge a=1 \wedge b=1$, and we apply S1 then T1 ; we reach thus $s=2 \wedge t=2 \wedge a=0 \wedge b=0$: all guards are then false and there is a deadlock. The iterative calculus yields

Rank 0 : all guards are true.

Rank 1 : the guards given in P.

Rank 2 : For S1 : $s=1 \wedge a=1 \wedge (b=1 \vee t=3)$
because $(2=1) \equiv (2=3) \equiv (0=1) \equiv$ false

For S2 : $s=2 \wedge b=1$ because $(3=3) \equiv$ true

For S3 : $s=3$

For T1 : $t=1 \wedge b=1 \wedge (a=1 \vee s=3)$

For T2 : $t=2 \wedge a=1$

For T3 : $t=3$

Rank 3 : same as Rank 2.

Here also we have rapid convergence. Only the guards for S1 and T1 are modified. The new program embodies the following policy : to request both resources at the same time $(a=1 \wedge b=1)$ or to request a first resource provided both resources have just been allocated to the other process ($a=1 \wedge t=3$ for S1, and $s=3 \wedge b=1$ for T1). In fact, the second case is redundant since both $a=1 \wedge t=3$ and $s=3 \wedge b=1$ are false ; this results from the invariance of $t=3 \supset a=0$ and of $s=3 \supset b=0$.

We shall later investigate approximation techniques to tackle in a constructive way the hard cases where induction is needed to compute the limits of the iterates C_i^n. Let us note that the proposed iterative derivations yield new guards which are the less restrictive ones : only the states causing deadlock are eliminated, nothing more. Finally, it should be recalled that the proposed calculus may serve to merely verify the absence of deadlocks : such a verification amounts to check that the solutions C_i are identical to the given B_i.

3. PREVENTING STARVATION

Starvation occurs for a component P_j when it remains blocked forever while the rest of the program continues. To prevent starvation means to ensure the guard B_j always becomes eventually true, after a finite number of executions of other components ; we assume the implementation ensures that if B_j becomes true an infinite number of times then P_j is executed an infinite number of times. We must thus strengthen the guards of the other components so as to ensure the considered guard is eventually established. We tackle this subproblem first, and then study starvation prevention.

3.1. Guards establishing a post-assertion.

Consider a post-assertion R and a program

$$\begin{aligned} P : \underline{do}\ & P1 : B1 \rightarrow S1 \\ & \| \ldots \\ & \| \ P_N : B_N \rightarrow S_N \\ \underline{od}& \end{aligned}$$

We want to derive new guards such that the new program is guaranteed to establish R after a finite number of steps. Let E_i be the new guards sought for. Again we propose to compute them by iteration :

$$E_i \equiv \lim_{n\to\infty} E_i^n \qquad (n \geqslant 1)$$

An iterate E_i^1 says R established after one application of $P_i : B_i \rightarrow S_i$. Hence

$$E_i^1 \equiv B_i \wedge wp(S_i,R)$$

Let us now consider the union of the iterates of rank n :

$$E^n \equiv E1^n \vee E2^n \vee \ldots \vee E_N^n$$

The idea is to consider E^n as the domain from which R is established after at most n transitions, where a transition is the application of one of the components P_i. Thus the next iterate E_i^{n+1} means that E_i^n holds or that one application of the guarded command P_i establishes the union E^n :

$$E_i^{n+1} \equiv E_i^n \vee B_i \wedge wp(S_i,E^n)$$

But this is dangerous : a state can be included in E_i^{n+1} whereas it was already included in some previous E_i^m with $m \leq n$; this could allow infinite computation cycles made of paths from E_i^{n+1} to E_j^m, which forever avoid R. A simple cure is then to add in E_i^{n+1} only new states which were not yet obtained in any of the $E_1^n, \ldots, E_N^n$:

$$E_i^{n+1} \equiv E_i^n \vee B_i \wedge wp(S_i, E^n) \wedge \neg E^n$$

Notice this safety restriction still preserves the property that all relevant states are included in E^{n+1} :

$$E^{n+1} \equiv E_1^{n+1} \vee E_2^{n+1} \vee \ldots \vee E_N^{n+1}$$

$$\equiv E^n \vee \bigvee_i B_i \wedge wp(S_i, E^n)$$

We do not illustrate this iterative calculus, because it is similar to the one proposed for deadlock prevention.

3.2. Application to starvation prevention

Assume we want to prevent the starvation of P1 in a program

$$P : \underline{do} \;\; P1 \parallel P2 \parallel \ldots \parallel P_N \;\; \underline{od}$$

$$P_i : B_i \rightarrow S_i$$

We apply the rules given in Section 3.1 for deriving new guards E_j, for j=2...N, which ensure $\underline{do}$ $E2 \rightarrow S2 \parallel \ldots \parallel E_N \rightarrow S_N$ $\underline{od}$ eventually establishes B1. Then the new program

$$P' : \underline{do} \; B1 \rightarrow S1 \parallel E2 \rightarrow S2 \parallel \ldots \parallel E_N \rightarrow S_N \; \underline{od}$$

guarantees P1 never starves.

To prevent the starvation of all components, one may apply the same method with respect to each component in turn, until one obtains a program where each guard is always eventually established by the other components. This procedure may appear rather formidable, but it corresponds to the intrinsic difficulty of the problem. Often, symmetry properties may simplify the work quite significantly as in the following example.

3.3. Dining philosophers

We refer to the classical problem as formulated in [3] ; it mirrors a fundamental case of competition for limited resources. The program is made of five parts, the philosophers. Each lives by being hungry, then eating, then thinking, and so on indefinitely. When eating, each philosopher sits at his reserved place at a round table ; he eats spaghetti using two forks ; but only one fork is available between any two neighbours. Clearly there is no deadlock. However starvation may well occur : a given philosopher may remain hungry forever without ever grabbing the two forks he needs for eating, because his two neighbours will behave so as to keep at least one of these forks in their hands, indefinitely.

The problem can expressed in the following program; the variables p_i (i=0,...4) record the state in which the i-th philosopher is, viz. hungry (p_i=1), eating (p_i=2), or thinking (p_i=3) :

P : <u>do</u>

‖ ...

‖ Pi1 : $p_i=1 \rightarrow p_i:=2$

‖ Pi2 : $p_i=2 \wedge p_{i-1} \neq 3 \wedge p_{i+1} \neq 3 \rightarrow p_i:=3$

‖ Pi3 : $p_i=3 \rightarrow p_i:=1$

‖ ...

<u>od</u>

The operations on indices are modulo 5. The guard in Pi2 expresses the requirement that the i-th philosopher P_i may start eating only when his two neighbours are not eating, given the scarcity of forks. Starvation of, say, the philosopher P2 corresponds to the following formal phenomenon: the guard of P22 may remain indefinitely false because $p1=3 \vee p3=3$ can be kept invariant by philosophers P1 and P3 : these may organize their lifes so that at any time at least one is eating.

Let us then prevent the starvation of P22 by applying the technique given in Section 3.2. We strengthen the guards in P0,P1,P3,P4 to ensure these components always eventually establish the guard of P22. The derivations, detailed in [2], yield the following new guards as results :

For P11 : $p1=1 \wedge (p2\neq 2 \vee p3\neq 3)$

For P12 : $p0\neq 3 \wedge p2\neq 3 \wedge (p2=1 \vee p3\neq 3)$

For P31 : $p3=1 \wedge (p2\neq 2 \vee p1\neq 3)$

For P32 : $p2\neq 3 \wedge p2\neq 3 \wedge (p2=1 \vee p1\neq 3)$

The strategy implied by this strengthening is sensible : when P2 is hungry and waits for eating, each of his neighbours may not start eating before the other neighbour finishes eating. This policy has been derived systematically by use of formal deductions, whereas usually such problems are tackled mentally. The resulting program is distributed in the sense that only the guards of the neighbours of P3 have been restricted ; but here this interesting property has been obtained by accident.

A remark concerning the applicability of these iterative calculi is in order here. We also made a number of experiments and exercises for deriving terminating programs with nontrivial algorithmic properties, using for instance the rules in Section 3.2. It appeared that for such applications the proposed iterative calculus is too hard : the inductions needed to express the limits and the simplifications required to obtain propositional terms are often difficult ; this is why weaker, approximated versions of the exact iterative rules are sought for, as shown in Section 6. However, as far as concurrent programming is concerned, the situation is better ; in this case, the sequential, algorithmic aspects are usually easy, and many classical case studies in the area have been tackled successfully with the proposed tools : often, the formal iterates rapidly converge to simple solutions. But even then, human intelligence remains necessary, and it is not clear how full automation could be achieved.

4. INCREASING THE INDEPENDENCE OF COMPONENTS

We now consider how to increase the efficiency of concurrent programs. Techniques for reducing the frequency of unnecessary reevaluations of guards have been proposed in [4,5] ; this is relevant e.g. in the context of "monitor" programs where signaling actions must be derived. Here we address another issue, relevant in distributed programs : how to increase the independence of components in a concurrent program. This improvement implies that the risk of waiting and the need for intercommunications diminish. In other words, the potentiality of parallelism is inversely proportional to the density of interactions such as concurrent use of shared variables or exchange of messages : local activities can be slowed down by too frequent or too long phases of synchronization or communication. We shall thus try to obtain components which compute locally as much as possible, by testing and operating on private variables, and which interact through shared variables as rarely as possible.

In our model, interpedence between components is expressed by accessing common variables ; for instance, a communication channel between two components is represented as a variable shared by these two components. We propose to reduce the interactions in two stages ; see [6] and also [7] . Firstly, we eliminate write-write conflicts caused by updatings of a common variable in two different parts. Secondly, we reduce the write-read interactions represented by the modification of a variable in one component and the reading of its value in another. A constraint on the methods to be proposed is that the order of magnitude of copying of data in the new programs be the same as that in the old programs ; this is why we did not follow the line of applicative programming, for which economical implementations are not yet really available.

4.1. Obtaining write-read conflicts only

A simple-minded method to eliminate the write-read conflicts is to permit the updating of a variable only by one specific component, and to give each other component a local copy of that variable. We thus transform a program P in a new one P' where each variable may be assigned to in only one component but where all components may still read **any** of the variables. Let us represent as follows the program to be transformed ; the identifier x represents the (vector of) program variables :

$$
\begin{array}{ll}
P : \underline{do} & P1 : B1(x) \rightarrow x:=f1(x) \\
 & \| \ P2 : B2(x) \rightarrow x:=f2(x) \\
 & \| \ \ldots \\
 & \| \ P_N : B_N(x) \rightarrow x:=f_N(x) \\
 & \underline{od}
\end{array}
$$

We then try to represent the value of x by the values of N new variables u1, u2,...,u_N, one per component. This requires the discovery of a function T such that

$$x = T(u1,u2,\ldots,u_N)$$

remains invariant. We aim at using this function for defining a transformation of P into a program P' with the new variables :

$$
\begin{array}{ll}
P' : \underline{do} & P1' : C1(u1,\ldots,u_N) \rightarrow u1:=f1(u1) \\
 & \| \ P2' : C2(u1,\ldots,u_N) \rightarrow u2:=f2(u2) \\
 & \| \ \ldots \\
 & \| \ P'_N : C_N(u1,\ldots,u_N) \rightarrow u_N:=f_N(u_N) \\
 & \underline{od}
\end{array}
$$

The intention is that the i-th guard C_i of the new program be satisfied only if the i-th guard B_i of the old program is satisfied, and that the effect on $x=T(u1,\ldots,u_N)$ of the corresponding command $u_i:=f_i(u_i)$ in P' be identical to the effect on x of the command $x:=f_i(x)$ in P. To fulfill these requirements, it is sufficient that

$$C_i(u1,\ldots,u_N) \equiv B_i(T(u1,\ldots,u_N)),$$

while the representation function verifies the following equations for i=1....N :

$$T(u1,\ldots,u_{i-1},f_i(u_i),u_{i+1},\ldots,u_n)=f_i(T(u1,\ldots,u_N))$$

or, in short,

$$T\,[f_i(u_i)/u_i] = f_i(T)$$

Let us illustrate the proposed method on a simple model of the bounded buffer :

```
P : do P1 : x < M → {production of datum ;
                     placement of datum in buffer ;}
                     x:=x+1
     ‖ P1 : x > 0 → {extraction of datum from buffer ;
                     consumption of datum  ;}
                     x:=x-1
    od
```

We need a representation T(u,v) of x such that

$$T(u+1,v) = T(u,v) + 1$$
$$T(u,v-1) = T(u,v) - 1$$

Obviously, T(u,v) = u+v is a satisfactory solution of these equations, and the new program is

```
P' : do P1' : u+v < M → u:=u+1
      ‖ P2' : u+v > 0 → v:=v-1
     od
```

We may interpret u as the total number of portions produced in the buffer, and −v as the total number of portions consumed from the buffer. Thus u+v=u−(−v) gives the number of portions which have been produced but not yet consumed, i.e. the number of portions within the buffer ; this number u+v=x must remain in the interval 0 to M.

We must also remark that, in each new component P_i', the command reads only the value of u_i : the new value of u_i depends thus on the previous value of u_i but on no other value whatsoever. All the interactions are confined in the guards : only there may outside variables be read. This property of P' is used in the next phase.

4.2. Reduction of write-read conflicts.

Whenever components of a concurrent program are evaluated in parallel, the actions on the program variables are combined in unpredictable ways but still within certain safe limits. In the example of the bounded buffer, this space of the game is defined by the size M of the buffer : for a buffer of only one portion, the production and consumption of data portions must strictly alternate, and there is no question anymore of unpredictable combinations of producer and consumer operations.

The idea of the next transformation is from time to time to distribute the available space of the game among the various program components in such a way that each component may be applied a number of times in complete freedom i.e. without any interaction with common variables. As soon as a component has used up his share of the space of the game, global interactions then take place for a redistribution of the space of the game. This idea will be formally expressed by use of auxiliary variables $u1', u2', \ldots, u_N'$, in addition to $u1, u2, \ldots, u_N$; they satisfy two requirements :

(A) Each u_i' represents a future value of u_i, in the sense that u_i' is obtained from u_i by applying $u_i:=f_i(u_i)$ a number of times ;

(B) In each component P_i, the command $u_i:=f(u_i)$ may be freely and safely performed until u_i reaches the value u_i'.

In other words, the future values u_i' must be chosen in such a way that the condition $u_i \neq u_i'$ be sufficient to guarantee the guard $C_i(u1, u2, \ldots, u_N)$ does hold, assuming of course that the other components P_k do not run past their own limits $u_k=u_k'$. As soon as a variable u_i reaches the value u_i' in a component P_i, a new set of future values $u1', u2', \ldots, u'_N$ verifying the requirements (A) and (B) must be computed, and then interactions are needed. Thus, the transformed programs roughly are on the form

$$
\begin{array}{lllll}
P' : \underline{\text{do}} & P1'' & : u1 \neq u1' & \rightarrow & u1:=f1(u1) \\
\| & Pa1 & : u1=u1' \wedge D1(u) & \rightarrow & \text{update } u \\
\| & \ldots & & & \\
\| & P_N'' & : u_N \neq u_N' & \rightarrow & u_N:=f_N(u_N) \\
\| & Pa_N & : u_N=u_N' \wedge D_N(u) & \rightarrow & \text{update } u \\
\underline{\text{od}} & & & &
\end{array}
$$

The variable u stands for the vector $u1, u2, \ldots, u_N$. The conditions

D_i ensure that, given the values $u1,u2,\ldots,u_N$, new future values $u1',u2',\ldots,u'_N$ can be found which verify (A) and (B) and which allow new and safe applications of the i-th component after the redistribution.

The stated requirements can be formally expressed as follows. We introduce a predicate $K(m1,m2,\ldots,m_N)$ telling each command $u_i:=f_i(u_i)$ may be safely applied m_i times :

$$K(m1,\ldots,m_N) \equiv \forall i: 1\leq i\leq N : \forall t_i : 0\leq t_i\leq m_i : [t_i\neq m_i \supset C_i(f_1^{t_1}(u1),\ldots,f_N^{t_N}(u_N))]$$

This definition means that the guard of the command $u_i:=f_i(u_i)$ holds as long as the number of applications of this command is strictly less than m_i whereas the number of applications of each other $u_j:=f_j(u_j)$ is less or equal to m_j. It remains now to explicit in the predicate K the future values $u1',u2',\ldots,u_N'$. These in fact result from the terms $f_i^{m_i}(u_i)$, defined by $f_i^0(u_i)=u_i$, $f_i^{m+1}(u_i)=f_i(f_i^m(u_i))$. The final expression of requirements (A) and (B) is thus the predicate $K'(u1,\ldots,u_N,u1',\ldots,u_N')$:

$$K'(u,u') \equiv \exists\ m1\ldots m_N : m1\geq 0,\ldots,m_N\geq 0 : \forall i : 1\leq i\leq N : [u_i'=f_i^{m_i}(u_i) \wedge K(m1,\ldots,m_N)]$$

In the example of the bounded buffer, the predicates K and K' take a simple and intuitive form :

$$K(m,n) \equiv \forall\ s,t : 0\leq s\leq m \wedge 0\leq t\leq n : [s\neq m \supset u+s+v-t < M] \wedge [t\neq n \supset u+s+v-t > 0]$$

Remark the expression u+s+v-t is maximized by the maximal value of s and the minimal value of t, and it is minimized by the minimal value of s and the maximal one for t. Hence we can simplify K to

$$K(m,n) \equiv [u+(m-1)+v-0<M] \wedge [u+0+v-(n-1)>0]$$
$$\equiv (u+v+m\leq M) \wedge (u+v-n\geq 0)$$

The first conjunct expresses the fact the producer may safely place m portions in the buffer provided m is less or equal to M-(u+v), the number of empty places in the buffer ; similarly the second conjunct says that the consumer may safely take n portions from the buffer where $n \leq u+v$, since u+v gives the number of occupied places in the buffer.

$$K'(u,v,u',v') \equiv \exists m,n : m\geq 0 \wedge n\geq 0 : (u'=u+m) \wedge (v'=v-n) \wedge K(m,n)$$
$$\equiv u'\geq u \wedge v'\leq v \wedge u'+v\leq M \wedge u+v'\geq 0$$

In other words, u may freely grow until it reaches the value u', which is not greater than the size of the buffer plus the number of portions already extracted from the buffer ; similarly, v may safely decrease until v' provided -v' is less than or equal to the number of portions already put in the buffer.

The guards and the commands of the auxiliary components Pa_i can now be precised in terms of K and K' :

$$Pa_i : u_i=u_i' \wedge K(0,\ldots,0,1,0,\ldots,0) \rightarrow$$
$$\text{update } u1',\ldots,u_N'$$
$$\text{so that } u_i\neq u_i' \wedge K'(u1',\ldots,u_N') \text{ holds}$$

We expressed D_i by K where $m_i=1$ and $m_j=0$ for all the other j : this means $u_i:=f_i(u_i)$ should be freely applicable at least once; otherwise it is of no use to update $u1'\ldots u_N'$ in Pa_i. Thus we finally obtain the following transformed program for the bounded buffer :

```
P" : do P1" : u≠u'                 → u:=u+1
      || Pa1 : u=u' ∧ u+v < M → u':=M-v
      || P2" : v≠v'                 → v:=v-1
      || Pa2 : v=v' ∧ u+v > 0  → v':= -u
     od
```

In the guard of Pa1, we have omitted the condition $u+v \geq 0$ because it is invariant. And in the command of Pa1, we could have assigned to u' any value from u+1 to M-v, given the requirements $u' > u$ and $u'+v \leq M$; we decided to choose the maximum M-v, and this does not necessitate an updating of v' which ranges from -u to v. By giving u' the greatest possible value, we give the producer a maximal share of the space of the game without penalizing the consumer ; and by not updating v', we disturb the consumer as little as possible. Thus note the method yields only boundary conditions ; within these boundaries, we still may choose u' and v' in strategically optimal ways.

4.3. Distributed reservation systems.

We develop the illustration of the bounded buffer by tackling two producers and one consumer. This extension can be interpreted as a rudimentary reservation system where two processes, the producers, may reserve places, such as hotel rooms or seats in a plane, and where one consumer process may free previously reserved places. We initially model this system by

$$\begin{aligned} P : \underline{do}\ & P1 : x < M \rightarrow x:=x+1 \\ \| \ & P2 : x < M \rightarrow x:=x+1 \\ \| \ & P3 : x > 0 \rightarrow x:=x-1 \\ \underline{od} & \end{aligned}$$

The variable x counts the places being reserved. The first transformation uses the representation function x=T(u1,u2,v)=u1+u2+v, and yields

$$\begin{aligned} P' : \underline{do}\ & P1' : u1+u2+v < M \rightarrow u1:=u1+1 \\ \| \ & P2' : u1+u2+v < M \rightarrow u2:=u2+1 \\ \| \ & P3' : u1+u2+v > 0 \rightarrow v:=v-1 \\ \underline{od} & \end{aligned}$$

To apply the second transformation, we need the predicates K and K'. We find without much ado

$$\begin{aligned} & K(m1,m2,n) \\ & \quad \equiv u1+m1+u2+m2+v \leqslant M \wedge u1+u2+v-n \geqslant 0 \end{aligned}$$

$$\begin{aligned} & K'(u1,u2,v,u1',u2',v') \\ & \quad \equiv u1' \geqslant u1 \wedge u2' \geqslant u2 \wedge v' \leqslant v \wedge u1+u2+v \geqslant 0 \\ & \qquad \wedge u1'+u2'+v \leqslant M \wedge u1+u2+v' \geqslant 0 \wedge u1+u2+v \leqslant M \end{aligned}$$

Consequently

$$\begin{aligned} & K(1,0,0) \equiv K(0,1,0) \equiv u1+u2+v < M \wedge u1+u2+v \geqslant 0 \\ & K(0,0,1) \equiv u1+u2+v > 0 \wedge u1+u2+v \leqslant M \end{aligned}$$

In Pa1, when u1=u1' and u1+u2+v < M, we must choose new values of u1' and u2' in such a way that u1' > u1, u2' ⩾ u2 and u1'+u2'+v ⩽ M all hold. To this end we divide the available space of the game, namely q=M-(u1+u2+v), as evenly as possible between u1' and u2'. Both producers will thus have the same amount of places available for reservation. When q is odd, in Pa1 the additional place is added to u1', in order to ensure u1' > u1 even if q=1. The command in Pa1 is thus

```
u1':=u1+(q+1)div 2 ;
u2':=u2+q div 2
```

Since q > 0, all the requirements are thereby satisfied. This strategic choice is optimal provided the mean frequencies of the requests by the two producers are equal. When q=1, that is when there remains only one non-reserved place, this place is taken from the second producer and given to the first one. The resulting program is

```
P" : do P1" : u≠u1'  → u1:=u1+1
     ‖ Pa1 : u1=u1' ∧ u1+u2+v < M
                     → u1':=(M+u1-u2-v+1)div 2 ;
                       u2' :=(M-u1+u2-v) div 2 ;
     ‖ P2" : u2≠u2'→ u2:=u2+1
     ‖ Pa2 : u2=u2' ∧ u1+u2+v < M
                     → u1':=(M+u1-u2-v) div 2 ;
                       u2':=(M-u1+u2-v+1) div 2
     ‖ P3" : v≠v'   → v:=v-1
     ‖ Pa2 : v=v' ∧ u1+u2+v > 0
                     → v':=-u1-u2
     od
```

Of course, this program does not appear too surprising once obtained. Yet it would not be much easier first to invent it without the help of design techniques and then to verify it using available proof rules for concurrent programs ; the necessary auxiliary assertions would inevitably play the role of the predicates K and K'. Moreover, the proposed method, albeit limited, does illustrate in precise ways the kind of reasoning which is used for the design of, e.g., distributed data bases : local, safe subspaces of the global games are identified on the basis of the system properties and then used to increase the independent activities or to decrease the interconnections. Last but not least, it goes without saying that rigorous tools are a must, especially in the area of concurrent programming, as once again shown in [8].

4.4. Distributed time.

Events in a distributed program can be synchronized by timestamps based on the ordering of local events [9]. We propose here the use of local dates (or time instants) which are approximated versions of global dates and which preserve certain useful logical constraints. The approximation is thereby kept under control, in the sense that a maximal delay between local dates can be defined. Yet, the updatings of the local dates are distributed. This case study is developed by following the spirit, but not the letter, of the method given in Sections 4.1 and 4.2.

Let be given a concurrent program do S1 ∥ S2 ∥ ... ∥ S_N od where a symmetric relation Next(i,j) between neighbours S_i and S_j is defined. Assume we want to order events in it by using a global date d. To this end we add a component Pd which updates d, and to each S_i we add operations to define the dates $begin_i$ and end_i of the begin and the end of S_i, respectively :

```
P : do Pd : true → d:=d+1
     ∥ P1 : (begin1:=d ; S1; end1:=d)
     ∥ ...
     ∥ P_N : (begin_N:=d ; S_N; end_N:=d)
    od
```

We recall that the additional sequencing within each P_i can be expressed by auxiliary counters if necessary. A sufficient condition ensuring an application E_j of S_j precedes and application E_h of S_h is that the corresponding end_j be strictly less than the corresponding $begin_h$; indeed, the only operation which may transform $d=end_j$ into $d'=begin_h$ such that $d < d'$, is the component Pd : in this case, E_j and E_h are separated by at least one application of Pd. Thus, if the date d is never incremented by Pd, there is no way to use d for ordering events. Conversely, the more often d is incremented, the finer can be the ordering between events.

As introduced, the date d is shared by all the components ; this may be unacceptable in a distributed system. We thus try to distribute that date among the components : first, we replace it by local versions, one per component, and second, we enforce logical properties to ensure some compatibility between these local dates. The result of the first step is

```
P' = do Pd1 : true → d1:=d1+1
      ∥ P1' : (begin1:=d1 ; S1 ; end1:=d1)
      ∥ ...
      ∥ Pd_N : true → d_N:=d_N+1
      ∥ P'_N : (begin_N:=d_N ; S_N ; end_N:=d_N)
     od
```

To define possible interactions between these local dates d_i, we impose the following requirement : the dates local to two neighbouring components are equal or differ by one. Formally,

$$\forall i,j : Next(i,j) : d_i = d_j \vee d_i = d_j \pm 1$$

This requirement is distributed in the sense that it restricts the value of each local date only on the basis of the local dates of the neighbours. Happily, from these local constraints, one readily derives a useful global invariant : the maximal difference between any two local dates is the diameter D of the graph defined by the neighbour relation, viz.

$$\forall i,j = 1 \ldots N : |d_i - d_j| \leqslant D$$

Because of this global property, a sufficient condition ensuring an application E_j of S_j precedes an application E_h of S_h, is

$$end_j + D < begin_h$$

where end_j and $begin_h$ are the local dates of the end of E_j and of the begin of E_h, respectively. To obtain the final program where the local dates are updated consistently, it remains to derive guards for the components Pd_i in order to preserve the invariance of the local requirements stated above. That invariance is secured if, for instance, each d_i is updated only when $d_i \leqslant d_j$ for each neighbour j of i. We obtain thus

P" : <u>do</u> ...
|| Pd_i' : $d_i \leqslant Min\{d_j | Next(i,j)\} \rightarrow d_i := d_i + 1$
|| P_i' : $(begin_i := d_i \; ; \; S_i \; ; \; end_i := d_i)$
|| ...
<u>od</u>

We do not develop here an example where certain specific events in a given application are ordered using the proposed local dates. To tackle this last issue, we should express each test on the ordering between events in terms of conditions $end_j + D < begin_h$.

A few remarks can be made about the proposed technique. If the graph of neighbours is hierarchically structured, the diameter D can be O(log(#vertices)) ; moreover, the local dates could interact through an auxiliary control graph, distinct from the graph of neighbours, and then the value of D can be tuned at will. A more specific optimization can be also considered: to take the distance D_{ij} between i and j as the maximal difference between d_i and d_j, instead of using the diameter everywhere.

The local dates d_i in fact approximate the global date with an "error" D : $d'-d > 0$ is replaced by $d'_h-d_j > D$. In other words, each local date d_i belongs to an interval $[d_i-D, d_i+D]$ of uncertainty : d_h and d_j express distinct global dates if d_h is outside $[d_j-D, d_j+D]$ or symmetrically. The technique can be extended to the case of composed events E and E', respectively defined as sets $\{E1,\ldots,E_M\}$ and $\{E1',\ldots,E'_N\}$ of more basic events. Then E precedes E' if each E_j precedes each E'_h, for j=1...M, h=1...N. A sufficient condition for this is

$$\begin{aligned}&\mathrm{Min}\{d'_h \mid d'_h \text{ is the local date of the begin of some } E'_h\}\\ &> D + \{\mathrm{Max}\ d_j \mid d_j \text{ is the local date of the end of some } E_j\}\end{aligned}$$

5. DESIGN DECISIONS AND DISTRIBUTED TERMINATION

5.1. Composing and specifying design decisions.

The methods for ensuring the absence of deadlock or starvation, or for increasing the independence of components become rather varied. A question then arises : how to organize the application of a number of given techniques in order to ensure the resulting programs do have the required properties. A typical difficulty is that the application of a given method to improve one aspect may invalidate properties obtained by previous design decisions. A possible scheme for composing and expressing program construction steps was presented [10], but still needs study. Instead of restating this tentative scheme here, we shall illustrate the underlying ideas in terms of a case study and thereby show why better definitions and formulations of design decisions do help in understanding program designs, especially in the case of concurrent programs.

5.2. Transformations for distributed termination.

One of the problems that every attempt to construct a distributed program has to face is that of "distributed termination". It is often relatively easy to distribute the global post-condition B(y) of a program P into a conjunction $B_0(y_0) \wedge \ldots \wedge B_N(y_N)$ of local post-conditions B_i on disjoint portions y_i of the global vector $y=(y_0,\ldots,y_N)$, in such a way that the following requirements are satisfied :

T1 : $[\bigwedge_i B_i(y_i)] \supset B(y)$

T2 : Program components $P_0,\ldots,P_N$ can be found so that by means of some finite sequence of communications the concurrent program P : do $P_0 \parallel \ldots \parallel P_N$ od reaches a globally stable state, in which each P_i is locally stable, namely $B_i(y_i)$ holds.

However, in a globally stable state, even though all processes are in their final local states, the whole program is blocked since no process "knows" that the other processes are in final states, and all wait for each other to initiate communications. Since the construction of a program verifying only T1 and T2 may be much easier than the construction of a terminating program with the same post-condition, the availability of a systematic technique which transforms every concurrent program with the described properties into a terminating program would be of much help, since such a transformation relieves the programmer from the possibly hard problem of combining the overall program design with the distribution of termination. Designing such a transformation in terms of some additional interactions, or control communications, is the problem of ensuring distributed termination.

A technique to achieve distributed termination has been developed in [11] . In that development, much attention was paid to the careful composition and explication of the design decisions because otherwise the resulting program schemes are hard to understand as well as to prove.

The first step in the transformation is to introduce so-called interval assertions : these are assertions such as B_i which hold during a given interval of time $[t_i:t_i']$. If there exist intervals of time $[t_i:t_i']$ such that $t_i \leq t \leq t' \leq t_i'$ for some t,t' and i=0,...N, and if $(\bigwedge_i B_i) \supset B$, then we know B must be true during the interval $[t:t']$, and the program may terminate. The further steps of the transformation are to represent the intervals $[t_i:t_i']$ by use of local times, and to express two main control phases : one for collecting the interval assertions as useful, and one for sending to all components either the confirmation or the disaffirmation of termination. Thanks to the reasoning with interval assertions, the "behind the back" phenomen is completely prevented : this logical failure occurs when a process is marked as locally stable and because of some interaction taking place after this marking, becomes locally active again, viz. not stable, but without notifying this change in state to the control process which implements distributed termination. The interested reader may study [11] ; we think better not to elaborate the details here, as they would require a disproportionate amount of space.

6. APPROXIMATED VERSIONS OF ITERATIVE CALCULI

This Section is devoted to an exploration of a difficult but essential research area. At the end of Section 3, we remarked that the iterative calculi proposed for deriving improved guards were in general not applicable outside the area of concurrent programs made of algorithmically simple components. We would like to present here new, yet not final, results in the search for weaker and safe versions of the foolproof, but Platonic, iterative calculi. Previous results were reported in [12].

6.1. Iterative calculus of guards to eliminate dead-ends.

We first present a technique to eliminate dead-ends in programs required to establish a given post-condition. This problem is intermediate between that of preventing deadlocks in non-terminating programs (Section 2) and that of ensuring termination on a given post-condition (Section 3.1).

Let be given a nondeterministic do-loop

```
P : do P1 : A1 → S1
    ‖  P2 : A2 → S2
    ‖  ...
    ‖  P_N : A_N → S_N
    od
```

and a post-assertion R such that $R \wedge A_i$ = false for each i=1...N.

We want to derive a new program

```
T : do T1 : C1 → S1
    ‖  ...
    ‖  T_N : C_N → S_N
    od
```

where each new guard is a possible strengthening of the old one, viz.

$$C_i \supset A_i$$

and such that T never reaches a dead-end, i.e. a state making $\neg R \wedge \bigwedge_i \neg C_i$ true. In other words, the predicate

$$R \vee \bigvee_i C_i$$

must be an invariant of T.

The new guards C_i are easily defined in terms of a natural iterative calculus akin to those given in Sections 2 and 3.1 :

$$C_i \equiv \lim_{n \to \infty} C_i^n$$

$$C_i^1 \equiv A_i \wedge wp(S_i, R)$$

$$C_i^{n+1} \equiv C_i^n \vee A_i \wedge wp(S_i, \bigvee_j C_j^n)$$

Clearly the solutions C_i verify the implications

$$C_i \supset A_i \wedge wp(S_i, R \vee \bigvee_j C_j)$$

which guarantee the invariance of $R \vee \bigvee_j C_j$. Note also the present calculus is the one given in Section 3.1 but without the conjunct $\neg \bigvee_j E_j^n$ introduced to exclude cycles.

6.2. Weaker criteria

The calculus proposed to derive the new guards C_i is, as the others, not suitable for effective computation. Our aim is then to replace these ideal C_i by weaker guards D_i :

$$C_i \supset D_i$$

Yet, these weaker guards should still prevent any dead-end. A simple way to express this requirement is

$$(\bigvee_j C_j) \wedge D_i \supset C_i$$

This means that when $\bigvee_j C_j$ holds each D_i implies the ideal new guard C_i ; thus the weaker guards D_i still preserve the invariance of $R \vee \bigvee_j C_j$, hence of $R \vee \bigvee_j D_j$. The heart of the method is a set of sufficient conditions ensuring the safety requirement given above.

In the following, we shall use the following abbreviations :

$S_j R$ for $wp(S_j, R)$

P^* for $P^0 \vee P^1 \vee P^2 \vee \ldots$

P^{n+1} for $P.P^n$

P^0 for the identity program

PR for $\bigvee_i A_i \wedge S_i R$, when $P \equiv$ do ... $\|$ $A_i \to S_i$ $\|$ ... od

$\tau \in P^n$ for : τ is a composed transition from P^n

We now give a useful lemma, then the main sufficient conditions, and finally a justification. In a first reading, the reader may proceed directly to Section 6.3.

Lemma : If $C_i \supset D_i$ and $D_i \supset A_i$
Then D_i may replace A_i in the derivation of C1, C2, ..., C_N.

The proof is omitted, but the idea is straightforward. Observe C_i itself may replace A_i in the given iterative calculus : if we apply the latter starting from

$$\text{do } C1 \to S1 \; \| \ldots \| \; C_N \to S_N \text{ od}$$

we obtain the same C1, ..., C_N as result. Thus any guard between C_i and A_i can be used as starting point.

Condition CR : For each i,j=1...N such that $i \neq j$: $C_i \supset D_i$, $D_i \supset A_i$:

Case 1 : $\exists k : D_i \wedge D_j \wedge S_j (R \vee TR \vee \ldots \vee T^k R \vee T^{k+1} \text{ true}) \equiv \text{false}$

or

Case 2 : $\exists k : \forall \tau \in T^k : \exists \pi \in P^* : \forall Q :$

$$[\, D_i \wedge D_j \wedge S_j \; \tau \; Q \supset Q \vee S_i \; \pi \; Q \,]$$
$$\wedge \forall n : n \leqslant k : D_i \wedge (\bigvee_j C_j^n) \supset C_i^n$$

Theorem : If Condition CR holds
Then $D_i \wedge (\bigvee_j C_j) \supset C_i$ holds for each i.

Proof : *Case 1* : Let us prove

$$\forall n \;\; \forall i,j : i \neq j : D_i \wedge C_j^n \equiv \text{false}$$

by induction on n.

Basis : $D_i \wedge C_j^1 \equiv D_i \wedge D_j \wedge S_j R$

$\supset D_i \wedge D_j \wedge S_j(R \vee TR \vee \ldots \vee T^k R \vee T^{k+1} \underline{true})$

$\supset \underline{false}$

Induction step : $n \geq 0$: Assume $C^n \equiv R \vee \bigvee_j C_j^n$;

$$D_i \wedge C_j^{n+1}$$

$$\equiv D_i \wedge D_j \wedge (S_j C^n \vee C_j^n)$$

$$\supset D_i \wedge D_j \wedge S_j(T^* R)$$

$$\supset D_i \wedge D_j \wedge S_j(R \vee TR \vee \ldots \vee T^k R \vee T^{k+1}(T^* R))$$

$$\supset D_i \wedge D_j \wedge S_j(R \vee TR \vee \ldots \vee T^k R \vee T^{k+1} \underline{true})$$

$$\supset \underline{false}$$

Case 2 : By induction too. Let us consider k.

Basis : Condition CR ensures $D_i \wedge C_j^n \supset C_i^n$ for $n \leq k$.

Induction Step : $n \geq k$: The induction hypothesis and thesis are respectively

$$\forall i,j : D_i \wedge C_j^n \supset C_i \qquad \text{(IH)}$$

$$\forall i,j : D_i \wedge C_j^{n+1} \supset C_i \qquad \text{(ITH)}$$

We deduce

$$D_i \wedge C_j^{n+1}$$

$$\supset D_i \wedge C_j^n \vee D_i \wedge D_j \wedge \neg C_j^n \wedge S_j(R \vee TR \vee \ldots \vee T^n R)$$

$$\supset C_i \vee D_i \wedge D_j \wedge S_j T^n R \qquad \text{by (IH)}$$

$$\supset C_i \vee D_i \wedge D_j \wedge S_j \tau T^m R \qquad \tau \in T^k,\ n = k+m$$

$$\supset C_i \vee D_i \wedge D_j \wedge S_j \tau C^m$$

$$\supset C_i \vee D_i \wedge D_j \wedge C^m \vee D_i \wedge D_j \wedge S_i \pi C^m \qquad \text{given condition CR, for } \pi \in P^*$$

$$\supset C_i \vee C_i \vee D_i \wedge D_j \wedge S_i C^{m+p} \qquad \text{by (IH)}$$

$$\supset C_i \qquad \text{by Dfn of } C_i$$

□

To ensure $C_i \supset D_i$, we check $A_i \wedge wp(S_i, R \vee \bigvee_j D_j) \supset D_i$

6.3. Summing up

The essence of Condition CR can be sketched as

$$(T_i\ \underline{\text{true}} \wedge T_j\ \underline{\text{true}}) \supset [\, T_j T^k C \supset C \vee S_i P^* C \,]$$

This says that if both T_i and T_j may be applied, then for each path in $T_j T^k$ establishing C there exists a path, possibly empty, in $S_i P^*$ establishing C as well. The theorem can then be pictured as follows, using conventions similar to those in [13] :

If

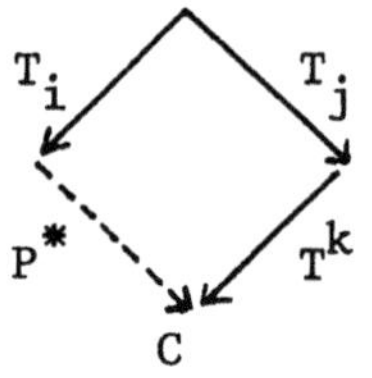

Then

It is in principle easier to derive guards D_i which verify Condition CR, than to compute the solutions C_i as in Section 6.1. However, we are still far from a fully constructive framework, comparable to, say, automatic generation of deterministic parsers for context-free languages. For instance, strategies are needed for finding small values of k for which Condition CR can be satisfied.

As already noted, there is a close relationship between the present approach and the many "Church-Rosser based" studies on rewriting systems. On the other hand, our attempts to discover schemes for feasible and safe approximations essentially differ from the many existing heuristic systems for controlling "knowledge-based" production rules ; it is not certain that such purely empirical techniques will remain acceptable in the long run.

7. DISCUSSION

The simplicity of the illustrations of program design given in the paper should not mislead the reader ; these simple illustrations are in fact classical abstractions of typical problems : simplicity is thus a desirable benefit obtained thanks to successful abstraction. Moreover the programs are derived systematically, and are thus foolproof by construction. They often embody original solutions, as in some of the illustrations and also in applications not developed here. Specific design techniques can be too limited or too hard for some problems, but then the underlying principles may still provide a valuable guidance. For instance, in Section 4.1, the equations which define the representation function from local data to global data may be difficult to solve ; yet, they amount to an algebraic definition of a representation function between datatypes, and this is a much studied area from which ideas can be borrowed. Conversely, the more design techniques become constructive, the more they can be automated. However, given the current state of the understanding in concurrent programming, it is not clear what could be automated, apart from the formal simplification of predicates and expressions or the application of some elementary program transformations.

Other research directions should be at least mentioned. The specification of properties required in concurrent programs may need more specific tools than, say, classical logic : an example is the use of temporal logic [14] where "always" and "eventually" are basic operators. Higher level languages are also studied in the context of concurrent programming ; applicative or stream-oriented languages attract the attention of many, as is well known. Much effort is also devoted to the search for good proof rules for distributed programs made of communicating components ; as in the case of sequential programming, such proof rules could help in the discovery of useful construction rules.

REFERENCES

[1] Dijkstra, A. W., A Discipline of Programming, Prentice-Hall, 1976.

[2] van Lamsweerde, A., and M. Sintzoff, Formal derivation of strongly concurrent programs, Acta Informatica 12(1979), pp. 1-31.

[3] Dijkstra, E. W., Hierarchical ordering of sequential processes, Acta Informatica 1(1971), pp. 115-138.

[4] Sintzoff, M., and A. van Lamsweerde, Constructing correct and efficient concurrent programs, Proc. ACM-IEEE Conf. on Reliable Software, Sigplan Notices 10(1975), N°6, pp. 319-326.

[5] Schmidt, H. A., On the efficient implementation of conditional critical regions and the construction of monitors, Acta Informatica 6(1976), pp. 227-250.

[6] Sintzoff, M., Principles for distributing programs, in : G. Kahn (ed.), Proc. Symp. Semantics of Concurrent Computation, LNCS 70, Springer , pp. 337-347, 1979.

[7] van Lamsweerde, A., Techniques de construction de programmes parallèles fortement corrects, Thèse de Doctorat, Fac. Sci., Univ. de Bruxelles, 1979.

[8] Mc Lean, G., Comments on SDD-1 concurrency control mechanisms, ACM Trans. on Database Systems 6(1981), pp. 347-350.

[9] Lamport, L., Time, clocks and the ordering of events in a distributed system, Comm. ACM 21(1978), pp. 558-565.

[10] Sintzoff, M., Suggestions for composing and specifying program design decisions, Proc. 4th Symp. on Programming, LNCS 83, Springer, 1980, pp. 311-326.

[11] Francez, N., M. Rodeh, and M. Sintzoff, Distributed termination with interval assertions, Proc. Conf. Formal Definition of Programming Concepts, LNCS 107, Springer, 1981.

[12] Sintzoff, M., Ensuring correctness by arbitrary postfixed-points, Proc. 7-th Symp. Math. Foundations of Computer Sci., LNCS 64, Springer, 1978, pp. 484-492.

[13] Huet, G., Confluent reductions : abstract properties and applications to term rewriting systems, J. ACM 27 (1980), pp. 797-821.

[14] Owicki, S., and L. Lamport, Proving liveness properties of concurrent programs, Unpublished document, 1980.

AN ANALYSIS OF CHOICE IN PROGRAM DESIGN

Roland BACKHOUSE

Department of Computer Science
Heriot-Watt University
Edinburgh, Scotland

An attempt is made to identify the most important decisions which arise in the processes of program synthesis and program transformation by analysing in detail the steps involved in designing two classical algorithms - binary search and Dijkstra's shortest path algorithm. The use of non-determinism in the early stages of the development of traditional, sequential programs is advocated as an aid to avoiding the pitfalls of premature decision making. Dynamic programming is then suggested as a technique for evaluating nondeterministic decisions and its use illustrated by the proof of the optimality of binary search in the worst and average cases and by developing an optimal gcd algorithm.

0. INTRODUCTION

Much of the challenge of algorithm design lies in the combination of correctness and efficiency. Over the last decade the work of, most prominently, Dijkstra and Hoare has subordinated the minutae of efficiency in algorithm design to the concerns of correctness; nevertheless efficiency remains an important attribute of any algorithm, and it is the desire to achieve efficiency whilst maintaining correctness which forms the backbone of recent work on program transformation and program synthesis.
The development of an efficient algorithm involves a multitude of decisions some of which are taken consciously, but many of which are not. If we are to be successful in automating the programmer's task we must carefully reexamine the development of classical algorithms looking not only at the alternatives posed and examined but also at the alternatives left untrod and perhaps even unseen by their authors. We must reinvent the wheel again

A. W. Biermann and G. Guiho (eds.), Computer Program Synthesis Methodologies, 79–124.

and again if we are ever to reproduce the decision process which lead to its design. In this series of lectures I want to reexamine two classical algorithms - binary search and Dijkstra's shortest path algorithm [Dijkstra, 1959] - with a view to identifying the many choices involved in their design. The development of binary search will offer no surprises but it does give me the opportunity of reviewing current ideas on program synthesis, the germs for which were sown by Hoare in his development of FIND [Hoare, 1971] and which have matured recently in the excellent texts of Dijkstra [Dijkstra, 1976] and Reynolds [Reynolds, 1981]. The design and implementation of Dijkstra's shortest path algorithm is, in my view, an excellent paradigm of decision making in program development. The aspect which I wish to emphasise most is the implicit decisions taken in its design which have important bearing on its efficient implementation but which went unchallenged for many years.

There is also an ulterior motive for using Dijkstra's algorithm as a paradigm of decision making: it allows me to introduce the basic principle of dynamic programming [Bellman, 1957], the so-called "principle of optimality". In Section 3 of these lectures I shall be exploiting this principle to evaluate some of the choices which arise in program transformation. Dynamic programming is already a well-known and well-tried technique used to solve a variety of problems [Aho, Hopcroft and Ullman, 1974], but my use of it is, I believe, novel - and also speculative. I call it "a calculus for the design of efficient algorithms?", the grandeur of which title is partly mollified by the question mark which indicates that the material is still formative and in need of further investigation.

The theme of these lectures is the role of choice in program design. My aim is to bring to the fore one aspect of Bauer's philosophy of program evolution [Bauer, 1979], namely the use of nondeterministic constructs in the postponement of design decisions. More specifically, I have three messages to convey:

1. Decisions should be taken explicitly, not implicitly.
2. Decisions should be postponed for as long as possible.
3. Nondeterminism should be used as much as possible in the early stages of program design.

My own contribution is an investigation into the feasibility of using a powerful, general technique - dynamic programming - for evaluating some of these decisions.

1. THE DEVELOPMENT OF BINARY SEARCH

1.1 The problem

To begin with let us consider the development of binary search, an algorithm with which you are all, no doubt, already familiar. The problem is this. Suppose one is given a fixed, indexed set of data items k_1 , k_2 , k_3 , ..., k_n and suppose they are ordered by the relation <. In other words,

$$k_1 < k_2 < \ldots < k_n . \tag{1.1}$$

Now, given an item x, the algorithm is to locate x, i.e. to return the index i (if any exists) such that $k_i = x$. Indication should also be made via a Boolean variable found if x cannot be located among K[†].

The main objective of discussing this problem is to review the ideas of program synthesis [Dijkstra, 1976; Reynolds, 1981]. A secondary objective is to highlight decisions which are made implicitly in traditional treatments of this problem.

1.2 Program Synthesis

The first step in synthesising a solution to this problem is a formal specification of its correctness which we can write down straightforwardly as returning the pair (found, i) where

$$\textit{if}\ \text{found}\ \textit{then}\ k_i = x\ \textit{else}\ (\forall j)\ 1 \le j \le n \Rightarrow k_j \ne x \tag{1.2}$$

Clearly, the occurence of ∀ in this specification necessitates a loop of some sort within the algorithm. Synthesising the loop's structure involves the invention of a loop invariant.

To get a handle on the latter problem let us consider a concrete instance of our searching problem, namely looking for a word in a dictionary. Now, in a dictionary each column of entries is normally headed by a single word, a repetition of the first word which appears in that column. By opening the dictionary at random and looking solely at the header it is possible to eliminate a large number of pages from further consideration. Similarly, some dictionaries have a thumb index labelled with a letter of the alphabet use of which allows the majority of the dictionary to be eliminated.

Abstractly, this idea of eliminating portions of the dictionary can be captured by the use of a variable range which is a subset of the set 1..n. At any instant we intend to eliminate from range those indices j for which we are certain that $k_j \ne x$.

Thus, range is the set of indices we have yet to try - if $x = k_i$ then we must have $i \in$ range.

More precisely, we propose the following loop invariant:

$$\textit{if}\ \text{found}\ \textit{then}\ k_i = x\ \textit{else}\ (\forall j)\ 1 \le j \le n\ \textit{and}\ j \notin \text{range} \Rightarrow k_j \ne x \tag{1.3}$$

Comparing (1.3) with (1.2) leads us to a skeleton of an algorithm.

---

† Here and elsewhere we shall use a capital letter to denote a vector of quantities. Thus K denotes (k_1 , k_2 ,..., k_n).

```
begin
{k1 < k2 < ... < kn }
"Achieve invariant";
{Invariant: if  found then ki=x
                       else (∀j) 1≤j≤n and  j ∉ range => kj ≠ x}
while (range ≠ ϕ) and (not found) do
  begin
  "Remove indices from range whilst maintaining invariant"
  end;
{Correctness: if found then ki=x
                       else (∀j) 1≤j≤n ⇒ kj≠x}.
end.
```

Algorithm 1.1 Skeleton

It is obvious how to "achieve invariant": found should be initialised to *false* and *range* to $\{1 \,..n\}$. The heart of the algorithm is, however, the strategy for removing indices from range. One strategy, which does not exploit the ordering on K, is simply to choose some arbitrary index i in range, compare k_i with x and if they are different remove i from range. More efficiently, if $k_i < x$ we know from (1.1) that $k_j \neq x$ for all indices j from 1 up to and including i. Equally, if $k_i > x$ then $k_j \neq x$ for all indices from i to n. Thus we may refine our skeleton algorithm to the following:

```
begin
found: = false; range: = {j | 1≤j≤n}
while (range ≠ ϕ ) and (not found) do
  begin CHOOSE i ε range;
  if ki = x then found : = true
  else if ki < x then remove {j | 1≤j≤i} from range
                 else remove {j | i≤j≤n} from range
  end
end.
```

Algorithm 1.2 Basic Algorithm

Note that in this algorithm we have our first explicit choice. The statement "CHOOSE i ε range" is intentionally nondeterministic because the correctness of the algorithm does not depend on how i is chosen. All that is required to guarantee correctness is that the subsequent statement either finds x or does not find x and removes at least one index from range, which it certainly does.

Although this is the first explicit choice it is certainly not the first choice that has been made. For one thing the chosen invariant is different to, say, that used by Reynolds [Reynolds, 1979]. (More about this later.) Secondly, the condition for terminating the loop is not as straightforward as it may seem. Strictly it is unnecessary to test found and if the line *while* (range ≠ φ) *and* (*not* found) were replaced by *while* (range ≠ φ) then the formal process of program synthesis might produce the following:

```
begin
range : = {j | 1≤j≤n}; found := false;
while range ≠ φ do
  begin CHOOSE i ε range;
  if k_i = x then begin found := true;
                        range := φ
                  end
  else if k_i <x then remove {j | 1≤j≤i} from range
                 else remove {j | i≤j≤n} from range
  end
end.
```

Algorithm 1.3 Alternative Basis

Such differences may perhaps have a miniscule effect upon the efficiency of the algorithm but they can affect its elegance quite significantly. If like Gries [Gries, 1980] (and me) you believe that elegance is important then the mechanics of such decisions is also an important consideration.
Let us now return to the choice of i. It's not difficult to see, firstly, that range is always a complete subrange of 1 ..n (i.e. takes the form $\{j \mid \ell \le j \le u\}$ for some ℓ and u) and, secondly, that the midpoint of range optimises the worst-case performance of the algorithm (again, we shall have more to say about this later). Thus the complete algorithm is shown below:

```
begin
range := {j | 1≤j≤n}; found := false;
while (range ≠ φ) and (not found) do
  begin i:= midpoint (range);
  if k_i = x then found := true
  else if k_i < x then remove {j | 1≤j≤i} from range
                  else remove {j | i≤j≤n} from range
  end
end.
```

Algorithm 1.4 Choice of i

1.3 Choosing Data Structures

We are now at the stage where we may choose data structures. Normally data structures should be chosen by examining the data items, the operations on those data items and the frequency with they occur. Here we need to choose a representation for range, but the choice is very straightforward. Clearly, two integers ℓ and u representing, respectively, the least and greatest elements of range will suffice. With this representation Table 1.1 shows the implementation of each of the operations on range.

Operation	Implementation
range := $\{j \mid 1 \le j \le n\}$	ℓ := 1; u := n;
range $\neq \phi$	$\ell \le u$
midpoint(range)	(ℓ+u) *div* 2
remove $\{j \mid 1 \le j \le i\}$ from range	ℓ := i+1
remove $\{j \mid 1 \le j \le n\}$ from range	u := i-1

Table 1.1 Implementation of range

Finally we show below the transformed algorithm complete with transformed assertions.

```
begin
{k1 < k2 <... < kn}
ℓ := 1; u:= n; found := false;
{Invariant: if found then ki = x
                      else (∀j) 1≤j<ℓ or u<j≤ℓ => kj ≠ x}
while (ℓ≤u) and (not found) do
   begin i := (ℓ+u) div 2;
   if ki = x then found := true
   else if ki < x then ℓ := i+1
                  else u := i-1
   end;
{Correctness: if found then ki = x
                       else (∀j) 1≤j≤n=> kj≠x}
end.
```

Algorithm 1.5 Final Implementation

1.4 Discussion

The foregoing development has been intended to review very briefly the ideas of program synthesis and program transformation. Program synthesis is a calculus for deducing program statements from assertions about the program's function. Beginning with the program specification and with a suitable choice of loop invariant the basic structure of the algorithm can be filled in. Program transformation begins with such an algorithm and transforms it in successive steps to an efficient implementation, each step preserving the correctness of the algorithm but, usually, complicating its structure.

Program transformation is most commonly associated with recursive programs [Burstall and Darlington, 1977; Bauer, 1979; Manna and Waldinger, 1979] but it need not be. The philosophy of beginning with a simple, obviously correct algorithm and transforming it in several easy stages to a complex but efficient version is

a very sound one and is relevant to all programming styles.
I mentioned earlier that my chosen (initial) loop invariant is different to, for example, that used by Reynolds [Reynolds, 1979]. Indeed, contrasting Reynolds' development with my own you will find that he introduces two integers ℓ and u from the outset whereas I have deliberately avoided taking this step until the very last.† This I believe to be important for two reasons. Firstly, the variables ℓ and u are not separable; they have meaning within the program only as a pair. So I think it is important to make this explicit within the development.
Secondly, it seems to me that Reynolds is making a choice prematurely. The more important choice is which index i to try next in the search and the representation of range should be dictated by this decision and not the other way around.
It is interesting to reflect on the number of decisions that have been made along the way in the development of this one simple program. The first was the choice of invariant and the decision to base the loop on removing indices from range on each iteration. The motivation for this decision was to guarantee termination, but Knuth [1973] presents two algorithms very similar to binary search (uniform binary search and Shar's method) which do not always remove indices from range (they may make redundant comparisons) and yet are guaranteed to terminate. A second decision, which I commented on earlier, was the condition for terminating the *while* statement. The third, the method of choosing i, was based on optimising worst-case performance. Equally valid alternatives are used in Fibonacci search and interpolation search [Knuth, 1973]. Finally, the fourth choice was the representation of range. These are the significant decisions; many more are hidden away.
Clearly there an inexhaustible number of branches in the tree of algorithms which stem from one basic algorithm. The question to be asked is "which decisions giving rise to this tree of possibilities can we automate, and how do we go about evaluating them?" However, before we can begin to answer this question we must develop a style of program development which exposes rather than hides the choices available to us. The pitfalls engendered by implicit decision making is the theme of the next section.

2. SHORTEST PATH ALGORITHMS

2.1 The problem

To illustrate in further detail the decision processes in algorithm design I wish to present a development of two shortest path algorithms. The presentation here is very much influenced by

† Reynolds is, of course, not alone; Knuth [1973], for example, does the same, as do many others. I have singled out Reynolds because his presentation of the algorithm is the best I have seen.

Johnson [Johnson, 1977].
The precise formulation of the single-source shortest path problem goes as follows. A labelled graph $G = (X,A,d)$ consists of a finite, non-empty set X of nodes, a set A of arcs and a length function d from A into the set of integers. Each arc a in A is an ordered pair (u,v) where $u,v \in X$.
A path through G from the node u to the node v is a sequence of arcs $(u_1, v_1), (u_2, v_2), \ldots, (u_k, v_k)$ where $u_1 = u$, $v_k = v$ and $v_i = u_{i+1}$ for each i, $1 \le i \le k$. Additionally there is by definition a path, consisting of the empty sequence of arcs, from each node u to itself.
The length function d is extended to paths by definining

$$d((u_1,v_1),\ldots,(u_k,v_k)) = d(u_1,v_1)+\ldots+d(u_k,v_k)$$

d applied to the empty sequence is zero.
A shortest path from u to v is defined to be a path p from u to v such that

$$d(p) = \min_{q \in P(u,v)} d(q)$$

where P(u,v) is the set of all paths from u to v. The single-source shortest path problem is to find the shortest path from a given source node s to each node v in a labelled graph G.

2.2 Computing paths from lengths

Almost invariably, shortest path algorithms work by evaluating shortest distances and then using these to deduce the paths. This process is perhaps best illustrated by a concrete example. Consider the graph in fig. 2.1. This figure uses conventional notation to represent a graph. The nodes are the numbered circles, and arc (u,v) is indicated by a line connecting u and v with the arrow pointing at v, and the length d(u,v) along the arc appears as a label on the line. The source node, in this case node 1, is indicated by an unlabelled arc pointing at the node.

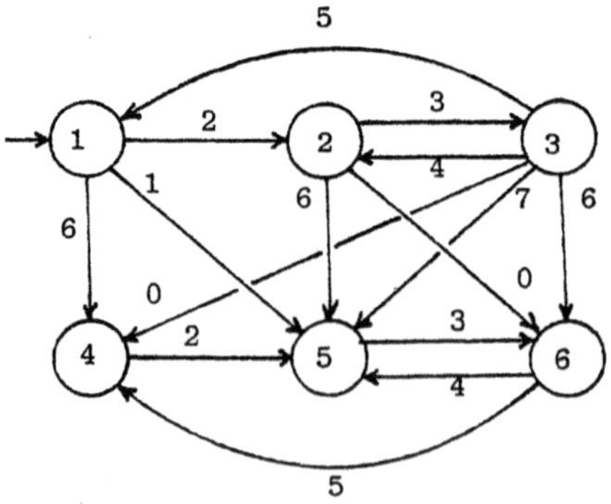

Fig. 2.1 A labelled graph

Let us use y^*_u to denote the length of a shortest path from node 1 to node u in the graph illustrated by fig. 2.1. Then we can write down the following system of simultaneous equations.

$$\begin{aligned}
y^*_1 &= \min\{y^*_3+5, && 0\} \\
y^*_2 &= \min\{y^*_1+2, y^*_3+4, && \infty\} \\
y^*_3 &= \min\{y^*_2+3, && \infty\} \\
y^*_4 &= \min\{y^*_1+6, y^*_3+0, y^*_6+5, && \infty\} \\
y^*_5 &= \min\{y^*_1+1, y^*_2+6, y^*_3+7, y^*_4+2, y^*_6+4, && \infty\} \\
y^*_6 &= \min\{y^*_2+0, y^*_3+6, y^*_5+3, && \infty\}
\end{aligned} \qquad (2.1)$$

The construction of this system of equations is based on the "principle of optimality" of dynamic programming [Bellman, 1957]. Each equation has a right-hand side which is the minimum of a set of terms, one of which is a constant (either 0 or ∞) and the remainder correspond to arcs in the graph. Look, for example, at the equation for y^*_4. This expresses the fact that in fig. 2.1 there are just three arcs to node 4, from nodes 1, 3 and 6. Thus a shortest path from the source to node 4 must consist of either a shortest path from the source to node 1 followed by the arc (1,4), or a shortest path from the source to node 3 followed by the arc (3,4) or a shortest path from the source to node 6 followed by the arc (6,4). The respective lengths of these paths is y^*_1+6, y^*_3+0 and y^*_6+5 and thus y^*_4 is their minimum. The term ∞ is added to be consistent with the term 0 in the equation for the source node, i.e. y^*_1. Here, by definition, there is a path of length 0 from node 1 to itself.
By inspection of fig. 2.1 one may verify that $y^*_1 = 0$, $y^*_2 = 2$, $y^*_3 = 5$, $y^*_4 = 5$, $y^*_5 = 1$ and $y^*_6 = 2$. Knowing these values we can retrace a shortest path from node 1 to each of the nodes. Consider, for example, node 4. We begin by asking which terms on the right-hand side of the equation for y^*_4 are equal to y^*_4.
In fact

$$y^*_4 = 5 = y^*_3 + 0$$

whilst $y^*_4 \neq y^*_1 + 6$ and $y^*_4 \neq y^*_6 + 5$.

We infer that the predecessor of node 4 on a shortest path from node 1 is node 3. Now, examining y^*_3 we have

$$y^*_3 = 5 = y^*_2 + 3$$

That is, the predecessor of node 3 is node 2. Finally,

$$y^*_2 = 2 = y^*_1 + 2$$

Therefore the predecessor of node 2 is node 1. Summarising, the shortest path from node 1 to 4 goes from 1 to 2 to 3 to 4.
What this example shows is that computing paths can be easily reduced to computing path lengths. Accordingly we shall now restrict ourselves to the latter problem.

2.3 Iterative techniques

There are two basic strategies for solving a system of simultaneous equations like (2.1) - we can use an elimination technique or an iterative technique [Carré, 1979]. We shall restrict our attention here to iterative techniques.
An iterative technique solves such equations by finding an approximation to each y_u^* and then repeatedly substituting the approximations into the right-hand sides in order to obtain new approximations. This process is repeated until the right-hand and left-hand sides become equal.
Let us use the notation y_u for an "approximation" to y_u^*.
(We shall give a precise meaning of "approximation" shortly).
Then, a first approximation to (2.1) is

$$y_1 = 0, \; y_2 = y_3 = y_4 = y_5 = y_6 = \infty$$

Substituting these values into the right-hand sides we obtain the new approximations:

$$y_1 = \min\{\infty+5, \qquad 0\} = 0$$

$$y_2 = \min\{0+2, \infty+4, \qquad \infty\} = 2$$

$$y_3 = \min\{\infty+3, \qquad \infty\} = \infty$$

$$y_4 = \min\{0+6, \infty+0, \infty+5, \qquad \infty\} = 6$$

$$y_5 = \min\{0+1, \infty+6, \infty+7, \infty+2, \infty+4, \qquad \infty\} = 1$$

$$y_6 = \min\{\infty+0, \infty+6, \infty+3, \qquad \infty\} = \infty$$

Various strategies can now be adopted for deciding when or when not to substitute a new approximation into the right-hand side of the equations. One strategy (the Jacobi method [Carré, 1979]) is to substitute <u>all</u> the new approximations into the right-hand sides. Other methods (for example the Gauss-Seidel method [Carré, 1979] and a method due to Yen [Yen, 1970]) are not so crude and more efficient. Each takes the general form given by algorithm 2.1.

```
begin
for all u ε X do y_u := ∞;   y_s := 0;
{Invariant: (∀u) y_u < ∞ => ∃ path p in G
            from s to u with d(p) = y_u}
while ∃ (u,v)εA s.t. y_v > y_u + d(u,v) do
   begin
   CHOOSE u ε X;
   for each arc (u,v) do y_v := min{y_v,y_u + d(u,v)}
   end;
{Correctness: (∀u) y_u = y*_u}
end.
```

Algorithm 2.1 Basic iterative technique

Strictly, of course, this is not an algorithm since it is not guaranteed to terminate. For one thing if the graph G contains cycles of negative length (a path p from some node u to itself with $d(p) < 0$) the problem is not well-formed and all the algorithms referred to earlier will fail. Additionally we need to be more specific about choosing u before we can claim termination. Nevertheless the algorithm is partially correct. The invariant added to it guarantees that the y-values are path lengths and not completely arbitrary. The invariant therefore formalises what we mean by saying that y_u is an "approximation" to y^*_u. It also has the important implication that if and when the algorithm terminates the y-values not only form a solution to the given equation set but are, indeed, lengths of shortest paths - it is well-known that equation sets like (2.1) may have more than one solution and we need to know that the obtained solution is the appropriate one!

2.4 A possible refinement

The condition for terminating the *while* statement in algorithm 2.1 gives a clue to a possible refinement of the algorithm. Consider the property R defined on nodes by

$$R(u) \equiv \exists\ (u,v) \in A \text{ such that } y_v > y_u + d(u,v)$$

Then it is clearly a waste of time to choose a node u not satisfying R(u) because for such nodes the subsequent *for* statement would have no effect. Moreover, the condition for terminating the *while* statement is equivalent to

$$\mathcal{R} \equiv \{u \mid R(u)\} = \phi$$

Our goal, therefore, is to be able to compute $\mathcal{R}$ or at least an approximation to $\mathcal{R}$.

Some headway can be made if we restrict our attention to the simpler requirement of computing a set U where

$R(u) \Rightarrow u \in U$

Specifically U can be initialised to {s}. A node v need only ever be added to U if y_v decreases and (assuming G contains no negative cycles) u can be removed from U once all arcs from it have been processed as shown in algorithm 2.2.

```
begin
{Assumption: G does not contain negative cycles}
for all u ε X do y_u := ∞; y_s := 0;
U := {s};
{Invariants: (1)  (∀u) y_u < ∞ => ∃ path p in G
                          from s to u with d(p) = y_u}

             (2)  ∃ (u,v)εA such that y_v>y_u+d(u,v) =>uεU}
while U ≠ φ do
   begin
   CHOOSE u ε U;
   for each arc (u,v) do
        if y_v>y_u+d(u,v) then begin y_v:= y_u+d(u,v);
                                U := U∪{v}
                                end;
   U := U-{u}
   end;
{Correctness: (∀u) y_u = y*_u}
end.
```

Algorithm 2.2 A choice of invariant

2.5 Dijkstra's Algorithm

Algorithm 2.2 is still not very satisfactory. Its total correctness can be established although termination is not at all obvious. However its running time can be exponential in the number of nodes in G [Johnson, 1977]. Real progress can only be made if we restrict our attention to graphs in which the arc lengths are all non-negative.

Consider again the graph in fig. 2.1. Having initially chosen s (node 1) and used it to obtain the new approximations $y_1 = 0$, $y_2 = 2$, $y_3 = \infty$, $y_4 = 6$, $y_5 = 1$ and $y_6 = \infty$ it seems reasonable to choose node 5 next. For, we may argue, y_1^* is certainly zero and $y_5 = y_5^* = 1$ because y_5 is the smallest of the new approximations, excluding y_1 - clearly, therefore, it can never decrease. In other words the values of y_1 and y_5 are permanent whilst the values of y_2 , y_3 , y_4 and y_6 may be temporary. The conjecture, which turns out to be correct when arc lengths are non-negative, is that any node u with the property that y_u is the least value among the temporary values can be chosen and declared permanent. This is the essence of the well-known shortest path algorithm due to Dijkstra [Dijkstra, 1959] which we show below.

In this algorithm permanent is a set of nodes u such that $y_u = y^*_u$. That is, the value y_u is no longer an "approximation" to the length of a shortest path from s to u but is the exact length. Temporary is always just the difference between X and permanent.

```
begin
{Assumption: (∀(u,v)εA) d(u,v)≥0}
for all uεX do y_u := ∞; y_s := 0;
permanent := φ; temporary := X;
{Invariants: (1) (∀u) y_u <∞ => ∃ path p in G
                    from s to u with d(p) = y_u

             (2) u ε permanent => y_u = y*_u.}
while temporary ≠ φ do
   begin
   choose u ε temporary such that w ε temporary
          => y_u ≤ y_w;

   permanent := permanent ∪ {u};
   temporary := temporary - {u};
   for each arc (u,v) do
          y_v := min{y_v,y_u+d(u,v)}
   end;
{Correctness: (∀u)  y_u = y*_u}
end.
```

Algorithm 2.3 Dijkstra's algorithm

One further refinement is added in many implementations of Dijkstra's algorithm. Specifically, it can be more efficient to compute two sets reached and frontier where frontier is the subset of temporary consisting only of those nodes u with $y_u < \infty$, and reached is the union of permanent and frontier. (The terminology here is intended to highlight the close correspondence between Dijkstra's algorithm and an elementary graph search [Backhouse, 1979, chap.1]). The computation of reached and frontier is straightforward as shown below. Note that permanent is now redundant since it equals reached - frontier; it has been retained for greater clarity.

```
                                                                   Frequency
      begin
      {Assumption: (∀(u,v) ε A) d(u,v) ≥ 0}
      for all u ε X do y_u := ∞; y_s := 0;                                  1
      permanent := φ; reached := frontier := {s};                           1
      {Invariants: (1) (∀u) uεreached => ∃ path p in G
                                from s to u with d(p) = y_u

                   (2) u ∉ reached => y_u = ∞
                   (3) u ε permanent => y_u = y*_u

                   (4) frontier = reached - permanent}
      while frontier ≠ φ do                                                ≤n+1
         begin
choose:  choose u ε frontier such that w ε frontier => y_u≤y_w;             ≤n
         permanent := permanent ∪ {u};                                      ≤n
         frontier := frontier - {u};                                        ≤n
         for each arc (u,v) do                                              ≤n
           if y_v > y_u + d(u,v)                                            ≤α
           then begin
                 if v ∉ reached                                             ≤α
                 then begin reached := reached ∪ {v};                      ≤n-1
                      frontier := frontier ∪ {v}                           ≤n-1
                      end;
update lengths:  y_v := y_u + d(u,v)                                        ≤α
                 end
         end;
      {Correctness: (∀u) y_u = y*_u}
      end.
```

Algorithm 2.4 A refinement of Dijkstra's algorithm and its worst-case analysis.

2.6 Choosing data structures

An analysis of Dijkstra's algorithm is very promising. Each statement of algorithm 2.4 has been annotated with an upper bound on the number of times it is executed using the notation

n = number of nodes in G, α = number of arcs in G.

The key property of the algorithm, which enables us to make this analysis, is that each node is added to reached at most once since no nodes are ever removed from reached and whenever one is added we check first that it is not already there. Now nodes are only ever added to frontier when they are added to reached; thus each node can be chosen at most once - i.e. the loop body is executed at most n times. The rest of the analysis follows easily from these observations.

The significance of this analysis is that Dijkstra's algorithm is basically linear time. Unfortunately this splendid property is destroyed when we begin to consider data structures which will

implement efficiently the frontier set. The problem is that there is a dichotomy between the efficient implementation of the statement labelled choose and that labelled update lengths.
One solution, adopted for example by Yen [Yen, 1972], is to use an elementary data structure like an array, stack or linked list to store elements of frontier. With such a choice the implementation of update lengths remains as it is in the algorithm; however, the location of the minimal value in the statement choose involves searching, and possibly rearranging a list of $O(n)$ elements. As a consequence, execution of the entire algorithm incurs $O(n^2)$ elementary operations. It is for this reason that Dijkstra's algorithm is often branded as an $O(n^2)$ algorithm.
A second, less attractive solution is to store frontier nodes in an ordered data structure, for example a linked list in which if u is next to v then $y_u \leq y_v$. The motivation for this is that choosing the minimal element involves only a constant number of elementary operations. Unfortunately this has a repercussion on update lengths in that each time a y-value is altered it becomes necessary to resort the data structure. Typically, frontier may have $O(n)$ elements and update lengths is executed $O(\alpha)$ times. Thus, if the graph is full - i.e. α is $O(n^2)$ - execution of the entire algorithm incurs $O(n.n^2) = O(n^3)$ operations. In any case it requires $O(n.\alpha)$ operations which will almost invariably be worse than the $O(n^2)$ operations required by an unsorted structure.
In fact, frontier is an instance of a commonly encountered object - namely, a priority queue - and on this account much effort has gone into analysing its implementation. A priority queue is a set of real numbers (the priorities) on which the following operations are performed

(i) insert new elements

(ii) find and delete an element with maximum (or minimum) priority in the queue

(iii) alter the priority of a specified element.

Johnson [Johnson, 1977] has suggested the use of a heap to store frontier. A heap (in the sense of J.W.J. Williams [Williams, 1964]) is a full binary tree ordered on each path from the root. Using it, choosing a node requires $O(\log_2 n)$ (the height of the tree) elementary operations and, also, update lengths requires $O(\log_2 n)$ elementary operations. Thus, in total, the worst-case analysis becomes $O(n.\log_2 n + \alpha.\log_2 n)$. Johnson [Johnson, 1975] also discusses the use of β-ary trees to form the heap structure: choosing a node then costs $O(\beta \log_\beta n)$ and update lengths costs $O(\log_\beta n)$ elementary operations. By setting $\beta = \lceil n^{1/k} \rceil$, for some constant k, the algorithm assumes a $O(n^{1+1/k} + \alpha)$ worst-case running time.
Finally Vuillemin has investigated the use of binomial forests to implement priority queues [Vuillemin, 1978]. His paper also includes a summary and bibliography of the many applications in which priority queues are to be found.
Table 2.1 summarises the discussion in this section. Its significance is that there is no "perfect" implementation of frontier.

Each has the effect of weighting either the statement choose or the statement update lengths and it seems impossible to reconcile the two.

Implementation Of Frontier	Number Of Elementary Operations In One Execution Of Choose	Number Of Elementary Operations In One Execution Of Update Lengths	Total Number Of Elementary Operations
linked list	$O(n)$	$O(1)$	$O(n^2 + \alpha)$
sorted linked list	$O(1)$	$O(n)$	$O(n + n \cdot \alpha)$
binary heap	$O(\log_2 n)$	$O(\log_2 n)$	$O(n \cdot \log_2 n + \alpha \cdot \log_2 n)$
$\lceil n^{1/k} \rceil$-ary heap	$O(n^{1/k})$	$O(1)$	$O(n^{1+1/k} + \alpha)$
Binomial forest	$O(\log_2 n)$	$O(\log_2 n)$	$O(n \cdot \log_2 n + \alpha \cdot \log_2 n)$

Table 2.1 Implementations of frontier

2.7 Starting again - Wagner's algorithm

We have seen in the previous section that Dijkstra's algorithm poses a problem in its final implementation. The only way to resolve this problem is to begin again our development of a shortest path algorithm, making use of our increased familiarity with the problem but looking for decisions we have made which can profitably be undone.

The algorithm I am about to describe is due to Wagner [Wagner, 1976] and, I believe, was suggested to him by his work on error correction [Wagner, 1974; Wagner and Fischer, 1974], for which application it is well suited. I shall, therefore, use a graph typical of those arising in spelling correction [Backhouse, 1979] to provide an intuitive understanding of the algorithm.

Consider, then, the graph in fig. 2.2. In this graph the nodes are the small dots, the source node being that in the upper left-hand corner. All the arcs have length 0 or 1; the dotted arcs have zero length and the remainder have unit length. Let us imagine that we apply Dijkstra's algorithm to this graph; in what order

we will choose nodes?
Initially (see fig. 2.3), beginning at the source node we will find just two nodes at distances zero from the source. Then our search will widen as we successively choose all nodes at distance 1 from the source. Next we will consider all nodes at distance 2, then at distance 3 and, finally, at distance 4 from the source.

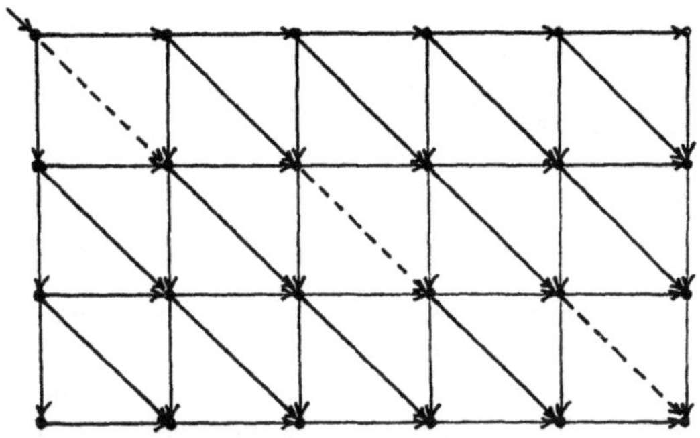

Fig. 2.2 A repair graph - solid arcs have length 1, dotted arcs have length 0.

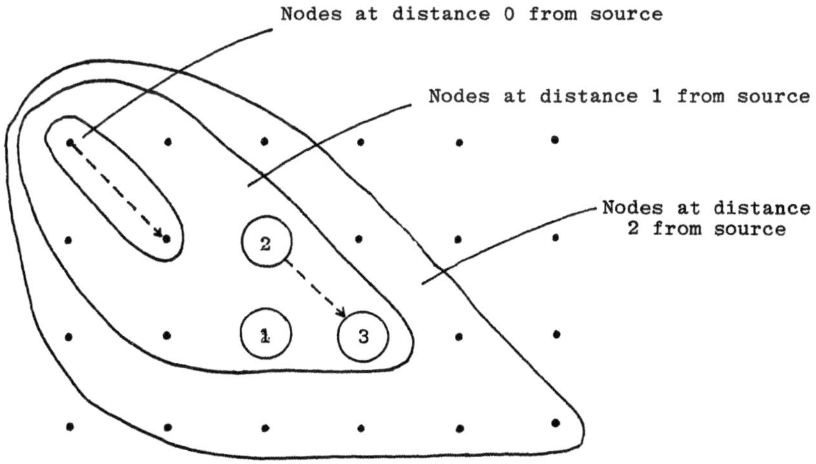

Fig. 2.3 Growth of frontier

This discussion suggests that we try to put nodes into a pigeon hole - the zeroth, 1st, 2nd etc. In other words, we should use radix sorting [Knuth, 1973] in our implementation of choose. This isn't easy, though, because nodes may be moved from one pigeon hole to another. Consider, for example, the node labelled 3 in fig. 2.3. If node 1 is chosen before node 2 then node 3 will at first be put into the 2nd pigeon hole (i.e. $y_3 := 2$) but later, when node 2 is chosen, it must be moved to the 1st pigeon hole (i.e. $y_3 := 1$). On the other hand, it is easy to use pigeon holes if we stop thinking and talking about nodes and think instead about arcs.
Let's look again at our most basic shortest path algorithm (algorithm 2.1 in section 2.3). The loop body in this algorithm consists of the two statements:

CHOOSE $u \in U$;
for each arc (u,v) *do* $y_v := \min\{y_v, y_u + d(u,v)\}$

Now the point I wish to emphasise, and emphasise most strongly, is that these two statements embody a decision which has been the cause of all our implementation difficulties. The decision is one of the worst of all; it has been made implicitly and unwittingly.
The decision that has been made is to process all the arcs from a given node u together. A more general and more flexible approach is to choose arcs independently, thus making the choice of arcs rather than the choice of nodes our central consideration (algorithm 2.5).

begin
for all $u \in X$ *do* $y_u := \infty$; $y_s := 0$;
{*Invariant*: $(\forall u)\ y_u < \infty \Rightarrow \exists$ path p in G from s to u with $d(p) = y_u$}
while $\exists\ (u,v) \in A$ s.t. $y_v > y_u + d(u,v)$ *do*
    begin
    CHOOSE $(u,v) \in A$;
    $y_v := \min\{y_v, y_u + d(u,v)\}$
    end;
{*Correctness*: $(\forall u)\ y_u = y_u^*$}
end.

Algorithm 2.5 Basis of Wagner's algorithm

The insight which is the basis of Wagner's algorithm is that a distance can be associated with each arc. This distance is called the head-distance of the arc and is defined by

$$\text{head-distance}((u,v)) = y_u + d(u,v)$$

Using the same reasoning as we did in the development of Dijkstra's algorithm we can argue that the chosen arc should be the one having minimum head-distance among the frontier arcs - those that have been reached but not yet chosen. Thus we arrive at the following algorithm.

```
begin
{Assumption:  (∀(u,v)εA) d(u,v) ≥ 0}
for each uεX do y_u := ∞; y_s := 0;
reached := {s}; frontier arcs := ϕ;
for each arc (s,u) from s do
     begin frontier arcs := frontier arcs ∪ {(s,u)};
     head distance((s,u)) = d(s,u)
     end;
{Invariants: (∀u) uε reached => ∃ path p from s to u
                              with d(p) = y_u
             (u,v) ε frontier arcs => y_u = y*_u and
                               head distance((u,v)) = y*_u + d(u,v)}
while frontier arcs ≠ ϕ do
   begin
   choose (u,v) ε frontier arcs such that (w,x) ε frontier arcs
         => head distance((u,v)) ≤ head distance((w,w));
   frontier arcs := frontier arcs - {(u,v)};
   y_v := min{y_v, head distance((u,v))};
   {Use new value of y_v to update head distances and frontier arcs}
   if v ∉ reached
   then begin reached := reached ∪ {v};
        for each arc (v,w) εA do
            begin frontier arcs := frontier arcs ∪ {(v,w)};
            head distance((v,w)) := y_v + d(v,w)
            end
        end
   end
{Correctness: (∀u)  y_u = y*_u}
end
```

Algorithm 2.6 The choice of arcs in Wagner's algorithm

The leap we have made in going from algorithm 2.5 to algorithm 2.6 is a large one but the details are not important to this discussion. Essentially the same steps have been used to develop algorithm 2.6 from algorithm 2.5 as were used to develop algorithm 2.4 from algorithm 2.1. The difference is that our attention is now concentrated on arcs rather than nodes. Thus in algorithm 2.4 we choose and remove nodes from the frontier set of nodes, now in algorithm 2.6 we choose and remove arcs from frontier arcs. In algorithm 2.4 a node is added to frontier at most once, in algorithm 2.6 an arc is added to frontier arcs at most once. Finally, in algorithm 2.4 the chosen node is that having minimal y value among the frontier nodes, in algorithm 2.6 the chosen arc is the one having minimal head distance among the frontier arcs.

One more observation needs to be made before we have fully described Wagner's algorithm. Returning to our discussion of the use of Dijkstra's algorithm on fig. 2.2 you will have noted that, if at any instant a node u is chosen, then all frontier nodes will be at a distance y_u or y_u+1 from the source. Thus, when a node u is chosen at zero distance from the source, all frontier nodes will be at distance 0 or 1 from the source. When we move on to choosing a node at distance 1 from the source all frontier nodes will either be also at distance 1 or at distance 2 from the source. And so it goes on.
The same observation can be applied to frontier arcs in Wagner's algorithm. When applied to our example graph, if an arc (u,v) is chosen the head distance of all other frontier arcs will be the same as or one more than the head distance of the arc (u,v).
Generally, if all arc lengths are non-negative integers and their maximum is maxd then the frontier arcs will all have head distance lying in the range curd to curd+maxd where curd is the minimum head distance of the frontier arcs.
The implication of the latter observation is that frontier arcs can be stored in one of maxd + 1 pigeon holes - if an arc has head distance h then it is stored in the (h *mod* (maxd + 1))th pigeon hole. Choosing an arc can then be achieved by looking successively in the curd *mod* (maxd + 1)th, (curd + 1) *mod* (maxd + 1)th, (curd + 2) *mod* (maxd + 1)th,.... holes until a non-empty one is found. Additionally, frontier arcs is empty if an only if all (maxd + 1) holes are empty.
This then is the complete description of Wagner's algorithm.

```
{Assumption: (∀(u,v) ε A) 0 ≤ d(u,v) ≤ maxd and d(u,v)
                                is integral}
var pigeon hole : array[0..maxd] of set of arc;
   procedure add((u,v): arc);
   begin {head distance((u,v)) := y_u + d(u,v)}
   h := (y_u +d(u,v)) mod (maxd + 1);
   pigeon hole [h] := pigeon hole[h] ∪ {(u,v)};
   end;
begin
for each u ε X do y_u := ∞; y_s := 0;
curd := 0; initialise all (maxd+1) pigeon holes to φ;
reached := {s};
for each arc (s,u) from s do add ((s,u));
k := 0; {number of times a pigeon hole has been found empty}
while k < maxd+1 do
   begin
   {choose a frontier arc}
   if pigeon hole[curd mod (maxd+1)] = φ
   then begin {try another pigeon hole}
        k := k+1; curd := curd +1;
        end
   else begin choose and remove arbitrary arc (u,v)
              in pigeon hole[curd mod (maxd+1)];
        {curd = head distance((u,v))}
        k := 0;
        y_v := min{y_v, head distance((u,v))};

        if v ∉ reached
        then begin reached := reached ∪ {v};
             for each arc (v,w)ε A do add((v,w))
             end
        end
   end
end.
```

Algorithm 2.7 Implementation of Wagner's algorithm

2.8 Analysis and summary

Wagner's shortest path algorithm is successful in avoiding the pitfall's of Dijkstra's algorithm provided that all arc lengths are small positive integers. A straightforward worst-case analysis of it [Wagner, 1976] shows that it runs in linear time; its drawback is that choosing an arc requires up to $O(\underline{maxd} + 1)$ elementary operations thus contributing a factor of $O(\overline{\alpha.\underline{maxd}} + 1))$ to the total running time. The term $(\underline{maxd} + 1)$ is, of course, negligible in cases like our example graph when all arc lengths are 0 or 1.

The moral of this development is that it is very easy to take a decision without realising it. Many, including myself, have failed to consider the possibility of choosing arcs rather than nodes when examining the deficiencies of existing shortest path algorithms. It only goes to show that sometimes one must begin one's design again from the very beginning if one is to overcome difficulties encountered along the way.

An analysis of Dijkstra's algorithm offers some clues which might suggest Wagner's approach. One such is that each arc is used at most once in update lengths (algorithm 2.5). Thus if we associate a distance with each arc it is set once only unlike the distances associated with nodes which are continually being decreased. But the most important clue to Wagner's algorithm comes from the application to which it is most suited - spelling correction - when all arc lengths are either 0 or 1. This should be taken as a warning to those who like to take abstraction to its limits. Very often a specific application can offer deep intuitive insights which it is foolish to ignore.

3. DYNAMIC PROGRAMMING: A CALCULUS FOR THE DESIGN OF EFFICIENT ALGORITHMS?

3.1 Introduction

The message of the two previous sections is one we all know well already: program design is a demanding and complex activity involving many decisions at every stage. It seems difficult to even begin to classify the decisions we have to make, let alone quantify and hence evaluate those decisions. Nevertheless, progress has been made in this direction in recent years. It is now well recognised that the data in programs are not characterised solely by the range of values they may take but also by the operations on those values and the frequency with which they occur [Hoare, 1972]; choosing data structures therefore proceeds accordingly. The development of program proving techniques [Hoare, 1971; Dijkstra, 1976; Reynolds, 1981] has also made it clear to us that the choice of loop invariant is of particular importance to the structure of any algorithm. In between these two there is a complex decision process involved in the transformation of a simple, correct but inefficient algorithm to a complex, correct and efficient algorithm, the science of which is still far from complete but which is under active development.

In this section I want to present some preliminary ideas I have on the use of dynamic programming to assist in evaluating decisions arising in program transformation. As with all such ideas they can only claim to be the tip of an iceberg; my ideas apply to situations in which a basic program structure has been developed which specifies nondeterministic choices of action - choices which do not affect the program's partial correctness but which do affect its efficiency and possibly its total correctness. The next section, which describes two problems of choice, will convey rather better the flavour of application I have in mind.

It's worth interposing a word on terminology at this point to clear up any confusion. The term "dynamic programming" was coined by Bellman [Bellman, 1957] to describe a technique for evaluating decisions in multi-stage processes (for example, producing olefines in a chemical plant). The use of "programming" in this term has nothing to do with computer programming and, nowadays, is somewhat unfortunate - a better term now would be, say, "dynamic planning".

3.2 Two problems

3.2.1. Binary search again

Let us return to the problem of section 1 again. That is, we are to develop an algorithm which given an ordered set of data items $k_1, k_2, \ldots, k_n$ and an additional item x will find that index i such that $k_i = x$.

Suppose that we have progressed somewhat in our development of the algorithm and have reached a point represented by algorithm 3.1 below. The feature of this algorithm which we wish to discuss is the statement:

$$\text{CHOOSE } i \in \{\ell..u\}$$

How should i be chosen?

```
begin ℓ := 1; u:= n; found := false;
     while (ℓ≤u) and (not found) do
        begin
        CHOOSE i ε {ℓ..u};
        if ki = x then found := true
        else if ki<x  then ℓ := i+1
                      else u := i-1
        end;
end.
```

Algorithm 3.1

3.2.2 Greatest common divisors

Our second problem concerns finding greatest common divisors (a well-worn problem in talks like this!) and our starting point is the first step in Dijkstra's development of a gcd algorithm [Dijkstra, 1976]. Basically, Dijkstra observes that the greatest common divisor of two integers obeys a number of simple identities. Specifically, we have

$$
\begin{aligned}
\gcd(x,y) &= \gcd(y,x) \\
&= \gcd(x+y,y) = \gcd(x,x+y) \\
&= \gcd(x-y,y) = \gcd(x,x-y) \\
&= \gcd(y-x,y) = \gcd(x,y-x) \\
&= \gcd(-x,y) = \gcd(x,-y)
\end{aligned}
$$

Dijkstra therefore proposed basing his algorithm on one similar to the following†. The problem is how to evaluate the choice of operation in this algorithm. Note that its successful termination depends on this choice (and so, strictly, it should not be called an algorithm). This is unlike algorithm 3.1 which always terminates whatever the choice of i.

```
begin  x := a;  y := b;
while (x≠0) and (y≠0) do
    begin
        CHOOSE:
           (interchange(x,y);
           x := x+y;
           y := x+y;
           x := x-y;
           y := x-y;
           x := y-x;
           y := y-x;
           x := -x;
           y := -y
          )
    end
{Correctness: (x≠0) => gcd(a,b) = |x| and
              (y≠0) => gcd(a,b) = |y| }
end.
```

Algorithm 3.2 Computing gcd(a,b)

3.3 Equational characterisation of the optimal choice

Before we can evaluate the choice of i in algorithm 3.1 or the choice of operation in algorithm 3.2, we must agree on suitable criteria. Usually, algorithms are evaluated according to their worst-case or average-case running times and so our intention

† To avoid minor details we assume $\gcd(a,0) = |a|$ for all a; in particular, $\gcd(0,0) = 0$.

is to show how dynamic programming can be used to optimise the performance of algorithms 3.1 and 3.2 with respect to these criteria. It is important to note, however, that the techniques we describe are not restricted in the choice of criterion. Indeed, very few algorithms are known which truly optimise worst-case or average-case running times and, in practice, it is necessary to resort to much cruder criteria. We shall have more to say about this in sec. 3.5.

Let us begin by considering the gcd algorithm (algorithm 3.2). In one respect the choice to be made in the gcd algorithm is simpler than the choice in the searching algorithm viz. there are no *if* statements in algorithm 3.2 and hence, unlike algorithm 3.1, there is no distinction between its worst-case and average-case performance. However, as we shall see, in another respect the choice in the gcd algorithm is much more difficult than that in algorithm 3.1.

Suppose we endeavour to choose operations from the list given in algorithm 3.2 so as to minimise the total number of such operations used in the calculation of gcd(a,b). Let t(x,y) denote the minimum number of operations required to find gcd(x,y). Then, referring to algorithm 3.2, we have:

$$t(x,0) = t(0,y) = 0 \text{ for all } x,y \tag{3.1}$$

$$\begin{aligned} t(x,y) = 1 + \min\{&t(y,x),\ t(x+y,y),\ t(x,x+y),\\ &t(x-y,y),\ t(x,x-y),\ t(y-x,y),\ t(x,y-x),\\ &t(-x,y),\ t(x,-y)\}\\ &\text{otherwise}\end{aligned} \tag{3.2}$$

The rationale behind Eqs. (3.1) and (3.2) is as follows. Eq. (3.1) is trivial: if x or y is zero the gcd of x and y can be output immediately. Now consider (3.2). If both x and y are non-zero algorithm 3.2 certainly needs at least one operation to evaluate their gcd - hence the rhs of (3.2) is 1 + <something>. After this operation the new values (x',y'), say, of x and y are either (y,x), (x+y,y), (x,x+y),...,(x,-y) depending on the choice of operation. Now, we argue that if the algorithm is to evaluate gcd(x,y) optimally then it must also evaluate gcd(x',y') optimally. (This is an important observation and is another instance of the <u>principle of optimality</u>.) Thus t(x,y) = 1 + t(x',y'). The final step is to observe that t(x',y') = min{t(y,x),...,t(x,y)}, i.e. the first operation is chosen to minimise the number of subsequent operations.

The same ideas can be applied to characterising the choice of i in algorithm 3.1. Consider the number of times the test $k_i = x$ (henceforth called a comparison) is performed in algorithm 3.1. Let us suppose, firstly, that the criterion for choosing i is to optimise the number of comparisons required, in the worst-case, for algorithm 3.1 to terminate.

Then, we have:

$$w(\ell,u,\text{found}) = 0 \text{ if } \ell>u, \text{ for all found } \varepsilon \{\mathit{true}, \mathit{false}\} \quad (3.3)$$

$$w(\ell,u,\mathit{true}) = 0 \text{ for all } \ell,u,\ 1\le\ell\le u\le n. \quad (3.4)$$

$$w(\ell,u,\mathit{false}) = \min_{\ell\le i\le u} \{1 + \max\{w(\ell,u,\mathit{true}),\ w(i+1,u,\mathit{false}),\ w(\ell,i-1,\mathit{false})\}\} \quad (3.5)$$

otherwise.

The construction of Eqs. (3.3) to (3.5) involves both the criterion function (the worst-case number of comparisons) and the structure of algorithm 3.1. The parameters of w determine the state of the computation when i is about to be chosen. Eqs. (3.3) and (3.4) are trivial; Eq. (3.5) is the crucial one. Consider its right-hand side and, for the moment, let us ignore the minimisation with respect to i. Then, if $\ell\le u$ and found is *false*, one comparison will certainly be made. Following this the value of ℓ,u or *found* will be changed and further comparisons initiated. In the worst case the number of further comparisons will maximise the number of comparisons made were *found* set to *true*, or ℓ set to i+1, or u set to i-1. Thus the rhs of (3.5) has the form 1 + max{<something>}. Now, to determine <something> we have invoked the principle of optimality once more. Applied to algorithm 3.1, the principle states that once a change has been made to ℓ,u or found an optimal algorithm will proceed optimally; that is, if found := *true* is executed then $w(\ell,u,\mathit{true})$ further comparisons will be made; if ℓ := i+1 is executed then w(i+1,u, *false*) further comparisons will be made etc.

The final item on the right-hand side is the minimisation over i. This simply reflects the requirement of choosing i to minimise the worst-case number of comparisons.

Now let us suppose that we wish to choose i to minimise the expected number of comparisons. Before we can do this we need to specify the "expected" input. For simplicity, we shall assume that x is indeed in the table (i.e. $\exists$ j such that $k_j = x$) and that each position is equally likely. Thus the probability that k_i is x is 1/n.

Let $e(\ell,u,\text{found})$ be the expected number of comparisons made by an algorithm which optimises the average-case performance.

Then, we have:

$$e(\ell,u,\text{found}) = 0 \quad \text{if } \ell>u \tag{3.6}$$
$$e(\ell,u,\mathit{true}) = 0 \quad \text{for all } \ell,u \;\; 1\leq\ell\leq u\leq n \tag{3.7}$$
$$e(\ell,u,\mathit{false}) = \min_{\ell\leq i\leq u} \{1 + \frac{1}{u-\ell+1} . e(\ell,u,\mathit{true}) + \frac{u-i}{u-\ell+1} . e(i+1,u,\mathit{false}) + \frac{i-\ell}{u-\ell+1} . e(\ell,i-1,\mathit{false})\} \tag{3.8}$$

Once again, the first two equations are trivial and the third is crucial. The justification for (3.8) goes like this. If $\ell\leq u$ and found is *false*, we certainly expect to make one comparison. Following this, found may be set to true with probability $1/(u-\ell+1)$; alternatively, ℓ may be set to i+1 with probability $(u-i)/(u-\ell+1)$; finally, u may be set to i-1 with probability $(i-\ell)/(u-\ell+1)$. These events are all disjoint and so the right-hand side of (3.8) follows immediately from the laws of probability and the principle of optimality.

3.4 Solving the equations

Constructing a system of equations like those in sec. 3.3 is easy; although we do not intend to give one, it is straightforward to provide a formal description of the process and its implementation in an automated system should present no problems. Solving the equations is by no means easy and requires a high degree of intuition about the problem in hand together with some skill in transliterating one's intuition into an equational form.

In this section we shall present the techniques which were used to solve eqs. (3.1) to (3.8). It would be wrong to claim, at this stage, that any is particularly important since I have only a modicum of experience in their use. We shall begin by considering algorithm 3.1 (the search algorithm), in particular its worst-case performance (eqs. (3.3), (3.4) and (3.5)).

3.4.1 The search algorithm - worst-case performance

In section 1.2 we made the claim that choosing the mid-point of the range $\ell..u$ optimises the worst-case performance of algorithm 3.1. The argument for this choice embodies two observations:

(i) the optimal number of comparisons depends only on $s = u-\ell+1$ i.e. the number of indices in the range $\ell..u$.

(ii) the optimal number of comparisons increases monotonically with s.

(These observations are rarely stated explicitly but are vital to a proof of optimality of binary search).

Now, if i is not the midpoint, the set $\{\ell..u\}$ will be split into two sets of unequal size, one of which has size greater than $\lceil s/2 \rceil$. Thus, in the worst-case, the search for x will continue in the larger set and, by observation (ii), the number of comparisons will be at least as many as if only half the set were being searched. In other words, a choice of i different from the midpoint will be suboptimal.

It is not uncommon for the running time of an algorithm to be monotonically increasing with the size of the input. This then is the first technique:

Exploit monotonicity

To see how this can be done we shall give a formal proof of the optimality of binary search. We begin by formalising observations (i) and (ii)

<u>Lemma 3.1</u>: (cf. observation (i)) $w(\ell,\ell+s-1,\underline{found}) = w(m,m+s-1,\underline{found})$, for all $s\geq 0$.

<u>Proof</u>: If s=0 or <u>found</u> = *true* the lemma follows immediately from eqs (3.3) and (3.4), respectively. Suppose s>0 and <u>found</u> = *false*. In this case eq. (3.5) applies and we proceed by induction on s. When s=1, there is no choice for i and we have:

$$w(\ell,\ell+s-1, \text{false}) = 1 + \max\{w(\ell,\ell+s-1,\mathit{true}), w(\ell+s,\ell+s-1, \mathit{false}), w(\ell,\ell-1, \mathit{false})\} = 1.$$

I.e. lemma 3.1 holds.
Now, suppose s>1 and, for all k<s, $w(\ell,\ell+k-1, \mathit{false}) = w(m,m+k-1, \mathit{false})$ for all ℓ,m. Then

$$\begin{aligned} w(\ell,\ell+s-1, \mathit{false}) &= \min_{0\leq k\leq s-1} \{1 + \max\{w(\ell,\ell+s-1, \mathit{true}), w(\ell+k+1,\ell+s-1, \mathit{false}), w(\ell,\ell+k-1, \mathit{false})\}\} \\ &= \min_{0\leq k\leq s-1} \{1 + \max\{w(m,m+s-1, \mathit{true}), w(m+k+1,m+s-1, \mathit{false}), w(m,m+k-1, \mathit{false})\}\} \\ &\quad \text{by the induction hypothesis} \\ &= w(m,m+s-1, \mathit{false}) \quad \text{by (3.5)} \end{aligned}$$

I.e. the induction hypothesis is true of s. □

In view of lemme 3.1 we shall rewrite eqs. (3.3) and (3.5) using v(s) in place of $w(\ell,\ell+s-1,$ *false*). We shall also substitute 0 for $w(\ell,u,$*true*). Thus:

$$v(0) = 0$$

$$v(s) = \min_{1\le i\le s} \{1 + \max\{0, v(s-i), v(i-1)\}\}$$

$$= \min_{1\le i\le s} \{1 + \max\{v(s-i), v(i-1)\}\}$$

Now we shall formalise observation (ii) and, as a by-product, we obtain the optimal policy for choosing i.

Lemma 3.2: $s \ge t \ge 0 \Rightarrow v(s) \ge v(t)$.

Proof: By induction on s.

Basis: when $s = 0$, $s \ge t \ge 0 \Rightarrow s = t \Rightarrow v(s) = v(t)$.

Induction step: Suppose $k > 0$ and lemma 3.2 holds for all $s \le k-1$. Consider v(k). Using the induction hypothesis we have:

$$v(k-i) \ge v(i-1) \text{ whenever } k-i \ge i-1 \text{ and } 1\le i\le k$$

and

$$v(k-i) \le v(i-1) \text{ whenever } k-i \le i-1 \text{ and } 1\le i\le k$$

Thus

$$v(k) = \min\{\min_{1\le i\le \lfloor (k+1)/2\rfloor} \{1+ v(k-i)\}, \min_{\lceil (k+1)/2\rceil \le i\le k} \{1+ v(i-1)\}\}$$

Now, using the induction hypothesis again,

$$v(k) = \min\{1+ v(k-\lfloor (k+1)/2\rfloor), 1+ v(\lceil (k+1)/2\rceil -1)\}$$

$$= 1 + v(\lfloor k/2\rfloor) \qquad (3.9)$$

(The last step is soon checked: by considering separately the case when k is odd and k is even.)
Similarly, if $t \le k$, $v(t) = 1 + v(\lfloor t/2\rfloor)$.
But $k\ge t \Rightarrow \lfloor k/2\rfloor \ge \lfloor t/2\rfloor$. Therefore, applying the induction hypothesis for the last time, $k\ge t \Rightarrow v(k) \ge v(t)$. □

Corollary: The optimal policy in algorithm 3.1 is to choose either $i = \lfloor (u+\ell)/2\rfloor$ or $i = \lceil (u+\ell)/2\rceil$.

Proof: Eq. (3.9) is what we are really after. It says that, of the terms being minimised on the rhs of the equation for v(k), the one which equals it is $1 + v(\lfloor k/2 \rfloor)$. Translating this back to w, we infer that:

$$w(\ell,\ell+k-1,\ false) = 1 + w(\ell,\ell+\lfloor k/2 \rfloor-1,\ false)$$

Setting $u = \ell+k-1$, we get

$$w(\ell,u,false) = 1 + w(\ell,\ \lceil(\ell+u)/2\rceil-1,\ false) \quad (3.10)$$

Alternatively, making use of lemma 3.1,

$$w(\ell,u,\ false) = 1 + w(\lfloor(\ell+u)/2\rfloor+1,u,false) \quad (3.11)$$

In other words the optimal algorithm is achieved by setting i to either $\lceil(\ell+u)/2\rceil$ (cf. (3.10)) or $\lfloor(\ell+u)/2\rfloor$ (cf. (3.11)). □

3.4.2 An optimal gcd algorithm

Let us now turn to the gcd algorithm of sec. 3.2.2 and show how eqs. (3.1) and (3.2) can be used to determine the optimal algorithm. Before we do so we wish to make an important observation about the analysis in sec. 3.4.1 of the optimal (worst-case) search algorithm. The observation is simply this: at no stage in the determination of the optimal policy did we ever calculate the value of $w(\ell,u,false)$ as an analytic function of ℓ and u. Generally, it is unnecessary to completely solve the equations in order to determine the optimal policy.
All we need to do is to establish equality of one term on the rhs of each equation with the ℓhs of that equation. Thus, in (3.9) we established the equality $v(k) = 1 + v(\lfloor k/2 \rfloor)$ from which it was straightforward to deduce the optimal choice of i.

Of course, an analytic solution for w is useful because it provides a worst-case analysis of the optimal algorithm. But, sometimes, it is possible to deduce the optimal policy from the equations whilst being quite unable to supply an analytic solution. The gcd algorithm is an example. What we are saying, therefore, is that we are able to construct an optimal gcd algorithm, and prove its optimality, but we cannot predict its running time! This is counter-intuitive and may explain why others have not attempted similar analyses.

Determining an optimal gcd algorithm is actually quite easy, but a long proof is required, primarily because there are a lot of operations among which to choose.

In the development we shall make an important assumption, namely that t(x,y) is finite for all x and y. Normally, before embarking on the sort of analysis we are about to perform, one would ascertain that there is at least one way of making the given choice which will ensure termination of one's algorithm. Thus, in algorithm 3.1 any choice of the index i will ensure its termination and so $w(\ell,u,\underline{found})$ and $e(\ell,u,\underline{found})$ are always finite. We shall not perform this preliminary analysis in the case of the gcd algorithm, but instead we will observe that the optimal algorithm does indeed terminate, hence validating our assumption. Formally, the assumption is this.

Assumption: $\forall x,y$ ($\exists$ k, such that $0 \le k < \infty$ and $t(x,y) = k$)

The reason for this assumption is simply to allow us to use induction on k.

Our first step is to observe that the interchange operation should never be used.

Lemma 3.3: $t(x,y) = t(y,x)$

Proof:
If x or y is zero then $t(x,y) = t(y,x) = 0$. So suppose they are both non-zero. Let a sequence of operations to evaluate gcd(x,y) be O_1, $O_2, \ldots O_k$. (Note: the assumption of finiteness of t(x,y) is being used here). Then, clearly from the symmetry of the operations, we can construct a sequence of operations $O_1', O_2', \ldots$ to compute gcd(y,x). (For example, if O_i is x := x+y, O_i' would be y := x+y). Thus $t(y,x) \le t(x,y)$. Hence, $t(x,y) \le t(y,x)$. I.e. $t(x,y) = t(y,x)$. □

Corollary: In an optimal algorithm the interchange operation is never used.

In the following lemmas we shall categorise the operations into three sets, namely:

(i) the additions: x := x+y and y := x+y.

(ii) the subtractions: x := x-y, x := y-x, y := x-y, and y := y-x.

(iii) the inverses: x := -x and y := -y.

Lemma 3.4: $t(x,y) = t(-x,y)$

Proof: This is proved by induction on t(x,y). The induction hypothesis is:

$$(\forall a,b)\ t(-a,b) \le k \Rightarrow t(a,b) \le t(-a,b)$$

Basis: k = 0. The hypothesis is clearly true.
Induction step. Let a and b be such that $t(-a,b) = k+1$ where $k \ge 0$. Let o be the first operation in an optimal sequence of

operations to evaluate gcd(-a,b). We perform a case analysis on o. Two relatively difficult cases are when o is x := x-y and when o is y := x-y. For these cases the induction hypothesis is established as follows.

(i) o is x := x-y.
We have $t(-a,b) = 1 + t(-(a+b),b)$.
Now $t(-(a+b),b) = k$
So, by the induction hypothesis,
$t(a+b,b) \leq k$
But $t(a,b) \leq 1 + t(a+b,b)$
$\therefore$ $t(a,b) \leq 1 + k = t(-a,b)$.

(ii) o is y := y-x.
We have $t(-a,b) = 1 + t(-a,-(a+b))$
$= 1 + t(-(a+b),-a)$ by lemma 3.3
Now $t(-(a+b),-a) = k$
So, by lemma 3.3 and the induction hypothesis,
$t(-a,a+b) = t(a+b,-a) \leq t(-(a+b),-a)$
Applying the induction hypothesis again we also have
$t(a,a+b) \leq t(-a,a+b)$
But $t(a,b) \leq 1 + t(a,a+b)$
$\therefore$ $t(a,b) \leq t(-a,b)$.

In the remaining cases the induction hypothesis is easily established. An illustration is the case when o is x := y-x:

$t(-a,b) = 1 + t(a+b,b)$
But $t(a,b) \leq 1 + t(a+b,b)$
$\therefore$ $t(a,b) \leq t(-a,b)$.

Thus by induction we have established that $t(a,b) \leq t(-a,b)$ for all a,b. Replacing a by -a we have the opposite inequality. Thus we conclude that $t(a,b) = t(-a,b)$ for all a,b. □

Corollary: $t(x,y) = t(x,-y)$

Proof: Immediate from lemma 3.3.

Corollary: In an optimal algorithm the inverse operations are never executed.

Lemmas 3.3 and 3.4 are useful in cutting down the number of operations we need to consider as well as assisting us in the proofs which follow. The next two lemmas add structure to the basic algorithm - they tell us that we must choose between performing additions or performing subtractions.

Lemma 3.5: In an optimal algorithm an addition is never followed by a subtraction.

Proof: This is proved by a case analysis.

(a) The sequences x := x+y; x := x−y and y := x+y; y := y−x are clearly suboptimal.

(b) Suppose x := x+y is followed by y := x−y in an optimal algorithm. Then, for some a,b,
$t(a,b) = 2 + t(a+b,a)$
$= 2 + t(a,a+b)$ by lemma 3.3
But $t(a,b) \leq 1 + t(a,a+b)$ which is a contradiction.

The case where y := x+y is followed by x := y−x is proved similarly.

(c) The sequence x := x+y; x := y−x is equivalent to x := −x and so is clearly suboptimal.

Similarly, the sequence y := x+y; y := x−y is suboptimal.

(d) Suppose x := x+y is followed by y := y−x in an optimal algorithm. Then, for some a,b,
$t(a,b) = 2 + t(a+b,-a)$
$= 2 + t(-a,a+b)$ by lemma 3.3
$= 2 + t(a,a+b)$ by lemma 3.4
But $t(a,b) \leq 1 + t(a,a+b)$ which is a contradiction.

The case where y := x+y is followed by x := x−y is proved similarly. □

Lemma 3.6: In a optimal algorithm a subtraction need never be followed by an addition.

Note our choice of words in the statement of this lemma. What is meant is that some optimal algorithms may involve a subtraction followed by an addition but there is one that doesn't. For example an optimal sequence of operations to evaluate gcd(3,2) is x := y−x; y := x+y; x := x+y which does involve a subtraction followed by an addition. However we can also evaluate gcd(3,2) optimally using x := x−y; y := y−x; y := y−x which involves no additions.

Proof: Again we perform a case analysis on the 8 possibilities.

(a) The sequences x := x−y; x := x+y and y := y−x; y := x+y are clearly suboptimal.

(b) Consider any pair a,b such that an optimal sequence S of operations to evaluate gcd(a,b) begins with the operation y := x−y followed by x := x+y.
Then $t(a,b) = 2 + t(2a-b,a-b)$.

Now we note

$$t(b-2a,b-a) = t(2a-b,a-b) \text{ by lemma 3.4.}$$

Consider, therefore, replacing the sequence S by the sequence S' composed of the subsequence y := y-x; x := y-x followed by an optimal sequence of operations to evaluate gcd(b-2a,b-a). S' will evaluate gcd(a,b) and has the same length as S. Thus S' is an optimal sequence which has one fewer addition than S.

By interchanging the roles of x and y, we can prove that any optimal sequence S which contains the operation x := y-x followed by y := x+y can be replaced by an optimal sequence S' in which the addition has been removed.

(c) The case where x := y-x is followed by x := x+y is similar. For suppose this sequence begins an optimal evaluation of gcd(a,b).
Then $t(a,b) = 2 + t(2b-a,b)$
$= 2 + t(a-2b,b)$ by lemma 3.4.
Thus by a similar argument an alternative sequence begins with x := x-y; x := x-y and continues with an optimal evaluation of gcd(a-2b,b).

Again, we can interchange the roles of x and y to show that one addition may be replaced by a subtraction in the case where y := x-y is followed by y := x+y.

(d) The sequences y := y-x; x := x+y and x := x-y; y := x+y can be shown to be suboptimal by an argument similar to case (b) in the proof of lemma 3.5.

Now the upshot of cases (a) - (d) is that any optimal sequence S to evaluate gcd(a,b) in which an addition follows a subtraction can always be replaced by an optimal sequence S' in which the first addition has been removed. Thus by repeatedly applying this process we can always transform S into an optimal sequence in which no addition follows a subtraction. □

The following theorem summarises the foregoing results.

Theorem 3.7: In an optimal algorithm only the additions or the subtractions are ever executed. Moreover, for given values of x and y, if additions are used to calculate gcd(x,y) then no subtractions are required and vice-versa.

An alternative way of stating theorem 3.7 is that the structure of an optimal algorithm is the following:

```
begin
CHOOSE: (add: while x ≠ 0 and y ≠ 0 do
                begin
                    CHOOSE: (x := x+y; y := x+y)
                end;
         sub: while x ≠ 0 and y ≠ 0 do
                begin
                    CHOOSE: (x := x-y; x := y-x;
                             y := x-y; y := y-x)
                end
        )
end.
```

Algorithm 3.3

Lemma 3.8: add should be chosen when x and y are of opposite sign. sub should be chosen when x and y have the same sign. The algorithm should ensure that the signs of x and y are preserved.

Proof: Clearly, if x and y are of opposite sign, no sequence of subtractions can ever make x or y become zero. Also, if x and y have the same sign, no sequence of additions can ever make x or y become zero.

Corollary:

```
y:= x+y  is chosen when sign(y) = sign(x+y)
x:= x+y  is chosen when sign(x) = sign(x+y)
x:= y-x  is not chosen when sign(x) ≠ sign(y-x)
x:= x-y  is not chosen when sign(x) ≠ sign(x-y)
y:= y-x  is not chosen when sign(y) ≠ sign(y-x)
y:= x-y  is not chosen when sign(y) ≠ sign(x-y)
```

Using the above corollary the structure of an optimal algorithm is:

```
begin
if sign(x) ≠ sign(y) → while (x≠0)∧(y≠0)
                          do begin z: = x+y;
                             if sign(z)=sign(x)→x:=x+y;
                                sign(z)=sign(y)→y:=x+y;
                             fi
                             end;
   (x≥0)∧(y≥0) → while (x≠0)∧(y≠0)
                     do begin
                        if x≥y→choose(x:=x-y; y:=x-y);
                           y≥x→choose(x:=y-x; y:=y-x);
                        fi
                        end;
   (x≤0)∧(y≤0) → while (x≠0)∧(y≠0)
                     do begin
                        if x≤y→choose(x:=x-y; y:=x-y);
                           y≤x→choose(x:=y-x; y:=y-x);
                        fi
                        end
fi;
end.
```

Algorithm 3.4

We shall now restrict our attention to the case $x>0 \wedge y>0 \wedge x \geq y$, and evaluate the choice of x:=x-y and y:=x-y.

Lemma 3.9: If $x>0$, $y>0$ and $x \geq y$, an optimal choice of operation is x:=x-y.

Proof: Suppose $x>0$, $y>0$ and $x \geq y$.
Then $t(x,y) = 1 + \min\{t(x-y,y),\ t(x,x-y)\}$.
Now if $x=y$, $t(x-y,y) = t(x,x-y) = 0$, so x:=x-y is an optimal choice. (Either choice would do.)
Suppose $x>y$. Then $x>x-y>0$
and $t(x,x-y) = 1 + \min\{t(y,x-y),\ t(x,y)\}$
$1 + \min\{t(x-y,y),\ t(x,y)\}$ by lemma 3.3

Hence $t(x,y) = \min\{1+t(x-y,y),\ 2+t(x-y,y),\ 2+t(x,y)\}$
$= 1 + t(x-y,y)$.
I.e. an optimal choice of operation is x:=x-y.
Similarly we have:

Lemma 3.10:

(a) If $x>0$, $y>0$ and $x>y$, an optimal choice of operation is y:=y-x.

(b) If $x<0$, $y<0$ and $x<y$, an optimal choice of operation is x:=x-y.

(c) If x<0,<y<0 and x≤0, an optimal choice of operation is y:=y-x.

Corollary: The following is an optimal gcd algorithm.

```
begin
if sign(x)≠sign(y) → while (x≠0)∧(y≠0)
                      do begin z:=x+y;
                         if sign(z)=sign(x)→x:=z;
                            sign(z)=sign(y)→y:=z;
                         fi
                         end;
   (x≥0)∧(y≥0) → while (x≠0)∧(y≠0)
                  do begin
                     if x≥y→x:=x-y;
                        y≥x→y:=y-x;
                     fi
                     end;
   (x≤0)∧(y≤0) → while (x≠0)∧(y≠0)
                  do begin
                     if x≤y→x:=x-y;
                        y≤x→y:=y-x;
                     fi
                     end;
fi
end.
```

Algorithm 3.5 An optimal gcd algorithm

3.4.3 The search algorithm - average performance

Equations (3.3), (3.4) and (3.5) were quite easy to solve because monotonicity allowed us to immediately evaluate the max operation and then the min operation in Eq. (3.5). Unfortunately, monotonicity does not seem to help in solving Eqs. (3.6), (3.7) and (3.8). Indeed, eqs. (3.6), (3.7) and (3.8) seem frighteningly difficult, yet one's intuition suggests that binary search should also be optimal in the average case. It is often the case that it is much easier to verify a hypothesis than it is to mechanically deduce the hypothesis and so the following technique is likely to prove extremely important.

> If you have an algorithm which you think is optimal then try it and see.

To see how this works in practice let us begin by simplifying Eqs. (3.6) and (3.8) defining e(ℓ,u,found). Specifically we claim, without proof, that e(ℓ,ℓ+s-1,found) = e(ℓ',ℓ'+s-1,found) for all ℓ,ℓ', s and found. (Supplying a detailed proof would be tedious.) Thus, abbreviating $e(\ell,\ell+s-1,\ \mathit{false})$ to $f(s)$ and eliminating $e(\ell,\ell+s-1,\ \mathit{true})$, Eqs. (3.6) and (3.8) become:

$$f(0) = 0$$

$$f(s) = \min_{1\le i\le s} \{1 + \frac{s-i}{s} f(s-i) + \frac{i-1}{s} f(i-1)\} \qquad (3.12)$$

Theorem 3.11: Binary search optimises the average number of comparisons in algorithm 3.1.

Proof: Suppose that binary search were optimal - what would Eq. (3.12) look like in that case? Well, we would be able to remove the min operation and claim that

$$f(s) = 1 + \frac{s-\lfloor s/2\rfloor}{s} f(s-\lfloor s/2\rfloor) + \frac{(\lfloor s/2\rfloor-1)}{s} f(\lfloor s/2\rfloor-1) \qquad (3.13)$$

Our proof, therefore, consists of verifying (3.13) by induction on s. More specifically we shall prove by induction on k that, for all j, $1\le j\le k$,

$$f(2j-1) = 1 + \frac{2(j-1)f(j-1)}{2j-1} \qquad (3.14)$$

$$f(2j) = 1 + \frac{jf(j) + (j-1)f(j-1)}{2j} \qquad (3.15)$$

Equations (3.14) and (3.15) are, of course, equivalent to (3.13) when s is odd and even, respectively.

It is easy to verify the basis of the induction since $f(0) = 0$, $f(1) = 1$ and $f(2) = 1 + \frac{1}{2}f(1)$ from Eq. (3.12).

To establish the induction step let us consider in turn $f(2k+1)$ and $f(2k+2)$.

Clearly $f(2k+1) \le 1 + \frac{2kf(k)}{2k+1}$

and

$$f(2k+2) \le 1 + \frac{(k+1)f(k+1) + kf(k)}{2k+2}$$

from Eq. (3.12). Thus it suffices to prove the opposite inequalities.

Now, from (3.12),

$$f(2k+1) = 1 + \frac{1}{2k+1} \min_{1 \le i \le 2k+1} \{(2k+1-i)f(2k+1-i)+(i-1)f(i-1)\}$$

Let us split the range of the min operation into two sets - when i is even and when i is odd. Thus

$$f(2k+1) = 1 + \frac{1}{2k+1} \min\{ \min_{1 \le j \le k} \{(2k+1-2j)f(2k+1-2j)+(2j-1)f(2j-1)\}$$

$$\min_{0 \le j \le k} \{2k-2j)f(2k-2j)+2jf(2j)\}\}$$

Now, $\min_{1 \le j \le k} \{(2k+1-2j)f(2k+1-2j)+(2j-1)f(2j-1)\}$

$= 2k + \min_{1 \le j \le k} \{2(k-j)f(k-j)+2(j-1)f(j-1)\}$

by the induction hypothesis

$= 2k + 2k(f(k)-1)$ by (3.12)

$= 2kf(k)$

Also, $\min_{0 \le j \le k} \{ (2k-2j)f(2k-2j) + 2jf(2j)\}$

$= \min\{ 2kf(2k) , \min_{1 \le j \le k-1} \{(k-j)f(k-j)+(k-j-1)f(k-j-1)$

$+ jf(j)+(j-1)f(j-1)\}\}$

by the induction hypothesis

$\ge \min\{2kf(2k), 2k + \min_{1 \le j \le k-1} \{ (k-j)f(k-j)+(j-1)f(j-1)\}$

$+ \min_{1 \le j \le k-1} \{ (k-j-1)f(k-j-1)+jf(j)\}\}$

$\ge \min\{2kf(2k), 2k + 2k(f(k)-1)\}$

by (3.12).

But $2kf(2k) = 2k + kf(k) + (k-1)f(k-1)$

by the induction hypothesis

and $f(k) \le 1 + \frac{k-1}{k} f(k-1)$ by (3.12)

$$\therefore \quad 2kf(2k) \ge 2k + kf(k) + k(f(k)-1)$$

$$= 2kf(k) + k$$

$$> 2kf(k)$$

Hence $\min_{0 \le j \le k} \{(2k-2j)f(2k-2j)+2jf(2j)\}$

$$\ge 2kf(k)$$

Thus $f(2k+1) = 1 + \frac{1}{2k+1} \min\{2kf(k), 2kf(k)\}$

$$= 1 + \frac{2kf(k)}{2k+1}$$

as required.

Now similar techniques to those used in the proof of (3.14) can be used to prove that

$\min_{1 \le j \le k+1} \{(2k+2-2j)f(2k+2-2j) + (2j-1)f(2j-1)\}$

$$= kf(k) + (k+1)f(k+1)$$

and

$\min_{0 \le j \le k} \{(2k+1-2j)f(2k+1-2j)\} + 2jf(2j)\}$

$$\ge kf(k) + (k+1)f(k+1)$$

Thus

$$f(2k+2) = 1 + \frac{kf(k) + (k+1)f(k+1)}{2k+2} \quad \text{as required.}$$

We have thus verified both parts of the induction hypothesis when $j = k+1$ so theorem 3.11 follows straightforwardly. □

3.5 Alternative "Optimality" criteria

The results presented in the previous sections have their attractions but there are good grounds for being skeptical of their ever being widely used. For one thing computer science can claim to have proved very few optimality results; the main thrust of complexity theory is to provide techniques for answering questions of the form "is there an algorithm which is better than ...?" but, so far, it has achieved few definitive results along these lines. Secondly, the examples I have given are of simple algorithms, well-known and well-understood by computer scientists. I can also point only to a few other articles where the principles of dynamic programming are evident (although not explicity) in the design of an algorithm - Pohl [1972] uses them in his design of a simultaneous min-max algorithm and the use of Fibonacci search in finding the maximum of a unimodal function is often justified by dynamic programming in textbooks on dynamic programming [Nemhauser, 1966]. But these are the only references I know where dynamic programming has been used to reason <u>about</u> the design of an algorithm. (Dynamic programming has, of course, found much use <u>within</u> algorithms, for example finding shortest paths, optimising straight-line code [Aho and Ullman, 1972], optimal binary search trees [Knuth, 1971], parsing context-free languages [Earley, 1970] and syntactic error correction [Wagner, 1974; Wagner and Fischer, 1974; Aho and Peterson, 1972]. This is a different use to that which I am advocating). Finally, even for these simple algorithms, the algebraic manipulation required may seem distinctly daunting to many; it certainly wouldn't appeal to the majority of programmers.

Nevertheless, I believe that dynamic programming can play an important role in algorithm design and the techniques I have presented are worth pursuing. In order to exploit them more fully we need to take a more simplistic approach. Worst-case and average-case analyses are not the sacred cows they often appear to be; it sometimes pays to adopt rather crude criteria when evaluating the efficiency of an algorithm.

Read Dijkstra's development of a gcd algorithm [Dijkstra, 1976] and you will find that he examines in turn two different possibilities. His aim is simply to get an algorithm which terminates and so he considers first always decreasing $|x-y|$ and secondly always decreasing $x+y$ (he assumes that x and y are maintained positive). A surprising observation is that it is possible to model his analysis as an application of dynamic programming.

Let's consider first the objective of always decreasing $|x-y|$. One way of modelling this is as finding an optimal gcd algorithm where our criterion of optimality is:

<u>Criterion A</u> Minimise the number of times $|x-y|$ does <u>not</u> decrease.

Using this criterion the choice of operations in algorithm 3.2 is easily converted into the definition of a criterion function:

```
A(x,0) = A(0,y) = 0                    for all x,y
A(x,y) = min{1 + A(y,x),
             if |x| < |x-y| then A(x+y,y)
                            else 1+A(x+y,y),
             if |y| < |x-y| then A(x,x+y)
                            else 1+A(x,x+y),
             if |x-2y| < |x-y| then A(x-y,y)
                               else 1+A(x-y,y),
             ⋮
            }           when x≠0 and y≠0.
```

The rationale for, for example, the second term in the equation for A(x,y) is as follows . If the operation x:=x+y is executed and |x-y| does decrease - i.e. |x| < |x-y| - then the number of times |x-y| does not decrease when evaluation gcd(x,y) equals the number of times |x-y| does not decrease when evaluating gcd(x+y,y). However, if |x-y| does not decrease then the number of times it does not decrease when evaluating gcd(x,y) is one more than the number of times it does not decrease when evaluating gcd(x+y,y). The first term, corresponding to the interchange operation, has been simplified to 1+A(y,x) because it is clear that it will never decrease |x-y|.

The conjecture, which Dijkstra shows to be unfounded, is that A(x,y) = 0 for all x,y. Thus Dijkstra dismisses this criterion and turns his attention to always decreasing x+y. We could, of course, model this as in criterion A and consider the minimum number of times x+y does not decrease. However, an equally valid criterion is as follows:

<u>Criterion B</u> Maximise <u>in a Boolean algebra</u> the proposition "x+y does not decrease".

This criterion is best clarified by presenting its defining equation:

```
B(x,0) = B(0,y) = true                         for all x,y
B(x,y) =       (if y+x< x+y  then true else false)
            or (if 2x+y<x+y  then true else false)
            or (if 2x+y<x+y  then true else false)
            or (if x<x+y     then true else false)
            or ....
                 when x≠0 and y≠0.
```

Here our conjecture is that B(x,y) = *true* for all positive x and y. The maximum operation referred to in the criterion is the Boolean *or* operation.

Criterion B does not quite capture all of Dijkstra's analysis.

Specifically, Dijkstra adds the constraint that x and y should be maintained positive. Likewise we can add this constraint to our criterion of optimality.

<u>Criterion C</u> Maximise in a Boolean algebra the proposition "x+y does not decrease and x>0 and y>0".

For positive x and y this criterion has the defining equation:

$$C(x,0) = C(0,y) = \mathit{true} \qquad \text{for all } x,y \geq 0.$$

$$\begin{aligned} C(x,y) = \; & (\mathit{if}\ y+x \not< x+y\ \mathit{and}\ y>0\ \mathit{and}\ x>0\ \mathit{then\ true\ else\ false}) \\ \mathit{or}\ & (\mathit{if}\ 2x+y<x+y\ \mathit{and}\ x+y>0\ \mathit{and}\ y>0\ \mathit{then\ true\ else\ false}) \\ \mathit{or}\ & (\mathit{if}\ 2x+y \not< x+y\ \mathit{and}\ x>0\ \mathit{and}\ x+y>0\ \mathit{then\ true\ else\ false}) \\ \mathit{or}\ & (\mathit{if}\ x<x+y\ \mathit{and}\ x-y>0\ \mathit{and}\ y>0\ \mathit{then\ true\ else\ false}) \\ & \ldots \\ & \quad \text{when } x>0 \text{ and } y>0. \end{aligned}$$

You may dismiss criteria A, B and C as tricks. But they do demonstrate the generality of dynamic programming. Provided one is imaginative in the choice of optimality criterion and one is willing to quantify the performance of an algorithm using values other than real numbers there is no limit to the technique.

4. CONCLUSION

Since the theme of this summer school is automatic program construction it behoves me to comment on the extent to which I foresee programming decisions being automated.

The decisions I have identified fall into a broad spectrum. At one end is the choice of loop invariants, at the other is the representation of data and in the middle are decisions arising in program transformation. To my mind there is no doubt that the initial choice of loop invariants cannot and should not be automated since they have a significant effect on the entire subsequent development. The role of the computer here should be purely a checking one - checking, for example, that the invariant and termination condition imply the postcondition. On the other hand, the choice of data representations offers much scope for automation and is a topic which is being actively pursued [Low, 1978; Schwartz, 1975]. Nevertheless, interaction with the programmer will, I believe, still be an important feature of an automated system since such choices depend on an analysis of one's algorithm - a task which is by no means easy for an automated system [Wegbreit, 1975].

I have attempted in Sec. 3 to show how dynamic programming may be used to evaluate decisions arising in program transformation with a view to its automation. But here again interaction with the programmer is essential because one's intuition about the problem is the greatest aid to solving the equations.

There are lots of decisions which I have not discussed - the tradeoff between storage and time is perhaps the most commonly occuring one. My main aim has been to remind you of the wealth of decisions that have to be made, the philosophy being that if you are aware of the possibilities then you are less likely to fall into the traps.

REFERENCES

(1) Aho, A.V. Hopcroft, J.E. and Ullman, J.D. (1974) "The design and analysis of computer algorithms", Addison-Wesley Publishing Co., Reading, Mass.

(2) Aho, A.V. and Peterson, T.G. (1972) "A minimum-distance error-correcting parser for context-free languages", SIAM Journal of Computing, 1, 4, pp. 305-312.

(3) Aho, A.V. and Ullman, J.D. (1972) "Optimization of straight-line code", SIAM J. Computing 1, 1, pp. 1-19.

(4) Backhouse, R.C. (1979) "Syntax of programming languages: Theory and practice", Prentice-Hall International, London.

(5) Bauer, F.L. (1979) "Program development by stepwise transformations - the project CIP", In Program construction", Eds F.L. Bauer and M. Broy, Springer-Verlag Lecture Notes in Computer Science 69, pp. 237-272.

(6) Bellman, R. (1957) "Dynamic programming", Princeton Univ. Press: Princeton, New Jersey.

(7) Burstall, R.M. and Darlington, J. (1977) "Some transformations for developing recursive programs", J. ACM 24, 1, pp. 44-67.

(8) Carré, B.A. (1979) "Graphs and networks", Oxford Applied Mathematics and Computing Science Series, Clarendon Press, Oxford.

(9) Dijkstra, E.W. (1959) "A note on two problems in connexion with graphs", Numerische Mathematik 1, pp. 269-271.

(10) Dijkstra, E.W. (1976) "A discipline of programming", Prentice Hall Inc., Englewood Cliffs, N.J.

(11) Earley, J. (1970) "An efficient context-free parsing algorithm", Comm. ACM 13, 2, pp. 94-102.

(12) Gries, D. (1980) "Educating the programmer: notation, proofs and the development of programs", Dept. of Computer Science, Cornell Univ., Tech. Rep. No. TR80-414.

(13) Hoare, C.A.R. (1971) "Proof of a program: FIND", Comm. ACM 14, 1, pp. 39-45.

(14) Hoare, C.A.R. (1972) "Notes on data structuring", In Structured programming by O.-J. Dahl, E.W. Dijkstra and C.A.R. Hoare, APIC Studies in Data Processing No.8, Academic Press, London.

(15) Johnson, D.B. (1973) "A note on Dijkstra's shortest path algorithm", J. ACM 20, 3, pp. 385-388.

(16) Johnson, D.B. (1975) "Priority queues with update and finding minimum spanning trees", Info. Proc. Letters 4, 3, pp. 53-57.

(17) Johnson, D.B. (1977) "Efficient algorithms for shortest paths in sparce networks", J. ACM 24, 1, pp. 1-13.

(18) Knuth, D.E. (1971) "Optimum binary search trees", Acta Informatica 1, pp. 14-25.

(19) Knuth, D.E. (1973) "The Art of Computer Programming , vol. 3, Sorting and Searching", Addison-Wesley Publ. Co., Reading, Mass.

(20) Low, J.R. (1978) "Automatic data structure selection: an example and overview", Comm. ACM 21, 5, pp. 376-385.

(21) Manna, Z. and Waldinger, R. (1979) "Synthesis: Dreams => programs", IEEE Trans. on Software Engineering, vol. SE-5, nº 4, pp. 294-328.

(22) Nemhauser, G.L. (1966) "Introduction to dynamic programming" J. Wiley and Sons, New York.

(23) Pohl, I. (1972) "A sorting problem and its complexity", Comm. ACM 15, 6 (June 72) pp. 462-464.

(24) Reynolds, J.C. (1979) "Reasoning about arrays", Comm. ACM 22, 5 (May 79) pp. 290-299.

(25) Reynolds, J.C. (1981) "The craft of programming", Prentice Hall International, London.

(26) Schwartz, J.T. (1975) "Automatic data structure choice in a language of very high level", Comm. ACM 18, 12 (Dec. 75) pp. 722-728.

(27) Sintzoff, M. (1976) "Eliminating blind alleys from backtrack programs" in Automata, Languages and programming, 3rd ICALP, Michaelson, S. and Milner, R. (Eds).

(28) Vuillemin, J. (1978) "A dat structure for manipulating priority queues", Comm. ACM 21, pp. 309-315.

(29) Wagner, R.A. (1974) "Order-n correction for regular languages", Comm. ACM 17, 5 (May 74) pp. 269-271.

(30) Wagner, R.A. (1976) "A shortest path algorithm for edge-sparse graphs", J. ACM 23, 1 (Jan. 76) pp. 50-57.

(31) Wagner, R.A. and Fischer, M.J. (1974) "The string-to-string correction problem", J. ACM 21, 1 (Jan. 74) pp. 168-173.

(32) Wegbreit, B. (1975) "Mechanical program analysis", Comm. ACM 18, 9 (Sept. 75) pp. 528-539.

(33) Williams, J.W.J. (1964) "Algorithm 232. Heapsort", Comm. ACM 7, 6 (June 64) pp. 347-348.

(34) Yen, J.Y. (1970) "An algorithm for finding shortest routes from all source nodes to a given destination in general networks", Quart. Appl. Maths. 27, pp. 526-530.

(35) Yen, J.Y. (1972) "Finding the lengths of all shortest paths in N-node non-negative distance-complete networks using $\frac{1}{2}N^3$ additions and N comparisons", J. ACM 19, 3 (July 72) pp. 423-424.

THE DESIGN OF VERY HIGH LEVEL LANGUAGES

Patricia C. Goldberg

Computer Sciences Department
IBM T. J. Watson Research Center
Yorktown Heights, NY

ABSTRACT: A characterization of very high level language is given that distinguishes them from merely high level languages. Three languages, APL, SETL, and BDL, are examined to justify this characterization.

I. INTRODUCTION

Although a number of characterizations have been given of very high level languages (VHLL's) [1], these have been presented largely with the simple aim of developing a taxonomy of languages. In this lecture, we shall attempt characterization of VHLL's in terms of the *programming language devices* they employ. To motivate and justify our characterization, we shall discuss in some detail three VHLL's, each designed from a rather distinct point of view. These are APL [2], SETL [3], and BDL [4].

These three languages, like VHLL's in general, treat aggregates as primitive objects, although they differ considerably in their actual choice of interesting aggregates. In each language the size and shape of aggregates may vary over time, so processors for the languages incorporate automatic storage management schemes.

A. W. Biermann and G. Guiho (eds.), Computer Program Synthesis Methodologies, 125–145.

Another general characterization of VHLL's that is often cited is the emphasis on *data flow* rather than *control flow* for describing program logic. Because of this characteristic, such languages are sometimes described as *non-procedural*, as prescribing *what* is to be computed rather than *how* the computation is accomplished. This description is technically incorrect, since each of these languages has a well-defined operational semantics. But such languages do suppress a considerable amount of detail that would be exposed in the equivalent program written in a conventional high level language (HLL).

In the sections that follow, we shall discuss the three languages in some detail. The discussion will be informal and incomplete, since our only goal will be to determine what devices these languages have incorporated that permit them to avoid explicit control flow.

II. APL

APL, the programming language, grew out of a notation developed during the 1960's by K. Iverson [5]. To this notation has been added a small number of programming language constructs. It is a programming language in which datatypes are not statically associated with variables; rather, variables are typed dynamically as a result of program execution.

In APL, the use of data flow to replace control flow has reached the level of an art form. APL "one-liners" are famous. Indeed, the only control flow mechanisms in APL are expression evaluation, function call, and the go to (→).

APL contains an unsurprising set of scalar datatypes: integers (these are also used to represent boolean quantities), reals, and characters. APL's primary data object is the *vector*. This is a collection of scalar objects of the same datatype. The individual elements of the collection are accessed by an index chosen from some prefix of the integers. Thus following the assignment

```
A←6 5 4 3 2 1
```

the value of A[3] will be 4.

Matrices and higher-order arrays are defined as *shaped vectors*. These can be introduced into a program using the dyadic shaping operator ρ. For example, the assignment

```
A←(2 3)ρ1 2 3 4 5 6
```

Causes A to hold a 2 x 3 matrix. A request to print A would then yield

```
1 2 3
4 5 6
```

Conversely, associated with every array A there is a vector, the *ravel* of A, written

```
,A
```

that enumerates the elements of A in row-major order.

The power of APL derives primarily from two sources: the choice of operators that have been made for the language and the systematic way these operators can be extended to higher order objects. In APL, it is generally true that

- *Scalar functions* can be applied, element by element, to vectors, matrices, and arrays of conforming shapes.

- *Vector operators and functions* can be applied to shape-conforming matrices and arrays, vector by vector.

- *Matrix operations* can be applied to general arrays by fixing (or specifying) axes of applicability.

Scalar Operators

The APL scalar operators are the usual arithmetic functions (+,×,-,÷,...), logical functions (∧,∨,~) and a set of comparators (<,≠,...).

The extension to matrices permits the computation

$$\begin{pmatrix}1 & 2\\3 & 4\end{pmatrix} \times \begin{pmatrix}2 & 3\\4 & 5\end{pmatrix}, \textit{ yielding } \begin{pmatrix}2 & 6\\12 & 20\end{pmatrix}$$

This multiplication is expressed in APL as

```
((2 2)ρ(1 2 3 4))×(2 2)ρ(2 3 4 5)
```

This basic mechanism can also be used, for example, to determine how two matrices compare:

$$\begin{pmatrix}5 & 7\\1 & 3\end{pmatrix} \leq \begin{pmatrix}8 & 2\\4 & 3\end{pmatrix} \textit{ yields } \begin{pmatrix}1 & 0\\1 & 1\end{pmatrix}$$

Similarly, bit vectoring operations can be expressed succinctly:

`(1 0 1 1)∧(0 1 1 0)` yields `(0 0 1 0)`

Vector operations

A key vector operation is *reduction* (/). This permits any scalar operation to be applied to the elements of a vector, yielding a scalar. Thus

`+/1 2 3 4 5` yields `15`

Reduction is particularly useful in combination with comparison and logical operations. For example

`+/P=Q`

determines the number of positions where `P(i)=Q(i)`, whereas

`∧/P=Q`

returns 1 exactly when `P=Q`.

Other operations are reverse (⌽):

`⌽1 2 3 4` yields `4 3 2 1`;

index generation (`ι`):

`ι5` yields `1 2 3 4 5`;

and take (↑), drop (↓) that accept or release the first n elements of a vector.

Matrix operations

The primitive matrix operations, matrix inverse and matrix division, are supported. More importantly, APL permits any two scalar operators to be combined to form a generalized "inner product" operator. For appropriately dimensioned arrays,

$$M + . \times N$$

represents ordinary matrix multiplication. That is,

$$a_{ij} = \sum_{k=1}^{(\rho N)[1]} m_{ik} \times n_{kj}$$

Similarly, if C represents the connectivity matrix for a graph, then

$$C \vee . \wedge C$$

represents paths of length 2; that is,

$$a_{ij} = \bigvee_{k=1}^{(\rho C)[1]} C_{ik} \wedge C_{kj}$$

General Array Operators

APL supports two very important general array operations: *transpose* (⍉) and outer *product* (∘ .).

The transpose operator is used both to re-arrange indices and to reduce the number of dimensions of an array. For example, if

```
B←(2 1) ⍉ A
```

then

```
B[i,j]= A [j,i].
```

Similarly, if

```
B←(1 2 1) ⍉ A
```

then

```
B[i,j] = A[i,j,i].
```

The outer product operator takes two arrays and a scalar function as input. It forms all possible pairs of scalar elements from the two arrays, then uses the scalar function to produce a scalar from each pair. The index of the scalar in the resulting array is the catenation of the indices of the input items. For example, if

```
ρA = 2 3
```

and

```
ρB = 1 5 3
```

and

```
C ← A ∘. x B,
```

then

```
ρC = 2 3 1 5 3
```

and

```
C [i;j;k;l;m] = A[i;j] x B [k;l;m].
```

The following two examples illustrate the APL programming paradigm.

Example: Transitive Closure

Let C be the connectivity matrix for a graph G. That is,

$$C_{ij} = 1$$

exactly if there is an *edge* in G from node i to node j. Compute and place in C the graph in which

$$C_{ij} = 1$$

exactly if there is a path in G from node i to node j.

The APL program to accomplish this is

```
L:→(~(∧/∧/C1=C←C1∨C1∨.∧C1←C))/L
```

To understand this program, remember that the APL expression evaluation rule is a single-precedence, right-to-left rule.

The whole program is written as a single conditional branch. As long as the condition expression evaluates to 1, L will be repeated. When the expression evaluates to 0, the program will fall through to the next statement. Thus we need to show that this computation will terminate, and that, when it does, C will contain the desired result. To understand that this is the case, let us step through an instance of the computation

```
L:→(~(∧/∧/C1=C←C1∨C1∨.∧C1←C))/L
```

Copy C into C1

```
L:→(~(∧/∧/C1=C←C1∨C1∨.∧C1←C))/L
```

Compute all "2-paths" to get paths of length 2n

```
L:→(~(∧/∧/C1=C←C1∨C1∨.∧C1←C))/L
```

Or in all old paths. This gives all paths ≤ 2n

```
L:→(~(∧/∧/C1=C←C1∨C1∨.∧C1←C))/L
```

Store result in C

```
L:→(~(∧/∧/C1=C←C1∨C1∨.∧C1←C))/L
```

Compute matrix of differences between last two iterations

```
L:→(~(∧/∧/C1 = C ← C1 ∨ C1 ∨.∧ C1 ← C))/L
```

And reduce along both dimensions to see if there are any differences. Compute 0 if there are.

We then negate the result to force the branch in the right direction.

This example is not precisely a "one-liner", since it depends for its effect on repeated execution of the single statement. Nonetheless, certain characteristics of APL can discerned from it.

An important APL principal is "compute first and test later." In this program, for example, only after all paths of length ≤ 2n have been computed is a test made to see if any new paths had been added. Even the computation of this test has the same form: compare all elements of the matrix and then look to see if at least one is different. It is the regularity of APL's data structures that permit this style; it is the exploitation of this style that permits detail to be suppressed.

Example: Table Look-up

Let TABLE be an array whose rows represent a set of items. Let ITEM be a particular item. Then the program to find the index of ITEM in TABLE is:

```
(∧/⍉(1 2 1)⍉ITEM ∘.= TABLE)/⍳(⍴TABLE)[1]
```

This program computes as follows

```
(∧/⍉(1 2 1)⍉ITEM ∘.= TABLE)/⍳(⍴TABLE)[1]
```

Form the vector 1,...,n where n is the number of items in TABLE

```
(∧/⍉(1 2 1)⍉[ITEM ∘.= TABLE])/⍳(⍴TABLE)[1]
```

Form array c(i,j,k) such that
C (i,j,k) = 1
exactly when ITEM (i) = TABLE (j,k)

```
(∧/⍉[(1 2 1)⍉ITEM ∘.= TABLE])/⍳(⍴TABLE)[1]
```

Form B(i,j) such that
B(i,j) = 1
exactly when C (i,j,i) = 1
i.e., when ITEM(i) = TABLE(j,i)

```
(∧/[⍉(1 2 1)⍉ITEM ∘.= TABLE])/⍳(⍴TABLE)[1]
```

Transpose, so that B and TABLE are in same form.

```
[(∧/⍉(1 2 1)⍉ITEM ∘.= TABLE)]/⍳(⍴TABLE)[1]
```

And reduce over rows, to get 0-1 vector,
where 1 = match with word.

```
[(∧/⍉(1 2 1)⍉ITEM ∘.= TABLE)/⍳(⍴TABLE)[1]]
```

Compress to select proper index.

In this example, the APL style is even more pronounced. The program begins by computing an array that tells us where a character in a given position of ITEM can be found in TABLE. Clearly, many of these matches are irrelevant to our search. Therefore, the generalized transpose is used to focus on those matches of interest. This is accomplished by computing for each word in TABLE, whether or not it matches ITEM -- even though only one match can succeed.

As before, this style is possible because of the regularity of arrays (as well as their dynamic nature) and the general way in which scalar operators, etc., have been introduced.

The APL Programming Paradigm

Exploit regularity of arrays to compute all possible values, then filter out those not of interest.

III. SETL

The SETL language [3], developed in the late '60's by J. Schwartz and his associates, is an attempt to use the basic axioms of set theory as the foundation for a programming language. So, in addition to integer, real, boolean, and string scalars, SETL supports finite sets. It also supports non-homogeneous tuples, and has language for special-case sets, called *maps*.

With the exception of reduction (/), SETL does not support the systematic application of scalar operators to aggregates. Rather, special attention has been paid to the definition of *iterators* that permit some filtering to be done with the iteration.

SETL Scalars

The SETL scalars are conventional, except possibly for strings, which admit of a substring notation:

```
a( 2 . . 5 ) = 'Xyzωq'
```

Unlike APL, none of the dyadic scalar operators extend to sets. Indeed many of the scalar operator symbols are given quite different meanings when their operands are sets. It is not surprising that extensional definitions have not been given, since there is no obvious mechanism, such as indexing, to indicate how the elements are to be paired.

SETL tuples

SETL tuples are quite different from APL vectors. They are non-homogeneous, e.g.,

```
X = [ 1, 2, 'abc', [ 5, 'X' ]].
```

Furthermore, the index set need not be a prefix of the integers. If the current value of X is that given in the above assignment, it is possible to make a further assignment of the form

```
X( 18 ) = [ 2, FALSE].
```

SETL sets

Sets can be constructed in a SETL program in a number of ways. One can simply enumerate the elements:

```
a = { 1, 2, 'abc' }.
```

Given two sets, one can form their union (∪), intersection (∩), and difference (-). WITH and FROM can be used to add or remove single elements from sets. An arbitrary element of a set can be accessed (arb).

A more powerful method of computing a set is the use of *set formers*. These have the general form

(expression:iterator)

Thus, one can define sets by

```
        (n: n = 1..100)
        ([X ** 2, X]: X = 1..5)
or      (a: a IN y | a > 5)
```

Sets of 2-tuples, e.g.,

```
{ [ 'a', 1], [ 'b', 2],...}
```

play a distinguished role in SETL and are referred to as *maps*. Maps play a role in SETL similar to that of *reference* variables in more conventional languages.

SETL Logic

As we indicated earlier, the non-homogeneous, non-indexed nature of SETL's aggregating methods make APL's device of using the shape (and, occasionally, the contents) of an aggregate as the basis for iteration relatively useless. Consequently, SETL has a much richer set of logic operators. On the surface these are rather conventional IF, CASE, and LOOP operators. For example, the syntax of LOOP is

```
LOOP iterator DO
     block
END LOOP
```

or

```
LOOP WHILE/UNTIL test DO
     block
END LOOP
```

What is *not* conventional is the form that the *iterators* and *tests* may take.

There are two basic iterators, FOR and FORALL. These typically take the form

```
   FOR m=1..n
   FOR m=2,4..
   FOR 1,2..n
or FORALL x IN s
```

Any of these forms may be combined with a filtering *test*. These tests may be conventional, e.g.,

```
   (FORALL x IN s | x > 5)
or (FOR i=1,2..10 | f(i) > 0)
```

or it may be one of the *quantified* test forms. These are the test forms EXISTS and FORALL. One may write as a test

EXISTS iterator | test

where the iterator is above, with the keyword omitted. If, during the iteration, the test succeeds, then EXISTS evaluates to TRUE. In addition the "iterator variable" if one is specified, is set to the value for which the test succeeded. Otherwise it returns FALSE.

Similarly, if one writes

FORALL iterator | test

FALSE is returned, with side effects, if the test ever fails; otherwise, TRUE is returned.

It is interesting to compare the SETL style of programming with that of APL.

Example: Table Look-up

In SETL, TABLE is represented as a map [item, index]. The SETL program to retrieve the index of ITEM is

```
INDEX = 0;
TRUTH = (EXISTS [INDEX,X] IN  TABLE | X = ITEM);
```

Here is a SETL "one-liner" that computes whether or not ITEM is in table. If it is, INDEX is set as a side effect of EXISTS.

Example: Transitive closure

Suppose G is encoded as the map [n1,n2], representing an edge from n1 to n2. Then the following SETL program refines the map to a "path map."

```
LOOP WHILE
      (EXISTS[N1,N2]IN G |
            (EXISTS[N3,N4]IN G |
                  N2=N3 AND[N1,N4]NOTIN G)))

DO
      G WITH[N1,N4];
END LOOP:
```

It is interesting to contrast this program with its APL equivalent. In the APL program, the complete connectivity matrix is computed in each iteration; only after that computation are filters and tests applied. In SETL, the emphasis is exactly opposite; all the filtering is done as part of the iteration condition, and an element added to the connectivity map if it meets all the tests. (In the absence of code optimization, neither computation is particularly efficient -- a subject to which we shall return in a subsequent lecture).

Although APL and SETL employ different aggregating primitives and encourage different styles of programming, certain basic devices can be discerned in both languages - in particular, a careful combination of aggregate creation and aggregate filtering operations.

But both APL and SETL are truly universal languages; that is, their primitive data objects and operations have not been specialized to any application area. It is therefore interesting to examine the semantics of a specialized language such as the Business Definition Language, BDL, to determine whether similar devices are employed.

IV. BDL

BDL was developed in the 1970's by Hammer et al [4]. Its intended exclusive domain of application is business computations, most particularly the kind of business computations that occur in an *office*.

The business characteristics that are reflected in the types of applications BDL aims at are:

- A business is usually divided into organizational units such as divisions and departments
- Formal communication and coordination among these units is managed via documents
- Document manipulation is an important function of these units
- This manipulation rarely involves complex mathematical computation
- Document creation depends in large part on information contained in other documents
- Operations are frequently performed on groups of documents
- Documents are typically stored for later use and reference.

BDL Scalars

BDL is a strongly typed language with a conventional set of scalars, plus a set specialized for business applications. The latter include times, dates, addresses, dollar amounts, etc. These are included both to accommodate input-output editing and to permit more careful type checking (it is nonsense to add two dates, for example).

BDL Aggregates

BDL admits of two levels of aggregating. The first, and basic, aggregate is a *document*. A document is a non-homogeneous, structured object, whose subparts may be either a structure or an indefinitely-sized, homogeneous, collection of structures. The leaves of the structures are, of course, the scalars.

Documents may be further aggregated into homogeneous *groups* of documents. Groups serve as input/output to a program as well as constituting the contents of a single *file*.

A file is simply a permanent repository for a group of documents that persists between different programs or different activations of a program.

BDL Control

BDL is a strict *data-flow* language. This has two implications:

(1) there are *no* control flow primitives

(2) any operation may compute whenever all of its inputs have been computed.

A program is divided into *steps*. A step corresponds to an algorithmic task or set of tasks associated with an organizational unit. To make the program more readable (it is the aim of BDL that every program should be understood, if not written, by its ultimate user), BDL steps are divided into two types: *composite* and *irreducible*. Composite steps are used only to indicate data-flow relationships; all computations are captured in irreducible steps.

Figure 1 shows a typical BDL composite step expressed in the graphic syntax (there is also a linear form). The boxes represent steps that may be either composite or irreducible. The edges represent data flow relationships: each is labelled with the name of the group of documents that flows over it. For example, Warehouse Order's are output from the Billing Step and are input to the Warehouse Processing step.

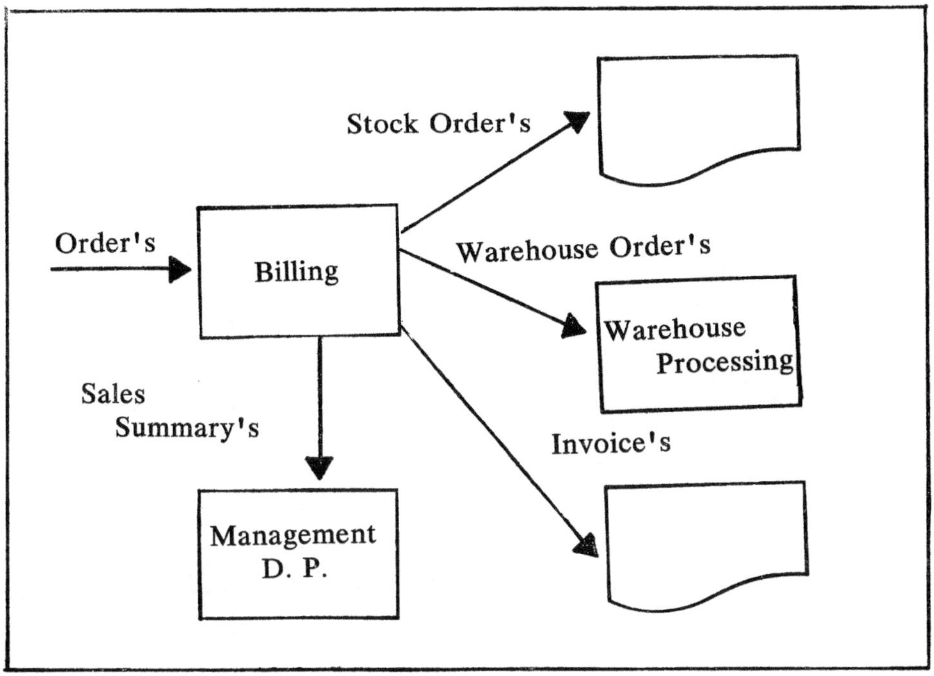

Figure 1. A BDL Composite Step

The Billing step is composite; its detail is shown in Figure 2. We see that the step has the same input-output as in Figure 1, but some internal components are shown. For example, two files, CM and IM, are internal to the Billing step. A special-purpose step, an accumulator, whose purpose is to accumulate Order's until a trigger causes it to output the accumulated group, is also used.

Our interest, however, is with the irreducible steps, since it is here that computational details are given. Figure 3 shows the BDL program for the Produce Invoice step.

The program for an irreducible step is written in terms of the group of documents it produces as output. The fields of these documents are displayed on the left. The convention is used that, when a homogeneous collection is named, it ends in "'s"; the name without the terminal "'s" refers to an individual in that collection.

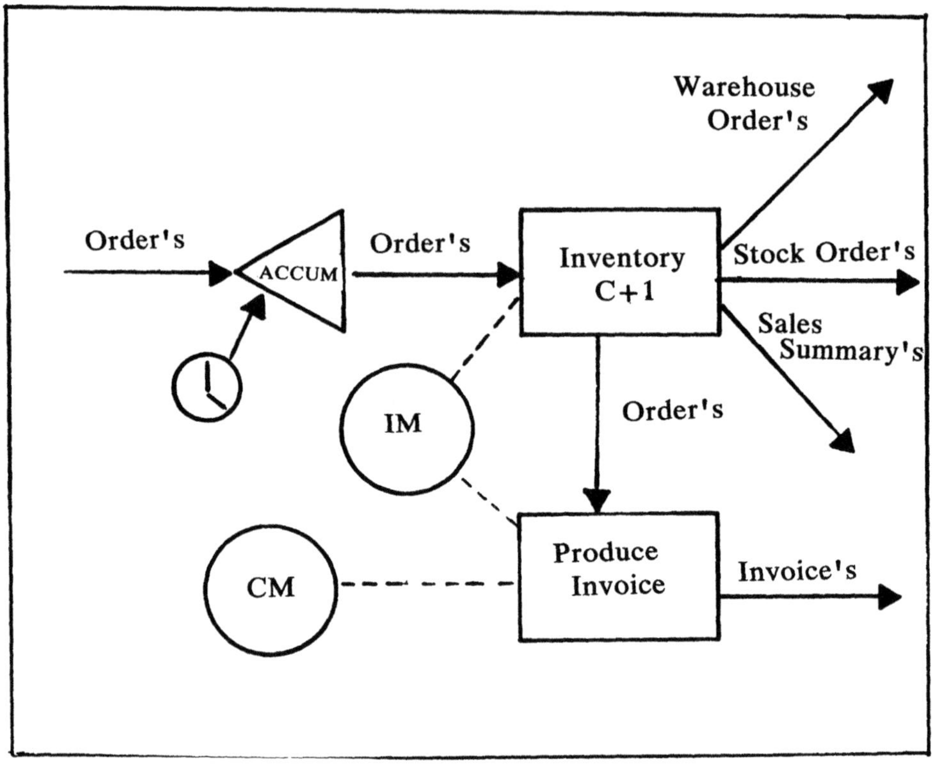

Figure 2. Details of the Billing Step.

The program is written as a series of "definitions", one for each field in the document. For groups, or homogeneous sub-collections, that definition is a *filter* (ONE PER) that describes a prototypical element of the collection. For a scalar, the definition is the computation for a single prototypical element.

Thus, Invoice's are described as ONE PER Order, where Order's are the input that caused the step to fire. This means that one invoice will be produced for each incoming order (later referred to as "CAUSE OF Invoice").

The customer number, called Cust#, is defined as the Incoming Cust#. The definitions associated with Cust# indicates this is to be computed by looking at the Order causing the invoice and finding the customer number there.

PRODUCE INVOICE PROGRAM	
Invoice's	ONE PER Order
Name	Customer Name
Address	Customer Address
Cust #	Incoming Cust#
Item's	ONE PER Same Item's
Item#	Incoming Item#
Quantity	SUM (Incoming Quantity's)
Price	Item Price
Gross	SUM (Quantity's × Price's)
Tax	.05 × Gross | STATE(Address)='NY' 0 | Otherwise
Amount	Gross + Tax

Definitions for Invoice's	
Order's	Input

Definitions for Cust	
Incoming Cust#	IN Order
Order	CAUSE of Invoice

Definitions for Item"s	
Same Item's	Order Item"s with common Ord Item#
Ord Item#	IN Order Item
Order Item's	In Order
Order	CAUSE of invoice

Definitions for Price	
Item Price	IN Assoc ItM
Assoc ItM	ItM with Item no = item#
Itemno	IN ItM
ItM's	INPUT

Figure 3. The Produce Invoice Irreducible Step

A subcollection, Items's, defined as ONE PER Same Item's, has a prototypical item associated with a *group* of items in the incoming order. Which subgroup? The subgroup with the same Item# in the Order.

It is not surprising that some of the computations for elements of a prototypical Item are aggregate computations, since each Item may correspond to a number of item lines in its causing Order. In particular, we see that Quantity is defined in terms of the reduction operator SUM, over the incoming quantities.

Another interesting computation is that for the Gross. This does not refer to the causing order, since it can be computed directly from the invoice, once all the fields for an Item have been computed. This computation involves both an extended scalar computation

Quantity's × Price's

that computes the *collection* of products, plus the reduction operator SUM, that sums the elements in these collections.

From this informal discussion of BDL, we can see that BDL combines some of the features of SETL and APL. Like SETL, it uses a repertoire of relatively powerful filters to reduce "inner loop" computations often to scalar computations. Where the filters do not reduce the computation to a computation over scalars, they do reduce it to a computation over homogeneous collections, so that APL-style aggregate operators can be used.

V. SUMMARY

We have examined three VHLL's quite difference in appearance. Each exploits a different primitive aggregate. Each of them, however, manages to avoid a great deal of explicit control flow. Furthermore, they accomplish this by appealing to a small number of basic ideas:

- the uniform application of lower level operators, both within and across aggregates;
- the inclusion of powerful aggregate formers;
- the definition of filtering operations that permit the intrinsic computation to be expressed either as a scalar computation or an extended scalar or array computation.

REFERENCES

[1] Leavenworth, B. M., and J. E. Sammet, Overview of Nonprocedural Languages. In Proceedings of a Symposium on Very High Level Languages, SIGPLAN Notices, Vol. 9, no.4, April 1974.

[2] APL/360 Primer. IBM Corp. Doc. No. GH20-0689.

[3] Schwartz, J., On Programming, An Interm Report on SETL Project, Comp. Sci. Dept., Courant Inst., New York Univ., 1973.

[4] Hammer, M.; W. G. Howe, V. J. Kruskal, I. Wladawsky. A Very High Level Programming Language for Data Processing Applications. *Comm. ACM*, Vol. 20, no. 11, Nov. 1977.

[5] K. E. Iverson, *A Programming Language*, John Wiley and Sons (New York), 1962.

IMPLEMENTATION CONSIDERATIONS IN VERY HIGH LEVEL LANGUAGES

Patricia C. Goldberg

Computer Sciences Department
IBM T. J. Watson Research Center
Yorktown Heights, NY

ABSTRACT: Although very high level languages offer advantages in expressivity, their naive implementations frequently suffer from run-time inefficiencies. In this paper we review some of the work aimed at improving the execution characteristics of such languages.

I. INTRODUCTION

In a previous paper [1] we categorized Very High Level Languages (VHLL's) as aggregate-oriented languages that rely largely on data flow rather than control flow for describing program logic. Using a small number of relatively straightforward devices, these languages permit a more succinct, usually simpler, description of a variety of algorithms. The language features on which this simplification depends include:

- the uniform application of lower level operators, both within and across aggregates;
- the inclusion of powerful aggregate formers;

A. W. Biermann and G. Guiho (eds.), Computer Program Synthesis Methodologies, 147–172.

- the definition of filtering operations that permit the intrinsic computation to be expressed as a scalar or an extended scalar computation.

The simplicity of expression is not without cost, however. If a VHLL is implemented by a direct appeal to its operational definition, the costs, both in execution time and in space, can be prohibitive, especially for large programs. Consequently, a great deal of intellectual effort has been put into developing "optimizing" translators for this class of languages.

To illustrate the problems of VHLL's, consider the APL [2] expression

```
(A+B)[I].
```

The operational semantics of APL suggest that it be computed by first evaluating the array expression

```
A+B
```

and then extracting the I'th element from it. Such an evaluation strategy requires `×/⍴A` additions and intermediate storage locations. But in the absence of run-time errors, it is sufficient to compute

```
A[I]+B[I]
```

because the other elements are ultimately discarded.

This example is not particularly compelling, since both the efficient and inefficient versions can be written as APL expressions. If, however, *I* is replaced by an arbitrarily complex scalar-yielding expression, a combination of the two forms is required in order to guarantee that the index expression is only evaluated once. This can be written as an APL expression only if embedded assignments are used.

Even when all of the elements of an expression participate in the final computation, there may be important savings of numbers in accesses as well as in storage from a non-obvious computation of the expression. Consider the expression

```
A ++/B ∘.×C .
```

The obvious evaluation requires the development of the storage consuming intermediate outer product,

```
B ∘.×C .
```

But since this outer product is to be sum-reduced, row by row, and the result added, element by element, to A, it is clear that only a single scalar quantity needs to be developed as an intermediate result.

Other VHLL's harbor similar "hidden inefficiencies". For example, the SETL [3] sequence

```
A=B+C
B = B WITH Q;
D=B+C;
```

can be converted to the more efficient sequence

```
A = B + C
B = B WITH Q ;
D = A WITH Q ;
```

A non-obvious transformation that cannot be trivially expressed in SETL occurs when the sequence

```
S={ };
LOOP(FORALL X IN S1 | X > 0) DO
S=S+{X};
END LOOP;
```

is implemented by adding {X} *destructively* to S, i.e., without copying S.

This set of examples suggests that there are rather large opportunities for improving the efficiency of VHLL programs. In the remainder of this paper we shall review some of the work that has been done in this area, especially for APL and SETL.

II. VHLL'S AND STYLES OF OPTIMIZATION

The design of optimization algorithms for VHLL's has followed closely the framework developed for high level languages (HLL's) of the FORTRAN, ALGOL, Pascal family. This work is thoroughly reviewed in [4], which should be consulted for details of algorithms mentioned below.

We shall assume that the source VHLL has been parsed and translated into straight-line intermediate code sequences, called basic blocks, and that the possible flows of control among these blocks have also been identified. The initial level of intermediate operations will be approximately that found in the source code. These are non-trivial assumptions; for example, the computed branch of APL makes it difficult to establish the control flow, although algorithms that begin to deal with this issue have been developed [5].

In VHLL optimization work, as in HLL's, there have been two important schools. One class of optimization algorithms restricts its attention to basic blocks; the other considers the global structure of a module (or, occasionally, the entire program). The argument for each of these frameworks, even for HLL's, is still being waged [6]; for VHLL's the situation is even less clear.

The localists argue that global algorithms are expensive; that it is difficult to obtain precise control flow information (*vide* APL), and hence the advantages of global algorithms are overstated; and that the kind of program improvements that can be obtained by global (as opposed to local) transformations could equally well have been done by the programmer in the first place. In any case, generating good local code gets rid of most of the inefficiency.

The globalists reply that the best global algorithms run in near-linear time; that optimal local algorithms are NP-complete, and hence heuristics have to be employed; that many important transformations cannot be performed without global understanding of the program structure. In any case, without global analysis, the quality of code generated will be very uneven, forcing the programmer to worry about issues that he should not have to care about.

In the case of VHLL's, there is evidence that both the local and global approaches have merit, and that any approach to code generation that aims at really good code will have to employ a mixture of the two. However, the present situation is that most of the APL work emphasizes local transformation; the SETL work, on the other hand, relies heavily on global techniques. Consequently, this review will illustrate both styles of code improvement.

III. APL OPTIMIZATION

Most of the work directed at improving the efficiency of APL has aimed at improving expression evaluation [7, 8, 9, 10]. There are several reasons for this:

- The APL computed branch makes it difficult to develop precise flow information.
- APL scoping rules for function calls make interprocedural analysis a necessary part of effective global analysis.
- APL performance can benefit considerably from improved expression evaluation.

Some of the earliest work in this area is contained in Abram's dissertation in the design of an APL machine [7]. He describes how certain computations can be avoided by manipulating address descriptors instead ("beating"). He also describes how lazy evaluation techniques [11] can be applied in the APL context. This work has formed the basis for the Hewlett-Packard APL compiler [12].

Guibas and Wyatt [10] have generalized Abrams work by considering complete expression trees. They have also extended it in several directions. In the discussion below, we shall largely follow their approach.

The operational semantics of APL suggest an evaluation strategy that is by no means lazy. For example, consider the APL expression

$$Z \leftarrow A \times B + C$$

where Z , A , B , C are conforming arrays of arbitrary rank.

APL semantics suggest the following evaluation sequence:

```
for   i = 1 to ×/ρZ
      t[i]←B[i] + C[i]
end;
for   i = 1 to ×/ρZ
      Z[i]←A[i] × t[i]
end;
```

This program requires the formation of a temporary array *t*. Furthermore, the elements of *t* are accessed twice: once to compute them and once to use them. Finally, the loop overhead is encountered twice.

The heart of APL evaluation strategy is contained in the following lemma: For all *scalar* operations f, g and all conforming arrays, A , B , C ,

$$A \, f \, B \, g \, C \Longleftrightarrow (\forall i) \, A[i] \, f \, B[i] \, g \, C[i]$$

From this, a lazy evaluation for expressions involving only scalar operations is clear: within a single loop develop the ith element of the result expression. Thus the example above becomes:

```
for   i=1 to ×/ρZ
          t ← B[i] + C[i]
      Z[i] ← A[i] + t
end;
```

This code requires only a scalar for intermediate storage. Indeed, this scalar may be realized in a register or on the top of a stack, according to the vagaries of the target machine and the scalar operation being performed.

The code produced by the lazy evaluation is correct, independent of the rank, etc., of the argument arrays, assuming that all the arrays are conforming. Since the expectation is that such an expression will be evaluated many times in the course of program execution, the following strategy is suggested.

On first attempt at execution:

(1) Check conformability of arrays.

(2) Compile code, using style outlined above.

(3) Execute code.

For subsequent executions, only steps(1) and (3) are required.

By elaborating the nature of the checks (step(1)) and of the code compiled (step(2)), this basic scheme can be extended to handle a broader class of operators. We shall describe one such extension from Guibas and Wyatt.

The Regular Selectors

The scalar APL operators are the primary set of computational operators. Scalar operators are sometimes combined with aggregate forming operators, e.g., reduction, inner and outer product. A set of operators totally unconcerned with calculation are the *selector* operators. The simplest selector is the subscript operator.

The algorithm outlined above can be elaborated to handle a particular class of selectors, the so-called *regular* selectors. These are drop (↓), take (↑), reverse(⌽), transpose (⍉), and $E[n_1;n_2;...;n_m]$, where each of the n_i are arithmetic progressions.

Regular selectors select some subarray from their principal argument, using the *control* argument to determine the subarray. They are called *regular* because if an APL atom *A* is treated as a rectangular grid of ⍴⍴*A* dimensions in a bounding box of size ⍴*A*, then the regular selectors designate a subgrid. That is, moving across any coordinate of the result array corresponds to moving along some set of coordinates of *A*, using equal size steps. This suggests that the ravel order of a subarrary named by a regular selector can be easily computed.

Regular selectors have two properties of interest:

(1) The composition of regular selectors is regular. That is, if f and g are regular selectors, then f(g*A*) can be computed by forming the composite selector fg and applying the result to *A*. The composite fg will also be regular (although not necessarily representable by a single APL operator).

(2) Regular selectors distribute over scalar operations. If f is regular, then

$$f(A+B) \equiv f(A)+f(B).$$

These properties suggest a method of extending the previously described technique for compiling scalar expressions to expression that also include regular selectors. On a pass through the expression tree from root to leaves, distribute regular selectors over scalars and compose regular selectors whenever they are encountered in a parent-child relation (Note. This tree-transforming requires that all the control operators associated with the regular selectors be previously evaluated. This can be accomplished by elaborating step (1) above.). The result of this pass is an expression tree that is essentially a scalar expression tree, since all regular selectors have been pushed to the leaves of the tree and are applied to the atoms. In this form, a modification of the code sequence compiled for scalar expressions can be generated.

This strategy requires a notation for composite regular selectors. If a selector is to be applied to a node of rank *n*, the selector can be represented by a data structure consisting of four vectors, defined over the interval [1, n]. These are:

q[i] an integer that indicates the coordinate of the result array to which the ith coordinate of the current node corresponds. Because of the transpose operator ⍉, several q[i] may have the same value.

s[i] the index along the ith coordinate of the origin element [0;0;...;0] d[i] the increment along the ith coordinate required to reach the next element along coordinate q[i].

ℓ[i] the size of the result along the $q[i]^{th}$ coordinate

The tree-transformation described above is accomplished by associating a selector data structure with every scalar node in a forward pass through the tree. Initially, a selector is associated with the root node *R*. There n=ρρ*R*, for all i, q[i]=1, s[i]=0, d[i]=0, d[i]=1, ℓ[i]=ρ*R*[i].

When an APL regular selector is encountered, its effect is "absorbed" into the selector and a new selector data structure is generated. Examples of the absorption rules are:

Reversal. To reverse along the kth coordinate

```
n'←n
for i in [1,n']
    do ℓ'[i]←ℓ[i]
       q'[i]=K
          if i=k then do
             s'[i]←r[i]Δs[i]+1
             d'[i]← -d[i]
             end
          else do:
             s[i]←s[i]
             d'[i]←d[i]
             end
end
```

Transpose. Let C[i], i in [1,p] be the control argument. Then

```
n'←p
    for i in [1,n']
        do
           q'[i]←q[c[i]]
           s'[i]←s[c[i]]
           d'[i]←d[c[i]]
           ℓ'[i]←ℓ[c[i]]
        end.
```

Absorption of the other selectors is equally straightforward.

At the end of this process, each reference to an atomic element A in the expansion has a composite selector associated with it. Because of the APL scalar evaluation rules, each of these must access an array of fixed rank. This rank can be computed from a stepper by taking max q[j].

For the purposes of compilation, it is convenient to apply the composite selector to its associated atom and to produce a *dope vector* that describes how to access the elements of the atoms in ravel order. For $1 \leq i \leq \max q[j]$ and an associated array A, this dope vector contains

α the relative offset in A of the origin element

γ_i the distance between two elements along the ith coordinate

δ_i the distance between two elements

$$[x_1, x_2, \ldots, x_i, \ell_{i+1}, \ldots, \ell_{max}] \text{ and } [x_1, x_2, \ldots, x_i + 1, \ldots, 1]$$

where the ℓ's represent limit values along their respective coordinates.

It is also convenient to keep in the dope vector π, the relative offset of the "current" element.

If the shape of A is $[k_1, k_2, \ldots, k_n]$, define

$$h(n) = 1 \,,$$
$$h(m) = k_{m+1} k_{m+2} \cdots k_n \qquad m<n.$$

Then

$$(\gamma = \sum [\,][\,]$$

$$\delta = \gamma - \sum_{<\leq}^{[\,]=} \gamma[\,][\,]$$

and

$$\alpha = \sum_{\leq\leq} [\,][\,].$$

The computation as described associates a dope vector with each atom. In fact, dope vectors can be shared by a number of atoms. All atoms with a common ancestor along a selector-free path through the tree can share the selector (and hence the dope vector) associated with that common node. This sharing cuts down both on the computation and storage overhead of this process.

With the dope vectors computed, code can be compiled for the expression. Since the original computation has been reduced to a scalar computation by the removal of the selectors from the tree, all of the dope vectors must represent arrays with a common storage, $[r_1,r_2,\ldots,r_m]$. It is useful in the evaluation routine to maintain, in addition to the vector r_i the vector

$$\text{count}\ [i],\ 1 \leq i \leq m,$$

to contain the current subscript value along the ith coordinate. The evaluation routine will also make use of the following functions:

initialize (D)	$D.\pi \leftarrow D.a$	(initialize a dope vector)
access (D,A)	origin of $A+D.\pi$	(get address of current element of A)
advance (D)	$D.\pi \leftarrow D.\pi + T.\delta$ [coord]	(locate next element in ravel order)

Using these functions, code for expressions continuing regular selectors can be compiled as shown in Figure 1.

Firther elaborations of this basic scheme are possible. Guibas and Wyatt show, for example, how to accommodate reduction and inner and outer products in this framework. Their paper should be consulted for details.

```
            coord ∪ 0 ;
            initialize ( D_1 ) ;
                .
                .
                .
            initialize ( D_N ) ;

Basic_Code:   do while coord < last_coord ;
                coord ∪ coord + 1 ;
                count [coor] ∪ 0 ;
            end ;
                .
                .      /* code for scalar computations, e.g. */
                .
            access ( D_1 , A )
                .
                .
                .
            access ( D_N ) X )

Advance: advance ( D_1 )
                .
                .
                .
            advance ( D_N )
            count [coord] ∪ count [coord] + 1 ;
            if count [coord] < r[coord]
                then go to Basic_code ;
                else if coord = 1
                    then exit ;
                    else do ;
                        coord ∪ coordΔ1 ;
                        go to Advance ;
                    end ;
```

Figure 1. Scalar code with Regular Selectors

IV. GLOBAL OPTIMIZATION AND SETL

As we have sketched above, the performance of APL can be greatly improved by exploiting the regularity of its data structure and simple algebraic properties of its operators. In particular, large intermediate results, costly in both time and space, can frequently be avoided. Because iterations are almost always expressed by operators embedded in an expression, it is expedient to focus on single-statement analysis as a way of improving APL performance. However, variants of the methods described below are also useful in the APL context (see [13] for an example), and may become more so if APL is extended to cover more complex data objects.

The language and basic data structures of SETL do not permit an APL-like approach to optimization. Iterations are not written as expression-yielding operators, but rather as control structures, albeit of a particularly powerful sort. Hence analysis of their effect cannot be accomplished by limiting one's attention to a single statement. Furthermore, the primary properties of sets that can be exploited to improve SETL performance are the relations of set inclusion and set membership. As the sequel will show, global analyses are required to establish these relationships precisely enough to exploit them.

The evolution and organization of the present SETL compiler has been described in a sequence of papers [14, 15, 16, 17]. We have freely borrowed from all of them. To motivate the structure and emphasis of the SETL compiler, it is necessary to understand a bit about the language and its naive implementation.

SETL is a weakly typed language with polymorphic operators. Although the type of a variable may be declared, generally the types of objects associated with program variables vary during the course of execution, depending on the most recent assignment. The meaning of operator symbols, furthermore, depends on the types of their operands; thus + variously represents ordinary addition (for scalars), set union, or string concatenation. Hence, SETL data objects carry run-time descriptors that are used both to determine the meaning of operators and the representation of the data object.

The default representation for a set is linked hash table. All elements can be located via a hash function; however, they are also maintained on a linear list to facilitate efficient iteration over a set. Figure 2 illustrates the default mapping for a set.

When a set is used as a map, this representation is converted (either statically or dynmaically) to a second representation. In this representation the hash function is applied only to the domain of the map. The value of the map is stored with the domain item, if the map is single-valued, or as a set of range items for multi-valued maps. This representation is illustrated in Figure 3.

Because of the size of the objects involved, set (and the analogous map) operations, such as assignment, union, insertion of an element, are accomplished by pointer manipulation. However, SETL has value, rather than pointer, semantics, and explicit copying must be carried out whenever pointer manipulation might violate the value semantics. In particular, in the absence of additional information

S WITH := X

requires S to be copied, rather than updated destructively, since additional pointers to the original value of S may exist. [*NOTE.* The current SETL implementation indicates that a data structure is shared by turning a "share-bit" on. Thus, once shared, a data structure cannot be "unshared". Although this could be remedied in an implementation that employed reference counts, this would not catch many important cases in which destructive updating may be safely employed, as we shall illustrate shortly.]

The structure of the SETL optimizer is that of a conventional optimizing compiler, except that register allocation is not performed. That is, the loop structure of the program is determined by *interval analysis* [18]. Then, using the interval structure, *use-definition information* is collected [19]. This information includes, for each use of a variable in the program, the set of definition points at which it may have obtained its value (a subset of all the definition points for that variable) and, conversely, for each definition, the set of use points it may affect. In SETL, this information is collected for a set of procedures, rather for the usual single procedure.

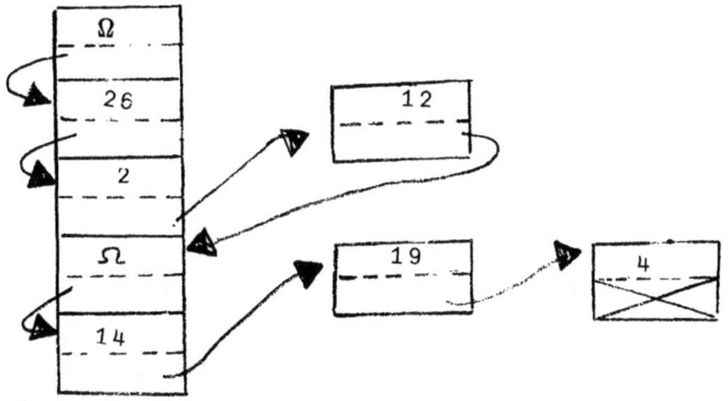

Figure 2. Default set representation for value
{2, 14, 26, 19, 12, 4}.
Hash function used is mod 5 arithmetic.

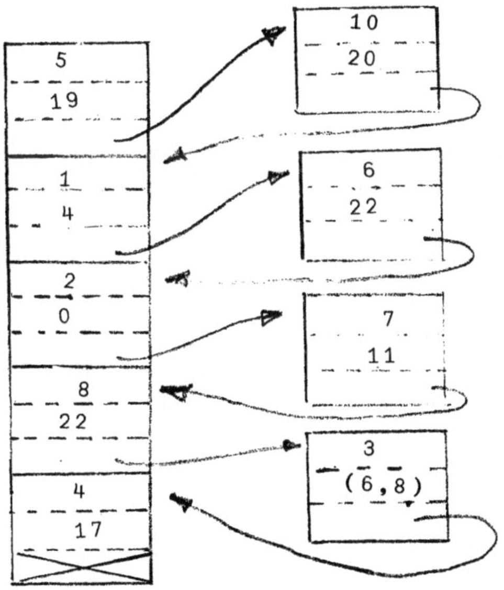

Figure 3. Default representation of the map
{[1,4], [4,17], [8,22], [5,19]
[6,22], [10,20], [2,0], [7,11], [3,6], [3,8]}.

Based on this, conventional optimizations: code motion, commoning, constant propogation, etc.; are performed. Then a set of three optimizations, structurally similar to these, that address problems peculiar to SETL and other VHLL's, are performed. These are:

- *Type determination.* Ascertain, for each use of a variable A, the smallest set of types that A might assume at that point. The result of this analysis is used both to map data more efficiently and to emit more efficient code sequences for operations on "simple" data objects. It also pinpoints the objects that represent sets (and maps) for the subsequent optimization.

- *Set inclusion and set membership determination.* For variables S_1, S_2 determined to represent sets, and for an arbitrary variable X, determine at what program points it can be safely asserted that

  $$S_1 \subseteq S_2 \text{ or } X \in S_2$$

  This information is used to select more efficient representations for sets.

- *Copy elimination.* Determine when expensive copying may be safely replaced by destructive updating.

In the subsequent sketches of these phases, we shall use the program of Figure 4, taken from [17], to illustrate and motivate the optimizations.

Type Determination. Type determination algorithms attempt to determine, from the text of a program, the "types" of variables at various program points. For example, in the sequence

```
a := b+1;
c := a+d;
```

it must be the case that d is an integer. To conclude this requires a series of deductions based on the fact that, in SETL, only integers can be added to integers. From this it follows that b, hence a, hence d, must be *integer*.

```
program HUFFCODE;
repr S: string; end repr;
read ( S );
FREQ = { }; CHARS = { };
( forall C in S )                    /*compute frequencies */
   if FREQ( C )=om then
            CHARS with: C; FREQ( C )=1;
   else FREQ( C )+; end if;
end forall;
HTREE = { };                         /*build Huffman tree */
( while #FREQ>1 )
   [C1,F1]: = GETMIN (FREQ); [C2,F2]:=GETMIN (FREQ )
   FREQ ( C=C1+C2 ):=F1 + F2;
   HTREE (C1): = [C,'O']; HTREE(C2):=[C,'1'];
end while;
HCODE = { }; /*compute actual Huffman codes */
( forall C in CHARS )
   HCD:=''; B:=C;
   ( while ( NDOE:=HTREE( B ) ) ≠( om )
            [B,LR]:=NODE; HCD:=LR+HCD;
   end while;
   HCODE( C ):=HCD;
end forall;
print ( Huffman Code is; HCODE );
procedure GETMIN ( rw G )/* remove min elf from F */
   C:=om; N:=om;
   ( forall M=F( A ) )
            if N=om or M<N then [ C,N ];=[A,M]; end if;
   end forall;
   F less f:=C;
   return [C,N]
end procedure GETMIN;
end program HUFFCODE;
```

Figure 4. SETL Program to Compute Huffman Codes

Type determinination algorithms attempt to develop a small, but conservative, set of types that may be associated with a variable at a given point. That is, if a set T of types is associated with a given variable use, it must be provably the case that, for any program execution that reaches that use, the type of the variable will be in set T. The type determination algorithm used in the SETL compiler is due to Tennenbaum. See [18] for details. A slightly more powerful algorithm can be found in [19].

Type determination algorithms rely on a finite lattice of types developed from a set of base types and a set of type generators. The base types used in SETL are: undefined atom, integer, real, character string, blank atom, label, boolean value, boolean string, null set, null type, error

Type generators are:

$t_1 \mid t_2$	the type t_1 or t_2
$\{t\}$	a set whose elements are of type t
$\langle t_1, t_2, \ldots, t_K \rangle$	a K-tuple whose components are of types $t_1, \ldots t_K$
and $[t]$	a tuple of unknown length whose components are of type t.

Although these type generators can be applied recursively to generate types of arbitrary complexity, almost all type-determination algorithms restrict them to types of finite complexity, lumping all others into the type *general* (but see (20) for an exception). By using the inclusion relation, the set of types can be treated as a lattice. Thus,

$$t_1 \le t_1 \mid t_2,$$

and

$$\{t\} \le \{t_1 \mid t_2\}.$$

Type determination algorithms exploit this lattice of types in a way now standard for global optimizers. First, based on entirely local considerations, a "safe" element of the lattice is assigned to each variable use. Then, global propogation algorithms are used to refine this information until the smallest safe lattice value is determined.

Thus, based purely on local considerations, we can only conclude that in

$$C := A + D \qquad (1)$$

each of the variables are of type *general.* However, in the statement

$$A := B + 1; \qquad (2)$$

we can conclude that A, B are of type *integer.* But then by global considerations, if (1) can be reached only from (2), and all parts from (2) to (1) are free of other assignments to A, we can conclude that, at (1), C, A, and D are of type *integer.*

Analysis of this sort allows us to conclude, for the Huffman codes program of Figure 4, that

CHARS	is a set,
FREQ	is a map from strings to integers,
C1, *C2*	are strings,
F1, *F2*	are integers,
C	is a string,
B	is a string,
HCD	is a string,
HCODE	is a map from string to string,
HTREE	is a map from string to string,
LR	is a string,
NODE	is a tuple (string, string),
etc.	

Set Inclusion and Membership Determination. The default representation for sets and maps is a linked hash table. Consequently, many simple operations on sets include relatively expensive hashing operations in their execution. The primary motivation of this optimization is to eliminate as many hashing operations as possible, substituting more direct access to values. For example, to perform the test

$$A \in X,$$

the value of A must be retrieved and hashed into X. On the other hand, if it could be established, for some set S, that

$$A \in S,\ X \subseteq S,$$

then membership in X could be represented by a flag in each element of S. Then

$$A \in X$$

would result in a test of a flag in a word indirectly addressed through A. Similarly, under favorable circumstances,

$$X\ \mathit{WITH} := A$$

could be implemented by an indirect setting of a flag.

The SETL language permits sets such as S, called *bases*, to be introduced into a program. A base cannot be manipulated in the executable part of a SETL program; it is used only as part of the declarative structure. If B has been declared a base, then if X is an ordinary program variable, one may declare

$$X \in B$$

or

X: [local/indexed/sparse] set ($\in B$)

or

X : [local/indexed/sparse] map ($\in B$)*

When such declarations are made, only B assumes the default set representation. Elements are hashed into B as required to maintain the truth of the declared relations.

Elements declared "based" are specially represented. For each set X declared

local set ($\in$ B)

a flag field is allocated in the elements of B which is on exactly when an element is in X. In the case that X is declared

indexed set ($\in B$)

an index into a bit string is stored in X, and the flags are kept remotely in that string. If a set X is declared

sparse set ($\in B$)

it is kept as a linked hash table of elements that point into B.

The meaning of the declarations for maps is similar. Rather than reserving space for a flag, space is reserved for the range element.

If used appropriately, basings can reduce the storage requirements and accessing cost of SETL algorithms to costs reasonably competitive with algorithms coded in more conventional languages (the primary additional cost being only that associated with maintenance of the based sets). However, the analysis of a SETL program required to use this basing language properly is extremely demanding and tedious. Hence, the SETL compiler contains a phase that attempts to determine automatically an appropriate set of basing statements.

There are two parts to this algorithm. First an appropriate set of bases is generated. The the issue of local vs. indexed vs. sparse is resolved.

The algorithmic strategy is familiar. First, each statement is examined in isolation to determine whether it would be profitable (in terms of avoiding hashing) to introduce one or more bases. If so, new bases are generated or introduced. For example

S WITH:= X;

causes a base b_j to be introduced, and the representations

S set ($\in b_j$), $X \in b_j$

to be established.

In some cases, a base is introduced that cannot be guaranteed to speed up executiion, but that certainly will not slow it down. Such bases are called *neutral* bases.

Finally, some statements do not lead to bases; e.g.,

```
X:=X+1.
```

Once basing information has been attached to each statement, global propogation techniques, exploiting use-use chaining, are employed to merge its bases into equivalence classes. Then these equivalence classes, representing maximal bases, are pruned to remove those that are ineffectual, i.e., those for which all basings are neutral.

This results in a set of basings attached to program points. Thus a program variable may have different representations in different parts of the program. This problem is overcome by inserting explicit representation conversion statements at low frequency program points where data flow analysis indicates that they are required.

Finally the basings are augmented by local, sparse, or indexed attributes according to program usage.

Application of this analyzer to the Huffman codes program results in the definition of one base *BASE*, and the following basings:

```
GETMIN  procedure ( indexed smap ( ∈BASE ) int ) tuple ( ∈BASE , int )
B∈BASE
C1∈BASE
    .
    .
    .
CHARS   indexed set ( ∈ BASE )
FREQ    indexed smap ( ∈BASE ) int
HCODE   local smap ( ∈ BASE ) string
HTREE   local smap ( ∈BASE ) tuple ( ∈ BASE , string )
NODE    tuple ( ∈ BASE , string )
```

Copy Elimination. SETL contains a number of operators, such as

```
S WITH: =A
S LESSF: =A.
```

For these operations, the most efficient implementation is to generate code that modifies the current value of *S*. Since, however, the object currently associated with *S* may be pointed to from other structures, a copy of *S* must be made before *S* is modified, if the SETL value semantices are to be preserved. But the paradigm

```
S = nl;
( forall X in Y | Z )
S  :=  S + {X};
end forall
```

is an important one that requires an efficient implementation. Clearly a copy is not required. However, the program

```
C = nl;
S = nl
( forall X in Y | Z )
    S := S + {X};
    C := C + {X}
end forall
```

does require *S* to be copied, since the old value of S is also pointed at by C.

The SETL optimizer therefore must determine, for every potential destructive use of a variable, whether any other variable that might hide a pointer to that data structure is live at the destructive use. If so, a copy is required; otherwise, the data structure can be modified directly.

The algorithm has two parts: an analysis of the program to identify possible variable occurrences that might contain a pointer to the data structure of interest, followed by a standard live-dead analysis to determine the need for copying.

To identify the variable occurrences that might contain a pointer an analysis reminiscent of constant proprogation is employed. First, using use-use chains, information gathered during the two previous analyses, and the nature of the SETL operators, a backward pass is made to determine the set of variable occurrences *from* which the pointer of interest might have *come*. Then, on a similar forward pass from each of these, the analysis determines another set of variable occurrences *to* which the pointer might have *gone*. The union of these two sets of variable occurrences are analyzed for·livliness at the point of interest. If none of them are live, a copy is not required.

Based on this analysis, the following copies can be eliminated from the Huffman codes program:

F	at statement	41
CHARS	at	9
FREQ	at	10
FREQ	at	11
FREQ	at	18
HTREE	at	19
HTREE	at	20
HCODE	at	30

V. SUMMARY

Straightforward implementations of VHLL's lead to inefficient programs, However, appropriate analysis of the program can lead to impressive improvements in execution times. It is therefore reasonable to believe that, with more research, both in the design and implementation of such languages, that efficient VHLL translation can be achieved.

BIBLIOGRAPHY

[1] Goldberg, P.C., The Design of Very High Level Languages. This volume.

[2] APL/360 Primer. IBM Corp. Doc. No. GH20-0689.

[3] Schwartz, J., On Programming, An Interim Report on the SETL Project. Comp. Sci. Dept., Courant Inst., New York Univ., 1973.

[4] Aho, A.V., and Ullman, J.D., *Principles of Compiler Design*, Addison-Wesley, Reading, Mass., 1977.

[5] Weihl, W.E., Interprocedural Data Flow Analysis in the Presence of Pointers, Procedure and Label Variables. Proc., Seventh Annual ACM Symposium on the Principles of Programming Languages, 1980.

[6 |)Harrison, W.H., and Johnson, S.C., Debate: The Role of Global Optimization in Compilers of the Future. Proc., Eigth Annual ACM Symposium on the Principles of Programming Languages, 1981.

[7]Abrams, P.S., An APL Machine. Tech. Report No. 3, Digital Systems Laboratory, Stanford Electronics, Stanford, Calif., Feb., 1970.

[8] Hassitt, A., and Lyon, L.E., Efficient Evaluation of Array Subscripts of Arrays, IBM J. of Research and Development, 16, 1, Jan., 1972.

[9] Perlis, A.J., Steps Toward an APL Compiler, Computer Science Research Report 24, Yale U., January, 1974.

[10] Guibas, L.J., and Wyatt, D.K., Compilation and Delayed Evaluation in APL. Proc. Fifth Annual ACM Symposium on Principles of Programming Languages, 1978.

[11] Henderson, P. and Morris, J.H., A Lazy Evaluator, Proc. Third ACM Symposium on Principles of Programming Languages, 1976.

[12] Van Dyke, E.J., A Dynamic Incremental Compiler for an Interpretive Language, Hewlett-Packard Journal 28, 11, July, 1977.

[13] Bauer, A.M., and Saal, H.J., Does APL Really Need Run-time Checking?, Software-Practice and Experience 4, 1974.

[14] Schwartz, J. T., Automatic Data Structure Choice in a Language of Very High Level, Comm. ACM 18, 12, 1975.

[15] Dewar, R.B.K., *et al*, Programming by Refinement as Exemplified by the SETL Representation Sublanguage, ACM Trans. on Prog. Languages and Systems 1, 1, 1979.

[16] Schonberg, E., *et al.*, An Automatic Technique of Selection of Data Representations in SETL Programs, ACM Trans. on Prog. Languages and Systems 3, 2, 1981.

[17] Freudenberger, S.F. *et al.*, Experience with the SETL Optimizer, to appear.

[18] Allen, F. E., Control Flow Analysis, Proc. Symp. Compiler Optimization, SIGPLAN Notices 5, 1970.

[19] Allen, F.E., and Cocke J., A Program Data Flow Analysis Procedure, Comm. ACM 19, 3, 1976.

[20] Tenenbaum, A., Automatic Type Analysis in a Very High Level Language, Ph.D. Dissertation, New York Univ., Oct., 1974.

[21] Kaplan, M.A., and Ullman, J.D., A General Scheme for the Inference of Variable Types, Proc. Fifth Annual ACM Symposium on Principles of Programming Languages, 1978.

[22] Suzuki, N., Inferring Types in Smalltalk, Proc. Eigth Annual ACM Symposium on Principles of Programming Languages, 1981.

A STUDY OF PROLOG.

Hervé GALLAIRE

Laboratoire de Marcoussis - C.G.E.
Route de Nozay
91460 MARCOUSSIS (FRANCE)

INTRODUCTION

This paper is intended to give an overview of PROLOG, both as a programming language and as a tool for some applications in automatic programming.

By now PROLOG is a well respected language, used mainly in Europe and catching up in United States where it is confronted to LISP and its many dialects. There are many reasons for such a situation, the most important being that LISP fulfills most of the expectations of its users. However when viewing PROLOG not just as a language but as a representative of the so-called "Logic Programming" languages, we believe this situation will change when it is realized in the AI community that PROLOG offers very desirable high-level features not belonging to LISP and its dialects.

The language was defined around 1971 at Marseille by A. COLMERAUER and Ph. ROUSSEL, for applications in the natural language analysis area (1) ; the first interpreter was written by Ph. ROUSSEL, soon followed by a portable FORTRAN implementation of G. BATTANI and H. MELONI (2) in 1973, reported in (3). Basically, the language has been stable up to now, although many different implementations have been seen, varying in minor aspects, varying in efficiency and in system interface capabilities.

A list of implementations is in appendix 1 of a PROLOG exercise book, by H. Coelho, J.C. Cotta and L.M. Pereira (4). Three main sources of PROLOG can be seen :

A. W. Biermann and G. Guiho (eds.), Computer Program Synthesis Methodologies, 173–212.

- The Marseille source, which as was said originated all PROLOG developments, and was most influential in France

- The Edinburgh source which, through the work of Warren (5) had a major influence on Prolog distribution in England and United States

- The Budapest source, which led to very many applications in Hungary (7)

Further implementations, still classical, are those of M. Bruynooghe (49), of G.M. Roberts (8). Newer implementations are underway, which sometimes involve different internal representations, and for some of them extensions to PROLOG ; we shall comment on them later in this paper.

But PROLOG is more than a language. The "logic programming" concept (9), (10), (11), (12) helped formalize many of the ideas which were basic in PROLOG and give better understanding of the whole PROLOG programming and interpretation process. Two recent workshops have been held, which give an up-to-date account of present research in this area (13), (14).

Although PROLOG, as a programming language can be presented only from programming concepts, we believe its most interesting features are drawn from it being embedded in logic. This is why we shall adopt a mixed presentation of the PROLOG language from both points of view.

The paper is divided in three parts. First PROLOG is presented and programming examples are given. Second we indicate some directions in which PROLOG can be enhanced. Finally we shall study applications of PROLOG to automatic programming.

Presenting a language is a difficult exercise. It will be presented by a user, without any idea of presenting it as would do PROLOG fathers or foremost PROLOG pushers. Any misconception will be mine. No claim of completeness will be made, and the syntax retained will be my own, differing from all current implementation in inessential ways.

Chapter 1

PROLOG - SYNTAX AND SEMANTICS

The main interest of PROLOG resides in its ability to let the programmer express definitions of functions, or rather of relations, with more emphasis on the definitions than on the ways to compute the relations. Of course it is no so simple to separate "what" from "how" and PROLOG goes only one step in that direction. Due to its high-level constructs, higher than in most other languages including LISP,PROLOG partly implements the declarative approach to programming. PROLOG was not the only trial in that direction but it is presently the most successful one.

In (15), Mc Dermott relates PROLOG to PLANNER, CONNIVER and like languages which aimed at similar goals. He points out several reasons for PROLOG success and for PLANNER failure which have to do with PROLOG simplicity, efficiency and more powerful features, such as pattern-matching, data-structuring and tools for massaging programs seen as data.

We shall indeed review these features. Most of them come from the original ideas which lead to PROLOG, namely that of "PROgramming in LOGic". However one does not need to be dogmatic about the logic aspects, as many PROLOG features do not borrow much from pure logic ; although we shall not be dogmatic, we shall insist on logic because it explains PROLOG unusual features.

1. AN EXAMPLE

Any PROLOG program can be read in at least two ways : a logic interpretation, and a procedural interpretation

1.1. A logical interpretation

A program is a set of clauses, restricted to Horn clauses. The following two clauses define the concatenation relation :

(1) CONCAT (nil, X,X) ←
(2) CONCAT (X.L, Y, X.Z) ← CONCAT (L, Y, Z)

Notation : Uppercase letters denote variables, while lower case denote constant values ; the '.'operator denotes concatenation (infix notation here, but prefix is standard) ; 'nil' is a constant defined as the empty list (i.e any list is terminated by nil, e.g a.b.nil is a list but a.b is not).

Clause (2) above can be read as :

List X.L, concatenated with list Y, yields list X.Z if list L, concatenated with Y, yields list Z.

Clause (1) reads as :

Concatenation of X to nil yields X

Such a program is a declarative definition of CONCAT rather than a procedural one, i.e it does not really say how CONCAT should be computed. Further, CONCAT being so defined is really a relation rather than a function, i.e it defines subsets of
{List} x {List} x {List}

instead of applications

{List} x {List} → {List} ;

with such an interpretation, queries about relation CONCAT can be asked. Given any triple of arguments do they belong to CONCAT ? given any couple of arguments does there exist a third argument such that the triple belongs to CONCAT ? and so on.

The logic interpretation of such queries and definitions is as follows : a query is a formula about which it must be shown that it is a logical consequence of the axioms corresponding to the definitions of CONCAT. Thus a proof procedure is to be used to prove the validity of the query ; this proof procedure, which in PROLOG case happens to be a refutation procedure (Resolution) when it succeeds, ought to yield values for variables which make the query a consequence of the axioms ; thus instead of computing values, using assignment, PROLOG relies upon the substitution of values to variables made in the proof procedure.

As an example, consider the query
← CONCAT (a.nil, b.nil, X)
whose semantics, according to the above discussion is :

Find the set of values v of variable X such that
concat (a.nil, b.nil, v)
holds i.e. is a logical consequence of the program (of clauses) given by clauses 1 and 2 above
The refutation proof is as described below interpreting A ← B, as the logical implication B → A, i.e as the clause ¬B V A and recalling that resolution is a cancellation process of negative literals with positive ones, computing substitutions or unifications to get the necessary matching (Fig. 1) :

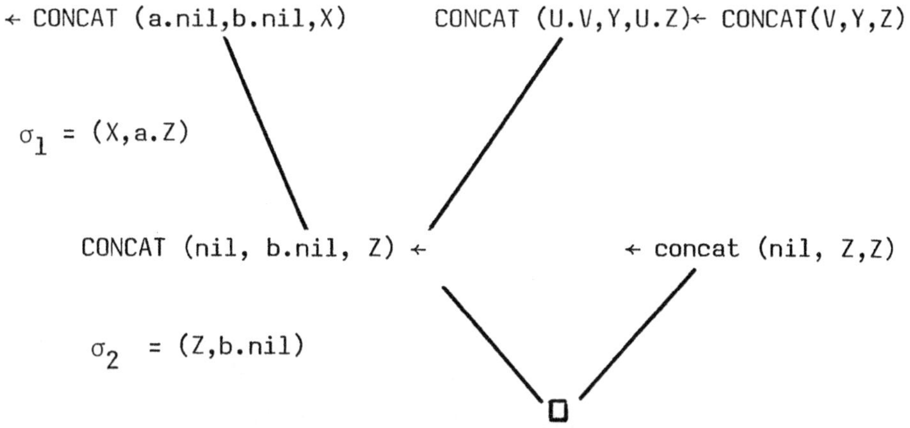

The consequence is established with substitution $\sigma_1 \quad \sigma_2$ i.e
(X,aZ) (Z,b.nil) hence (X,a.b.nil)

Fig. 1

As another query consider CONCAT (X, Y, a.b.nil) i.e
← CONCAT (X,Y,a.b.nil)
One should hope for 3 triples satisfying this relation, hence 3 couples of values X, Y ; namely ((X,nil), (Y,a.b.nil)), ((X, a.nil),(Y, b.nil)), ((X, a.b.nil), (Y, nil)).
The refutation trees are given in Fig. 2.

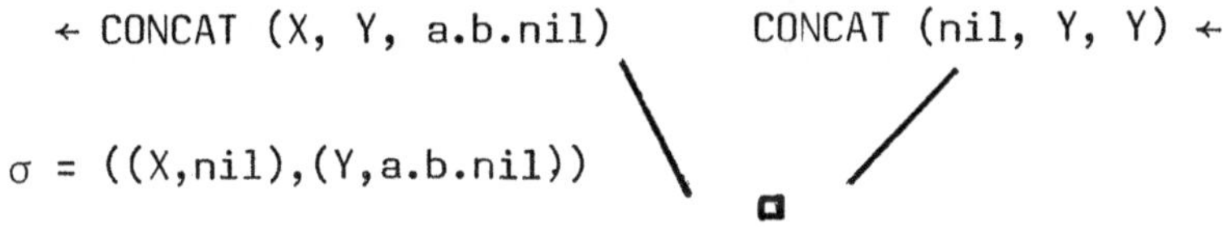

But also

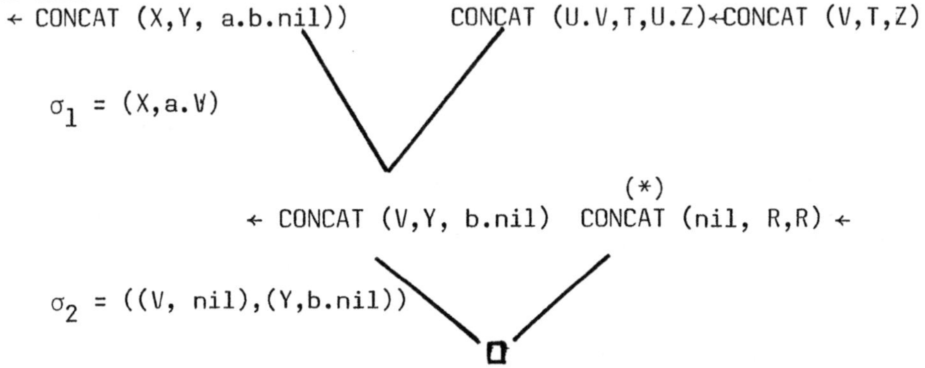

thus $\sigma_1 \quad \sigma_2$ yields
((X,a.nil), (Y,b.nil))

Fig. 2

Finally replacing node (*) above by the second clause of the program yields a new bottom of proof tree shown in Fig. 3.

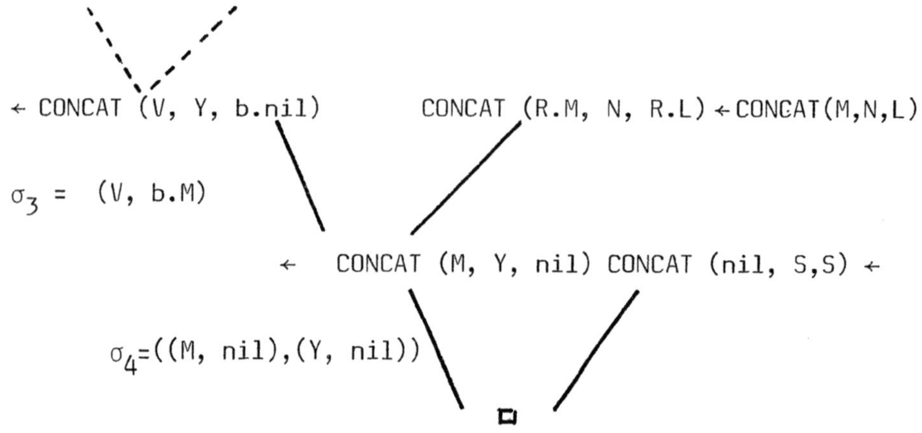

thus $\sigma_1\ \sigma_3\ \sigma_4$ yields ((X, a.b.nil), (Y, nil))

Fig.3

Of course the order in which solutions are found, trials and errors organization are up to the theorem prover.

Before coming to the procedural interpretation let us remark that what we called up to now the "definition" of relation CONCAT, is at best half a definition ; i.e nowhere did we check that the definition was complete and that each part of the definition was an equivalence rather than an implication. We shall comment later on these problems.

1.2. A procedural interpretation

Let us now see clauses (1) and (2)

(1) CONCAT (nil, X, X) ←
(2) CONCAT (X.L, Y, X.Z) ← CONCAT (L, Y, Z)

as rewriting rules, with left-hand sides being rewritten in their right hand sides whenever the left hand side matches an occurrence of a query (or of a problem to be solved) ; equivalently one can see the left hand side as a procedure head, with the right hand side as a body of the procedure. In order to make this interpretation more evident, A. Colmerauer now proposes a new implementation in which clauses are written as follows :

(1) CONCAT (nil, X, X) →
(2) CONCAT (X.L, Y,X.Z) → CONCAT (L, Y, Z)

It is interesting to see this change of syntax which to me reflects a change of attitude of PROLOG inventors.

Although (1) and (2) are procedure calls their parameter specification is not a standard one in procedure handling. The interpreter of procedure calls will take care of parameter passing, depending on which variables values are known at call time.

The interpreter builds an and-or tree (16) whith OR nodes corresponding to alternative procedure calls, solving the same problem, and AND nodes corresponding to the bodies of procedures. In the current example, there is no AND node, as the body of each procedure is made of but one literal. In fact the OR node is not built as the PROLOG interpreter implements a backtracking algorithm which will be presented later. Fig. 4 shows an and-or tree for CONCAT.

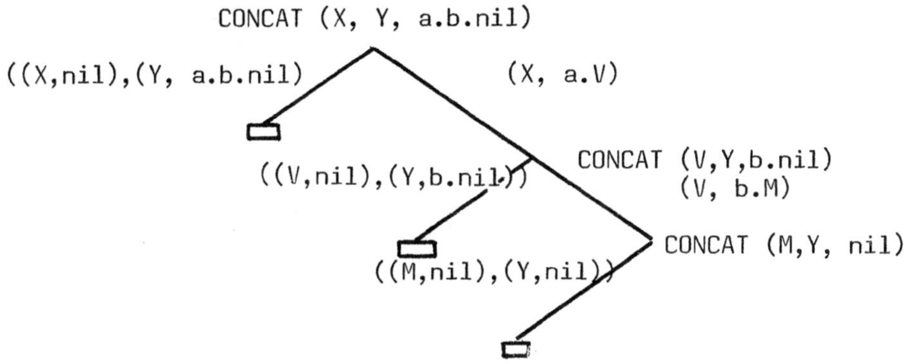

Fig. 4

An example including true AND nodes would be made of the following clauses :

```
      GF (X,Y) ← F(X,Z), F (Z,Y)          (a)
      GF (X,Y) ← F (X,Z), M(Z,Y)
       M (Z,Y) ← M(Z,K), B(K,Y)
  F(john, mary) ←
  M(mary, paul) ←
  B(paul, peter)←
```

For instance clause (a), under the logical interpretation would have been read as :
X is GF of Y if X is F of some Z and Z is F of Y
or X is GF of Y if there is Z such that X is F of Z and Z is F of Y.
In the procedural interpretation clause (a) becomes a procedure with head GF (X,Y) and body F (X,Z) F (Z,Y) which is to replace the head upon a call of GF.

The and-or tree corresponding to call GF (john, peter) is drawn Fig. 5.

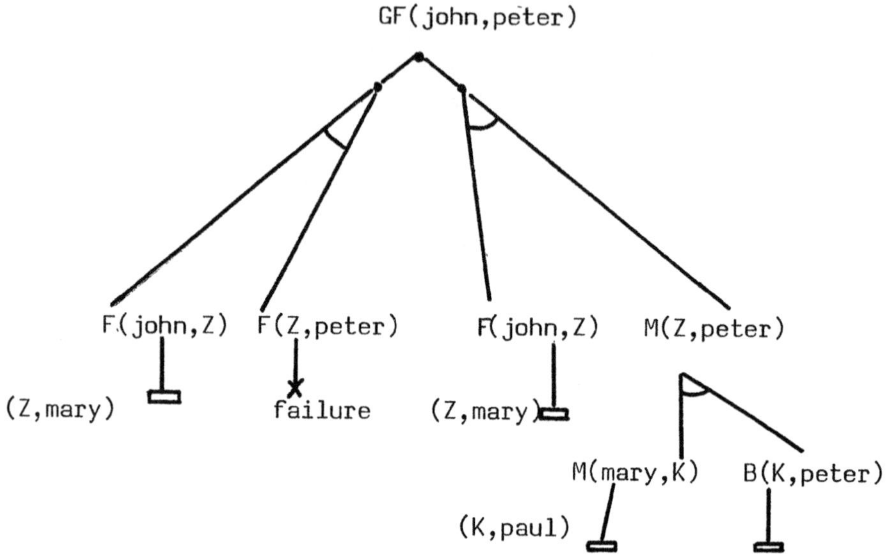

Fig. 5

The node called "failure" in Fig. 5 indicates that no match is possible with F (Z,peter) when Z is instantiated as "mary". This call succeeds (john is GF of peter).

In fact Prolog build this tree in a specific way, namely one alternative at a time, working on the leftmost literal and backtracking on failure. Fig. 6 and Fig. 7 illustrate two PROLOG trees corresponding to the above proof tree ; nodes have been numbered in the order in which the PROLOG interpreter visits them for expansion.

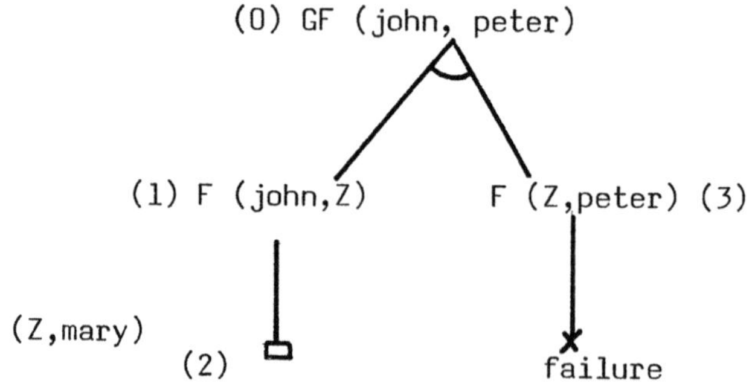

Fig. 6

Nodes in Fig. 6 are explored in order (1), (2), (3) - (3) is a failure node ; backtracking takes back to node (0) from which the alternative is taken, giving the tree of Fig. 7. as no more choice is left in nodes 2 and 1.

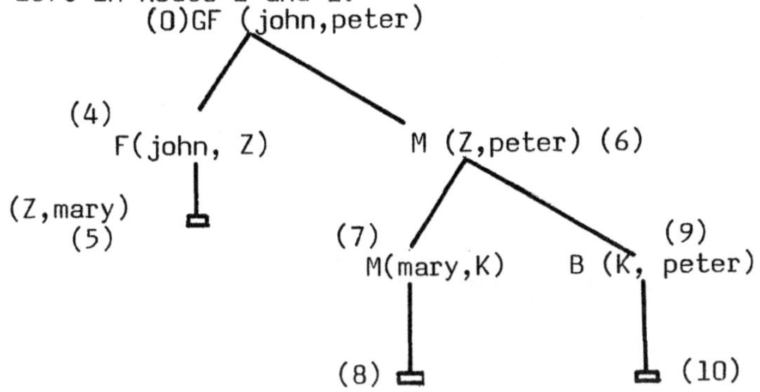

Fig. 7

Our first example illustrated one of the nice characteristics of PROLOG i.e the non-determinism of clauses ; several types of non-determinism are possible ; they all participate of the declarative approach and bring something to it ; it is often said that such features are difficult to implement efficiently (i.e as efficiently in each case as would be the corresponding deterministic algorithm) ; another important characteristic shown in the example was that of manipulations of terms (i.e lists) and the progressive "computation" of the values of variables.

Before turning to the basics of PROLOG let us mention that a third interpretation of clauses can be given : a head of a clause is a problem to be solved ; its body is then the conjunction of subproblems to be solved for solving the given problem. This is the problem reduction formulation of problem resolution systems. It will not be pursued here, but it is common to use it in applications such as e.g plan generation (in robotics).

2. PROLOG : A BASIS

Because a major part of PROLOG rests upon a formal basis, i.e logic, we shall, in this paragraph, present PROLOG from this formal viewpoint. Truly, important aspects of PROLOG are not in logic, and can better be explained without any subtle and twisted argument from outside of logic ; the '/' (cut) example is the best known example ; input and output belongs to that family too. Because we are not dogmatic about PROLOG we do not bother too much about this ; we prefer to explain with a neat basis what can be so explained. In what follows a knowledge of logic at the level of Chang and Lee (17) is assumed ; in particular the resolution proof procedure and its basic variations are freely discussed.

2.1. Horn clauses

PROLOG admits Horn clauses only. A clause is a first order well formed formula (wff), universally quantified, which is a disjunction of negative and positive literals :

$$A1 \vee \ldots \vee Am \vee \neg B1 \ldots \vee \neg Bn \qquad n, m > 0$$

Where Bj's, Ai's are positive literals or atomic formulas.
Thus $\neg$ Bj is a negative literal.
The Bj's are the conditions of the clause, and Ai's are its conclusions : an equivalent formulation of the above is :

$$B1 \wedge \ldots \wedge Bn \quad A1 \vee \ldots \vee Am$$

which can be read or noted :

$$A1 \vee \ldots \vee Am \text{ if } B1 \wedge \ldots \wedge Bn$$

$$A1 \vee \ldots \vee Am \leftarrow B1 \wedge \ldots \wedge Bn$$

Variables being universally quantified, if this formula contains variables X1, ... Xk it becomes

$$\forall X1 \ldots \forall Xk \; (A1 \vee \ldots \vee Am \leftarrow B1 \wedge \ldots \wedge Bn)$$

If n = 0 it is an unconditional assertion

$$\forall X1 \ldots \forall Xk \; (A1 \vee \ldots \vee Am)$$

while if m = 0 it is a refutation $\forall X1 \ldots \forall Xk \; \neg (B1 \wedge \ldots \wedge Bn)$

Horn clauses are a restriction of the clausal form containing but one positive literal, at most, per clause : $0 \leqslant n \leqslant 1$. Thus they have one of the following three forms :

(1) $A \leftarrow B1,\ldots,Bn$
(2) $A \leftarrow$
(3) $\leftarrow B1,\ldots,Bn$

where A, Bi's are positive literals. If X1,...,Xk are all the variable of (1) appearing both in A and in some (at least one) Bj, and if Y1,...,Yr are variables appearing in some Bj and not in A, then (1) can be written (under non clausal form)

(1') $\forall X1 \ldots \forall Xk \; (A \leftarrow \exists Y1 \ldots \exists Yr \quad B1 \wedge \ldots \wedge Bn)$

Similarly (3) becomes

(3') $\leftarrow \exists Y1. \; \exists Yr \quad B1 \wedge \ldots \wedge Bn$
(i.e $\neg \exists Y1 \ldots \exists Yr \,.\, B1 \wedge \ldots \wedge Bn$)

clauses (1), or (1') express procedures (heads, bodies) ; clauses (2) express assertions ; clauses (3) or (3') express questions, refuted problems, or procedure calls : if it must be shown that there exist values of Y1,... Yr such that B1 and B2 ... and Bn are all simultaneously satisfied (true) then the logical statement (3) or (3') is the refutation of such a statement.

The proof procedure is thus a refutation procedure called resolution which will try to derive the empty clause from such wffs (3) ; if it succeeds, as the empty clause cannot be true, and as the derivation process (resolution which is a process which cancels positive literals (A) with the negative ones ($\neg$ Bi)) preserves logical consistency, this means that the initial set of clauses is not consistant ; thus $\neg\exists$Y1 ... $\exists$Yr B1 $\wedge$ ... $\wedge$ Bn is not consistant with the other clauses ; finally this shows that $\exists$Y1 ... $\exists$Yr B1 $\wedge$ ... $\wedge$ Bn is consistant with the other clauses, i.e it has been shown that there exists values of Y1,..., Yr such that B1,...,Bn are simultaneously satisfied, and these values must be extracted from the proof procedure.

2.2. More about the proof procedure

Only logic can provide us with a satisfactory discussion of what is or what is not a correct implementation of the proof procedure. More explicitly one would like to answer the following questions :

- is there a complete proof procedure for Horn clauses ?
- if so, what freedom do we have in implementing it ?

The first question is positively answered ; i.e if a wff of type (3) is not consistant with a given set of clauses, then the proof procedure known as resolution will find it.

The second question is more subtle. What choices are offered in the various steps of the proof :

- given a (so called resolvent) wff of type (3) say $\leftarrow$ C1,...,Cs, when resolving it with another clause, of type (1) or (2), one of the Cj must be chosen to start with : this is the choice of literal ;

- given a choice of literal to be resolved upon, there may be several A, heads of clauses of type (1) or (2) to unify with the chosen literal : this is the choice of clause.

Given these two choices at each step of the proof, the refutation will proceed until it fails or succeeds.

The fact that resolution is complete, as recalled above, does not tell you which choices of literal and clause at each step make it complete. It assumes exhaustive combinatorial search. Several implementations of resolution have been proved complete : SL resolution, linear with set of support, ... PROLOG implements the so-called LUSH resolution (48) which is complete for Horn

clauses ; this allows one to
- take an initial type (3) clause : the goal G
- take a program (i.e a set of clauses of type (2) and (1))
- use linear resolution with G as set of support

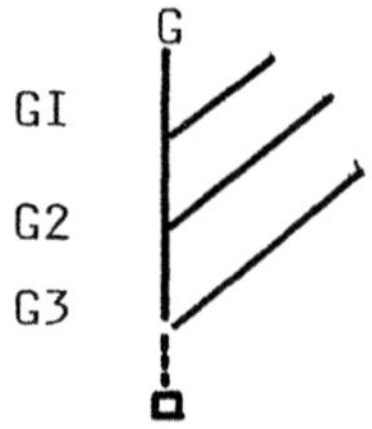

where Gi is C1, ..., Cs

- use any literal selection function.

The result, which is intuitive, but established formally only in logic (and not in programming terms), says that any literal selection criterium is as good as any other as far as completeness is concerned.
Clearly the choice of clause is unimportant for completeness too as long as we can guarantee that all possible candidates will be tried in some order.

Completeness is not enough ; freedom of choice is not enough either. One would like efficient proof procedure. The problem is to find out how to make the most efficient choice at each step. Let us illustrate this problem which is known as the control problem :

using the above clause GF (X,Y) ← F (X,Z), F (Z,Y) with variables X, Y free or bound at call time one will get as resolvent ←F (X,Z), F (Z,Y) with X, Y free or bound.
It should be clear that depending on values of X and of Y, depending on whether X or Y is free, one will be better advised to start resolving F (X,Z) first rather than F (Z,Y) first or conversely.

If F (U,V) is relation "U is father of V", then it is clear that since a child has but one father while a father may have several children, it is more efficent to resolve (i.e to "evaluate" or to "call" in terms of the procedural interpretation) F (Z,Y) first when Y has a value ; similarly if Y is free and X is known it is clearly more efficient to call F (X,Z) first. Because not much is known on this subject most systems implement a fixed strategy of choice. We shall come back to this point later ; the fixed choices made by PROLOG are the following : leftmost literal of the resolvent first,

clauses are tried in the order in which they have been input, and backtracking is chronological

Up to now we have discussed the top-down interpretation of logic only.

We could have discussed the bottom-up interpretation, in which one derives new assertions of type (2) from assertions of type (2) and clauses of type (1) until an assertion unifies with a clause of type (3) to give the empty clause. Fig. 8 describes the refutation tree corresponding to bottom-up interpretation of resolution.

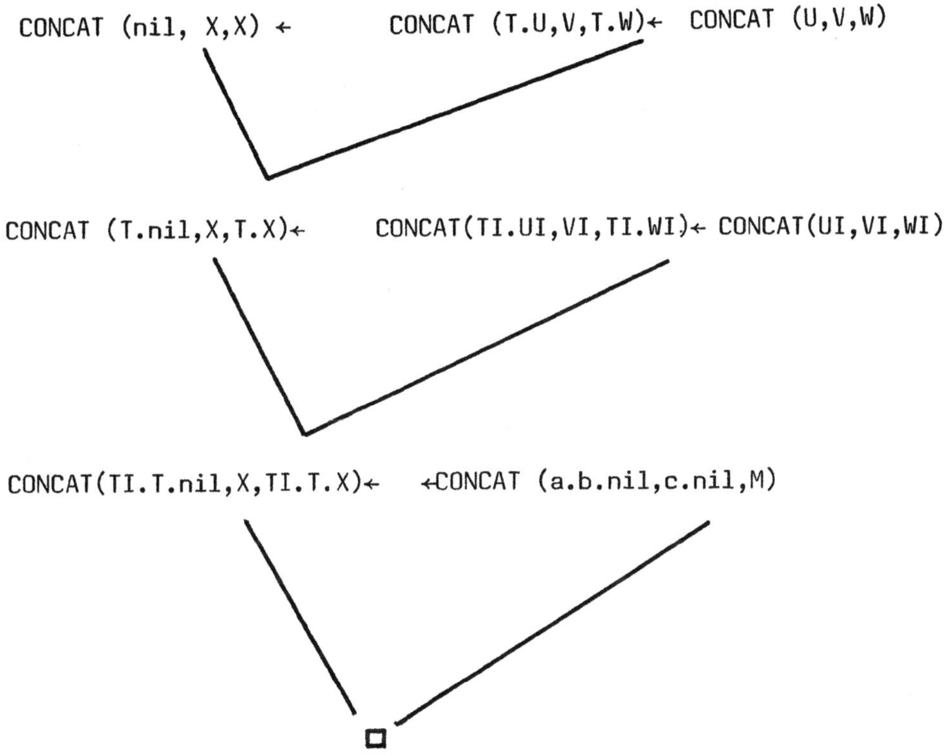

Fig.8

The problem to be solved was represented in Fig. 8 as

← CONCAT (a.b.nil, c.nil, M)

2.3. Limitations of clausal logic

Is clausal logic too much of a constraint in expressing programs ? is Horn clause logic really even more limitative ? It is known that clausal logic is as powerful as first order logic, but it is clear that it is not as natural as is "natural" logic (i.e first order logic). For example, consider

Some men are animals

$$\exists X\ (M\ (X) \wedge A\ (X)) \leftarrow$$

Such an assertion is represented by the set of clauses (i.e the conjunction of clauses)

$M\ (a)\ \leftarrow$
$A\ (a)\ \leftarrow$

where "a" is a Skolem function introduced by eliminating the existential quantifier. Which one conveys more semantics ? Also forms (3') and (1') above are often clearer than their respective clausal forms (3) and (1). Compare :

$$Anc\ (X,Y) \leftarrow Par\ (X,Y) \vee \exists Z\ (Anc\ (X,Z) \wedge Anc\ (Z,Y))$$

and its equivalent clausal form (10)

$Anc\ (X,Y) \leftarrow Par\ (X,Y)$
$Anc\ (X,Y) \leftarrow Anc\ (X,Z),\ Anc\ (Z,Y)$

which one is clearer and for which reasons ?
Theoretically Horn clauses are not restrictive ; it can be shown that logic, when restricted to Horn clauses can still represent the provability relation of first order logic. This answer which is obtained via a coding à la Gödel, is not satisfying. In reality it is not possible to express disjunctions as conclusions of a clause :

$A \vee B \leftarrow C$

is not expressible as is $A \leftarrow \neg B,\ C$
and thus $A \leftarrow (B \leftarrow C)$
is not expressible either.
Note that disjunction of hypotheses can be expressed in Horn-clause formalism. $A \leftarrow B \vee C$ is equivalent to the set

$A \leftarrow B$
$A \leftarrow C$

since $A \leftarrow B \vee C \equiv A \vee \neg(B \vee C) \equiv (A \vee \neg B) \wedge (A \vee \neg C)$.

Another Horn-clause restriction is the impossibility to handle, in the general case, definitions by equivalence (if and only if definitions). The "only if" part is quite often expressed by disjunction in the conclusion part of the clause. Consider a data base example in which one wishes to express that $p_1,\ldots p_n$ are all the persons belonging to a given set Obj ; it must be expressed as :

(a) $X = p_1 \vee X = p_2 \ldots \vee X = p_n \leftarrow Obj\ (X)$

(b) $\left\{ \begin{array}{l} Obj\ (X) \leftarrow X = p_1 \\ \text{------------------} \\ Obj\ (x) \leftarrow X = p_n \end{array} \right.$

(b) gives the if part of the definition of Obj (sufficiency) while (a) gives its only if part (necessity) and is not equivalent to a Horn form. Of course (b) is the Horn form of

$Obj (X) \leftarrow X = p_1 \vee \ldots \vee X = p_n.$

Where is an information such as (a) needed ? Any time one wants to establish a property dealing with all elements of set Obj it will be needed ; for instance if one wanted to know whether <u>all</u> elements of Obj are Male the wff $\forall X\ (Obj\ (X)) \rightarrow Male\ (X)$ cannot be proved in logic from (b) and the set of wffs $\{Male\ (p_1) \leftarrow\ , \ldots ,\ Male\ (p_n) \leftarrow\}$ as one does not know from (b) that $\{p_1, \ldots, p_n\}$ are the only elements of Obj.

This problem has been studied recently (18), (19), (20). Deductive systems which, as Prolog, do not allow to express informations such as (a) have been enhanced in that the interpreter can be told to assert " ¬P" when P is not provable from what is known (this is <u>negation by failure</u>). Three remarks are in order here :

i) such an extension of the interpreter is not dangerous as long as no use of ¬P is made in further deductions

ii) this extension is not sufficient to give sufficient power to Horn clauses :

assuming that negations are allowed in the condition part (or disjunctions allowed in the conclusions) "P← ¬ P" becomes allowed ; assuming a query about P, namely "P← " then the enhanced Horn clause interpreter would loop :

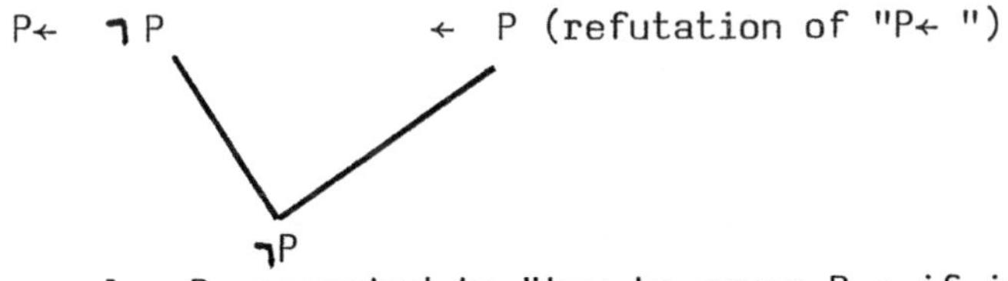

this new goal P converted to "try to prove P ; if it fails, assert ¬P" yields new goal ← P i.e a loop. Instead standard logic derives easily "P ← " from P← ¬P as $P \leftarrow \neg P \equiv P \vee P \equiv P$

iii) when used outside of Horn logic this enhancement becomes a dangerous one ; given :

B (X) V N (X) ← C (X)

C (felix) ←

then as "B (felix)← " cannot be proved it is asserted " ¬B (felix)← ". But similarly as "N (felix)← " cannot be proved " ¬N (felix) ← " can be asserted ; altogether

{C (felix), ¬B (felix), ¬N(felix), B (X) V N (X) ← C (X)}

is inconsistent.

One way out of this problem has been to use terms instead of assertions ; by having a predicate Obj whose argument is the list of its elements, one obviates the need for such conventions and discussions ; for instance the above problem about Obj and Male can now be expressed as :

← Obj (X), Test (X)

where predicate Test is defined as follows :

Test (nil)←
Test (U.L) ← Male (U), Test (L)

To be more general while stepping outside of first order logic (although Warren proposes to see this as syntactic sugar (21)), one would want to express this as (use of UNIV is necessary See 3.2.vi) :

← Obj (X), Test (Male, X)
Test (P, nil)←
Test (P, U.L) ← P (U), Test (P,L)

Making explicit the list of elements (X) to which the query applies one has avoided all theoretical problem about "only if". J.Mc Dermott, in (15) objects about the translation of the query as we have just done ; the reason he gives is that the formulation of the query "Does there exist an element in Obj which is Male" would be

← Obj (X), Member (A,X), Male (A)

and that the superficially identical queries (where "all" is replaced by "there exist")now become quite foreign formulations. Although this observation is correct, it is not as simple as it appears to use "natural" logic ; the existential query, expressed naturally is :

$$\exists X\ [\text{Obj}\ (X)\ \wedge \exists A\ (\text{Member}\ (A,X)\ \wedge\ \text{Male}\ (A))]$$

while the universal one is :

$$\exists X\ [\text{Obj}\ (X) \wedge \forall\ A\ (\text{Member}\ (A,X) \rightarrow\ \text{Male}\ (A))]$$

thus it is not sufficient to replace " $\exists$ A" by " $\forall$ A", a connector must be changed in order to correctly express the semantics of the query. The comparison then is to be made to natural language not to a formalism.

We are now ready to study PROLOG itself.

3. PROLOG - HOW TO USE IT

We shall distinguish between PROLOG basic features, and the system defined ones which are often non logical features ;

we must insist on the second type of features as they are more complex to comprehend and to use. As said earlier we do not follow systematically the syntax of an implementation ; but as far as semantics is concerned, we stay as close as possible to the Marseilles implementation (3)

3.1. Basic constructs in PROLOG

Basic constructs in PROLOG are Horn clauses, using terms as arguments of literals, constants and variables, goals assertions and rules. All of Horn logic is allowed and the user is free to define any set of predicates he wishes ; some technical restrictions are enforced, as usual in any language, such as the fact that in principle all clauses with identical head predicate name must be given sequentially and all at once. There is no declaration;the terms, arguments of predicates, are interpreted in infix notation.

PROLOG works on such constructs as we defined it in earlier sections i.e

- from the goal (the first resolvent)

- selecting the leftmost literal in the resolvent (the left most node in the derivation tree)

- matching it with clauses heads, selecting the first match (in the order in which clauses have been given), replacing the head by the body, propagating the instantiations made by matching both in the body and in the other literals of the resolvent, obtaining the new resolvent and a new set of nodes in the tree which are offsprings of the node corresponding to the selected literal ; if the body is empty, □ i.e the empty node, is attached as an offspring, to the selected node ; the resolvent is the set of the leaves of the tree;

- remembering the chosen clause, i.e remembering at that node of the derivation tree corresponding to the chosen literal, the list of clauses not yet selected for matching it ;

- backtracking, when a literal cannot be matched, to the node visited last in the tree and for which at least one more alternative is attached : chronological backtracking ;

- succeeding when the resolvent is empty i.e when no more nodes are left to visit, i.e when all leaves are □ ;

- failing partially when a node fails to be matched, and totally when the failing node is the root of the tree

- finding all solutions to the given goal by simulating failure after each global success of the initial goal

CAUTION : PROLOG distinguishes integers from other constants (128 is an integer, while a128 is a constant), and constants from variables (X is a variable)

CAUTION : as opposed to LISP, functional symbols are never evaluated ; they do not bear any operational semantic.

NOTE : as argued earlier, predicate symbols are relations, not functions ; non-determinism is a fact of life in PROLOG and the default option ; enjoy it (but see 3.2.).

3.2. System-defined features

CAUTION : system defined predicates are not subject to backtrack ; they are skipped on backtrack.

i) Some of these features are just shortcuts for the PROLOG programmer ; there is a standard prologue in which system defined predicates are common to all PROLOG users. Among them we have :

- arithmetic predicates : (Minus, Plus, Div, Mult, ...)
- comparison predicates : (Inf, ...)

Some restrictions are imposed on the use of such predicates : their arguments must be integers, and the first two arguments of basic mathematical operators (Minus, ...) must have a value at call time (the system would not want to evaluate Minus (X,Y,25)) ; also notice that in basic PROLOG only positive integers are handled ; user-defined extensions give easily negative integers, reals, complex numbers.

ii) Other features deal with input-output. Truly, original PROLOG was lacking in this area, but it was theoretically possible for the user to define his own I/O. Classical features are LIGNE, SORT (term) (i.e output a term), ENT (term) (i.e input a term), SORM ("text") (i.e output a message).

. CAUTION : These predicates are the first of a long list of non logical features ; by this we mean that although the syntax of such predicates is first order, their semantics is not explained with logic ; they act upon the environment of the program ; others will act on the interpretation process. Remember that we have decided not to try any logic interpretations of these obviously necessary features.

iii) Other system-defined allow for a better syntax of terms, or at least one chosen, within some limits by the user who can choose between prefix (standard) and infix, priorities of operators in terms, directions of associativity of operators. For example up to now in all previous manipulations of lists we used an infix "." for concatenation, although it is a standard prefix operator (see 3.1.). It would have been defined as follows, calling system-defined predicate AJOP

← AJOP (".", 3,"X↑ (X ↑X")

to be of priority 3, infix with right to left associativity, thus interpreting a list like a.b.c.nil as the tree

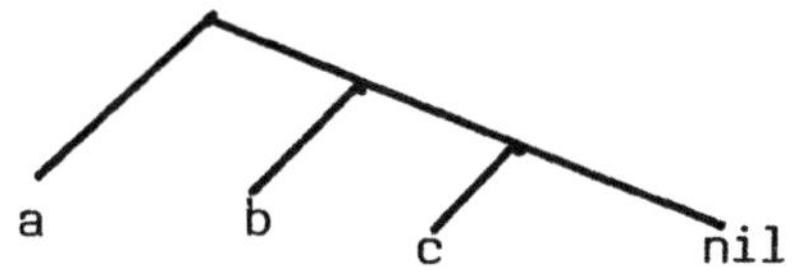

or as . (a,. (b,. (c,nil))) in prefix notation.
Similarly if one wants to define an operator SUM, unary the following will do :

← AJOP ("sum", 1, " ↑X")

iv) A very useful feature, already illustrated in 2 above is that operators names are not distinguished from predicates names, i.e the same name can be used in a term and as a predicate (X in the example)

```
    ← Test (Male, X)
← Select (X,L)
  Select (X,X.L) ← Priority (X,Y) Greater (Y,3), X
  Select (Y,X.L) ← Select (X,L)
```

This feature is very useful for writing interpreters.

v) There are system predicates to add and suppress new clauses to the current program ; this feature is also very odd for logic because it amounts to changing, on the fly, theories ; the syntax is awkward in Marseilles implementation and we do not give it in detail ; for example :

← AJOUT (term standing for the list representation of the clause)

with variations on predicate to express where to add the clause (at the head or bottom of the already existing list of clauses with same head predicate).

vi) The above predicates are useful for modifying a program ; other predicates can handle clauses and give ways to massage them so as to modify them, at the user's will, taking advantage of the fact that clauses are represented internally as terms :

← UNIV (X,T)

tries to unify terms X and T ; if X is a free variable it assigns to X term T; if X is a term it matches it against T. For instance, given the following clause :

```
CONCTERM (X1, X2, X3)←  UNIV (X1,.(N1, LARG 1)),
                        UNIV (X2,.(N2, LARG 2)),
                        CONCAT (N1, N2, N)
                        CONCAT (LARG1, LARG2, LARG)
                        UNIV (X3,.(N,LARG))
```

a call to CONCTERM with X1 and X2 assigned two terms say t1 (p,q) and s2 (r) as values at call time, will build a new term X3 whose value will be t1s2 (p, q,r). The sequence of terms and unifications is given in Fig. 9.

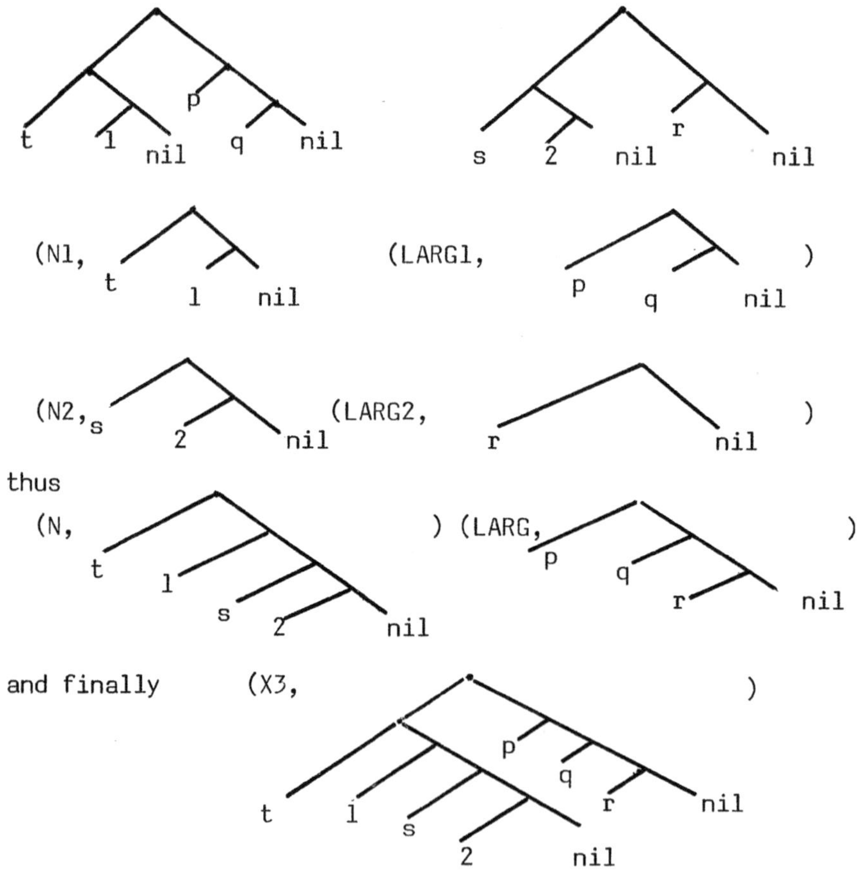

Fig. 9

vii) Some predicates are defined which give rise to testing of the status of variables, terms, ... For instance " ← VAR (X)" will succeed iff X is a true free variable not yet assigned (even to another free variable) ; other such predicates are ATOM (X), ... Predicate EGALF (X,Y) succeeds when X and Y are formally the same terms (i.e it does not test that they have the same value : this could be tested by calling an EQ predicate, where "EQ (X,X)← " is its definition.

viii) Last and very important, not only for efficiency reasons but because it influences very much the programming style of PROLOG users, pushing them away from the "what" spectrum down to the "how" spectrum (see Introduction), there is a system -defined predicate, the cut, which will be noted here as "/" which acts directly on the interpreter as follows : when traversing first '/' in the resolvent (hence going forward, from its left), the interpreter succeeds i.e acts as if it had the definition " /←" ; whenever the interpreter encounters it as a backtrack position, i.e from the right, it fails and it automatically fails its father node. See Fig. 10.

Ex :

```
P ← K,A,B
A ← D,/,C
A ← G
K ←
K ← E
K ← F
E ←
D ←
  ← P
```

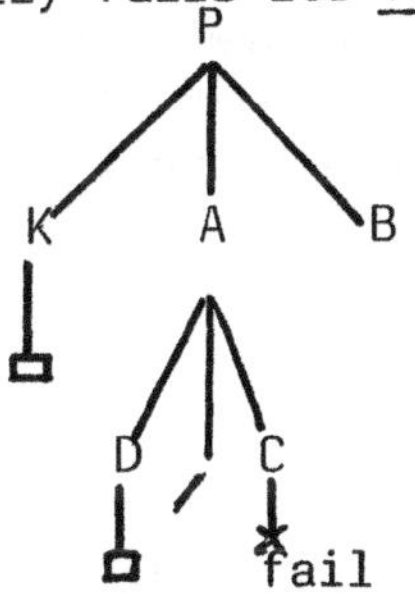

When C fails normal backtrack would examine alternatives for D ; if there were none it would try alternatives for A ; if there were none it would try those for K and finally those for P ; here, due to the "/", when C fails, failure propagates to A and the first backtracking node to be examined is K, all subnodes of A being deleted yielding

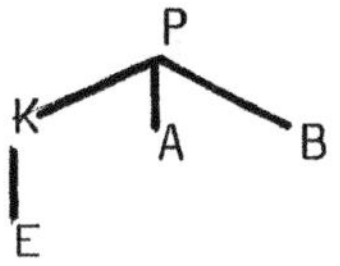

and proceeding from then on.

Fig. 10

The "/" predicate can be used in more subtle ways as it has the option of directly indicating to the interpretor which node to backtrack to. It takes the form "/(name)" where name is the literal name to which one wants to backtrack. We shall see outstanding examples of the '/' construct, whose meaning is the expression of determinism of the head literal of the clause in which it appears.

Before we turn to specific programming examples let us remark that :

- there are also predicates to access the derivation tree(Ancestor, State) whose use is rare in classical applications

- PROLOG fails in its implementation of unification to make the "occur check" ; thus when one tries to unify f (X, g (X)) with f (Y,Y) PROLOG will succeed (it should not as it is trying to match X and g (X)) ; there are good efficiency reasons to behave in this way. Moreover A. Colmerauer in (22) has investigated the semantics of such constructs and shown that the infinite trees which they synthesize are but a representation of context free grammars (restricted), and A. Colmerauer uses this feature in many applications dealing with properties of programs.

4. EXAMPLES

It is difficult to give a representative set of examples of a programming language in a restricted space. Most of the examples given below will manipulate lists ; the reader is referred to (4) for a thorough set, from which some of the classical examples have been borrowed. They mostly illustrate the functional aspects of PROLOG, not its non deterministic aspects.

4.1. Lists and sets manipulations

Remember that lists (and terms) are represented as :

The empty list is nil.

i) Write a program to find an element in a list which has property P

```
Elem (.(X,Y),X)← P (X)
Elem (.(X,Y),Z)← Elem (Y,Z)
```

due to the order of the 2 clauses this will work ; in different order the interpreter would loop. This definition of Elem is complete, and this program will stop in all cases. Notice that since Elem (nil) is not defined one could question the completeness of the definition ; but in that case, Elem (nil) not being defined fails, which is exactly what is needed.
Because it is simpler we will assume in most examples to come that '.' i.e the concatenation operator has been externally defined as infix and right to left associative ; the example becomes (exchanging arguments)

```
Elem (X, X.Y) ← P (X)
Elem (Z, X.Y) ← Elem (Z,Y)
```

Finally if one wants to express determinism (so as to really find the first element <u>only</u>, counteracting PROLOG trial to find all solutions), one should write :

```
First (X,X.Y) ← P (X), /
First (Z, X.Y)← First (Z,Y)
```

The '/' predicate does not recover all space when there is success. Further predicates have been introduced to take advantage of this opportunity.
Note that nil is not taken to be an element of any list ; the list made of nil is the empty list (see the above schema).

ii) Compute the intersection of two lists

```
Intersect (H.T, L, H.U) ← Elem (H,L), Intersect (T,L,U)
Intersect (X.T, L, U)   ← Intersect (T,L,U)
Intersect (nil, L, nil) ←
```

In this example it is easy to show completeness of cases ; one should see that the last clause could be replaced, without any change as far as the answer is concerned by

```
Intersect (X,L,nil) ←
```

because if the interpreter is attempting to match with clause 3,it implies that clauses (1) and (2) failed ; the only possibility left, if the initial argument was a list, a true one, is nil ; thus using Intersect (nil, L, nil) makes a syntactic test on the argument (and if it was not a true list ending with nil, this would fail). Only the first solution found by Prolog is correct ; one should have written, avoiding backtrack :

```
Intersect (H.T, L, H.U) ← Elem (H,L) , /, Intersect (T, L, U)
```
Note that a different order of clauses would work the same because nil can be tested any time ; there is no potential redundancy with the other matching : H.T ; X.L cannot match nil. A different program can be written :

```
Intersect 1 (nil, X, nil) ←
Intersect 1 (X.R, Y, X.Z) ← Elem (X,Y),/, Intersect 1 (R, Y, Z)
Intersect 1 (X.R, Y, Z)   ← Intersect1(R, Y, Z)
```

iii)Replace intersection by union in ii) :

```
Union (nil, X, X) ←
Union (X.R, Y, Z) ← Elem (X,Y),/, Union (R, Y, Z)
Union (X.R, Y, X.Z) ← Union (R,Y, Z)
```

iv)Find whether a list is a sublist of another one ; find whether it is disjoint of another one : .

```
Subset (A.X, Y) ←  Elem (A,Y), Subset (X,Y)
Subset (nil, Y) ←

Disjoint (A.X, Y)← Elem (A, Y) -/- Fail
Disjoint (A.X, Y)← Disjoint (X,Y)
```

v) Split tail T of a list whose head is H in two lists U1, U2, such that all elements of U1 will be ⩽ H, all those of U2 will be > H and initial order in T is preserved in U1 and U2.

```
Split (H, H1.T1, H1.U1, U2)← Infeq (H1,H), Split (H, T1, U1, U2)
Split (H, H1.T1, U1, H1.U2)← Gt (H1,H), Split (H, T1, U1, U2)
Split (H, nil, nil) nil) ←
```

Again it would be better to write, for efficiency

```
Split (H, H1.T1, H1.U1, U2)← Infeq (H1,H),/,Split (H,T1,U1,U2)
Split (H, H1.T1, U1, H1.U2)← Split (H, T1, U1, U2)
Split (H, nil, nil, nil)←
```

This last program, without /, would work too, i.e the first solution found would be correct (and only that one). The Gt (H1,H) test is not necessary because the test Infeq (H1,H) has already failed when considering the second clause. The solutions found by backtracking, (with / omitted) are not correct (as in the Intersect example).

<u>Problem</u> : Given Inf (X,Y), write Infeq (X,Y) which succeeds if X ⩽ Y ; then write Gt (X,Y) which succeeds if X > Y.

vi) Write a Quicksort program, using Split, to sort a list of elements.

```
Quicksort (H.T, S) ← Split (H, T, U1, U2)
                     Quicksort (U1, V1)
                     Quicksort (U2, V2)
                     CONCAT (V1, H.V2, S)
Quicksort (nil, nil)←
```

It is possible to insert one / in the above program ; where ? if there are several possibilities how do the behaviors vary ?

<u>Problem</u> : write a Permutations program. If you have several versions compare their efficiency.

4.2. Negation

Write a program NOT (X) which succeeds if and only if program X fails (see discussion 2.2.)

```
NOT (X)← X, /, FAIL
NOT (X)←
```

Here the cut i.e / is used differently from most of the above uses ; it is not a matter of determinism. We used it this way in Disjoint above. We could define now :

```
Diff (X,Y) ← NOT (Subset (X,Y)), NOT (Subset (Y,X))
```

Turning back to the negation discussion we see that PROLOG, enhanced with NOT, implements the negation by failure principle ; to support discussion (ii) in 2.2., PROLOG would fail with this enhancement, to derive P from P ← NOT (P) ; indeed we get the following derivation :

```
         P  ←  NOT (P)         )
        NOT (X) ← X,/,Fail     }  Given
        NOT (X) ←              )
                ← P             Goal
then            ← NOT (P)
then            ← P, /, Fail
```

and the interpreter loops.

4.3. Trees

Let us handle the binary property only of balanced binary trees (b.b.t) ; if such a tree has root K, all elements in the left subtree are smaller than K, all elements in the right subtree are bigger than K (no duplicates). We would like to write a PROLOG program which deletes the root element and re-balances the tree (keeps only its binary property).
What cases are to be considered ?

i)

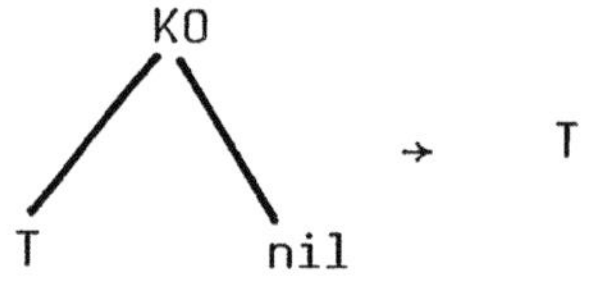

in case K0 is the root and the right subtree is empty, the new tree is T

ii)

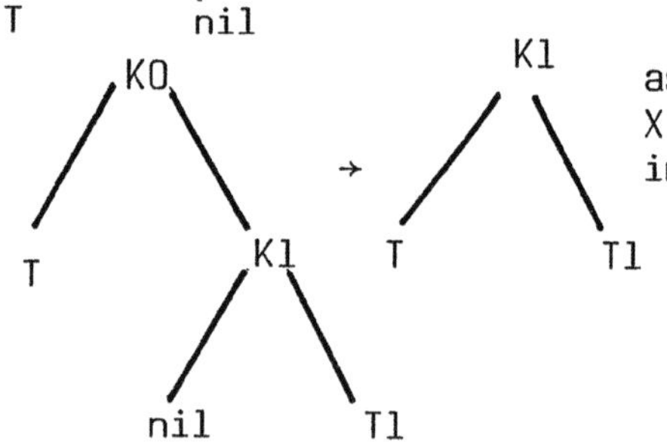

as for all X ∈ T, Z ∈ T1
X < K0 < K1 < Z
in the initial tree, the new tree is balanced.

iii) the only remaining case is (T1 ≠ nil) :

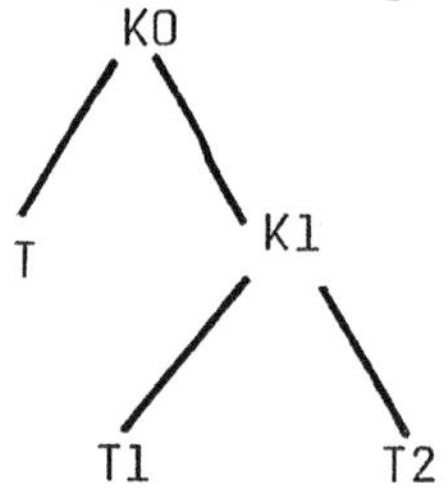

in this case the new root is to be extracted from T1 : it is its smallest element, i.e its leftmost element which has no left descendent.

If K2 is the smallest element of T1, and if "T1-K2" is rebalanced as T3, then the resulting tree in this case is

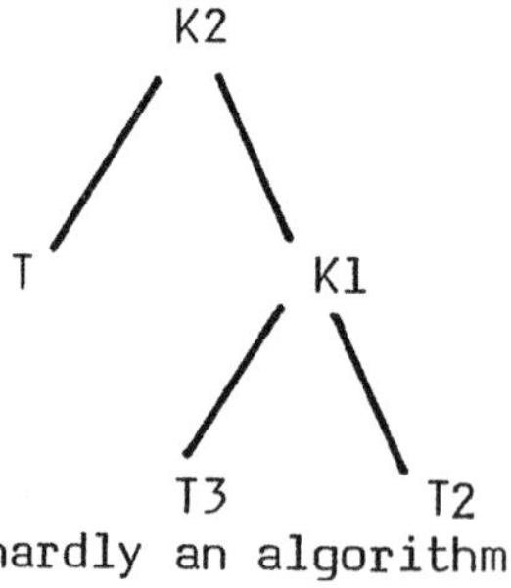

This is hardly an algorithm ; it is merely a definition by cases ;

The PROLOG program is a simple translation of that definition taking into account the tree representation : assume a node to be coded as follows :

T = 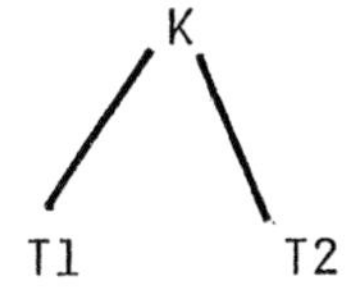

coded as T' = t (T'1, K, T'2) where T'1 and T'2 are coded versions of T1, T2 ;

nil coded as nil ;
thus a <u>leaf</u> K or 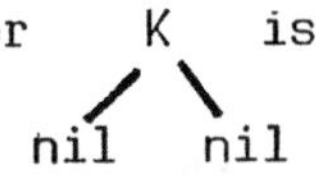

is coded as t(nil, K, nil).

Then, we get for each case (eliminating the ' notation)

```
(i)   Del (K, t (T1, K, nil), T1)  ←
(ii) Del (K,t (T,K,t (nil, K1, T1)), t (T,K1,T1)) ←
(iii) Del (K,t (T,K,t (T1,K1,T2)), t (T,K2,t (T3, K1, T2))) ←

          Smallest (T1, K2), Bal (T1, K2, T3)  (*)
```

at this point, Smallest and Bal predicates could be coded ; however it does not seem to be the right way to <u>implement</u> an algorithm, because looking for the smallest element (and calling it) K2, some work is done which ought not be done again when knowing K2 value, looking for it in T1, and rebalancing "T1-K2" as T3. Both operations should be done in one step, i.e the (*) line is to be replaced by

```
          Elim-small-bal (T1, K2, T3)
```

Elim-small-bal can also be defined by cases :

if T1 is a tree with no left subtree, K2 is its root, and if not : T1 is 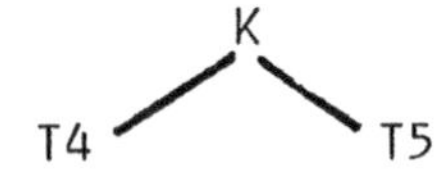

, K2 must be the smallest element in T4 and thus T3 is 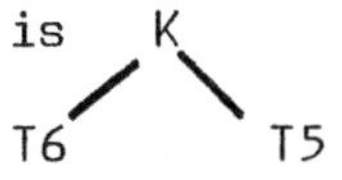

if elim-small-bal (T4,K2,T6).

This gives after coding of the trees :

```
elim-small-bal (t (nil, K2, T3), K2, T3) ←
elim-small-bal (t (T4, K, T5), K2, t (T6, K, T5)) ←
                        elim-small-bal(T4, K2, T6)
```

<u>Problem</u> : Modify the above example so as to write a program which searches a balanced tree for a given value and,if it finds it, deletes it and re-balances the whole tree.

Problem : Write a program to search a balanced tree, looking for a given value, and if it does not find it, add the value to the tree and"rebalances"it (there are several ways to do so for the binary property).

4.4. Miscellaneous

Problems :

i) Write an eight-queen program
ii) Write a program to color a planar map so that two adjacent regions have a different color.

Example : Natural Language Analysis (11), (26), (38)
Finding a predicate logic representation of the parsing problem does not have a unique solution.

Given a context-free grammar such as suggested by the following excerpt :

```
Sent →  Np   Vp  (a Sentence is a Noun phrase, followed by a
                  Verb phrase)
Np   →  Det Adj Noun
Vp   →  Aux  Verb
Det  →  The
Det  →  a
...
```

it is easy to imagine a PROLOG program whose execution would act just as the top-down parser directly derived from that grammar. A trivial solution would be :

```
Sent (X,Y) ← Np (Y,Z)  Vp (Z,Y)
Np (X,Y) ← Det (X,Z) Adj (Z,U) Noun (U,Y) there is a sentence from
          X to Y, if there is a noun-phrase from X to Z and a Verb
          phrase from Z to Y.
Det (X,Y) ←  the (X,Y)
Det (X,Y) ←   a (X,Y)
...
```

Usually a top-down interpreter (such as PROLOG) will correspond to a top-down parser.
An equivalent formulation is

```
Sent (X) ←  Np (Y)  Vp (Z)   Concat (Y, Z, X)
Np (X) ← Det (Z) Adj (U) Noun(V) Concat (Z, U, T) Concat (T, V, Y)
```

Le us come back to the first formulation ; if we want the terminals to appear as data, i.e as terms, rather than as literals, i.e if we want to use terminals in rules, we can use an alternative formulation such as :

```
Det (X, Y) ← Connects (X, the, Y)
Det (X, Y) ← Connects (X, a, Y)
```

this will work if the input phrase is tagged with integers at word frontiers and represented as a set of unit clauses ; such as :

```
Connects (1, a, 2) ←
Connects (2, man, 3) ←
```

This new representation is now independent of the actual terminals of the grammar, which is a nice feature as it corresponds to using a dictionary. Instead of tagging each point in the input string with an integer it is possible to tag each point with the sublist of the input which occurs after that point in the string ; doing so, Connects can now be defined in a single clause:

```
Connects (W.S, W, S) ←
```

and the list to be parsed can be represented as :

```
← Sent (a. man. who. lives. nil, nil)
```

One can then preprocess all calls to Connects in the grammar clauses, getting for instance instead of the clause

```
Relative (X, Y) ← Connects (X, that, Z)  Vp (Z,Y)
```

a clause `Relative (that. Z, Y)  ← Vp (Z,Y)`
eliminating altogether the need for the Connects predicate. This is the <u>metamorphosis grammar</u> form of Colmerauer (11).

This formulation has been extended by F. Pereira and D. Warren into the <u>definite clause grammars</u> which are to correspond to the augmented transition networks, and which extend them in interesting ways as argued in (38) ; definite clause grammars provide for :

- building of parse trees (as necessary in any parser)
- imposition of extra conditions on the constituents of a phrase
- general treatment of context dependency

This is not much of a surprise as we know PROLOG can easily handle tree manipulation (see I.4.3.), and that extra predicates can be appended to each clause. A thorough discussion could then take us to extraposition grammars (39) which go steps further than context-free grammars in expressiveness ; of course this is not a well-settled matter and we shall not go into such debated details. The point to be made is that, at this time, it seems that logic provides a quite efficient and powerful tool to attack such problems and front-ends to data bases can and have been produced using it.

Chapter 2

DIRECTIONS FOR IMPROVEMENTS

PROLOG has been used in many applications ; refer again to (4) for a sketch of some of them, and to (13) for papers describing others. These applications range from natural language analysis as seen above to house planning, provers, program synthesis systems, plan generators, drug synthesis systems etc ; many more are now being developed. Logic itself has had influence on whole fields in computer science, eg data bases (55).

PROLOG is used in these applications either as a general programming language (with unusual features), or specially for its unusual features (deduction, non determinism, ...) i.e for its interpreter capabilities as a theorem prover-put it differently:PROLOG can be seen as but another programming language or as a problem solver.

From these applications, a lot of experience has been gained in PROLOG usage and it is now possible to summarize this experience either through implementations which already incorporate modifications to the original PROLOG or by initiating research proposals.

Let us first mention syntax evolution ; as we said no two PROLOG systems have an identical syntax ; but their correspondances are usually obvious. We shall not deal here with such syntactic changes, but rather concentrate on deeper ones.

1. PROLOG'S EXTENSIONS THROUGH BUILT-IN PREDICATES

The Edinburgh's PROLOG is a major PROLOG implementation which provides very interesting features and extensions to previous PROLOG. See (24) for an account of it. We summarize below what we believe to be essential ones :

i) Input-output and stream management : several predicates allow the user to read files, overridding previous clauses already read-in or not overriding them, to open or close files for I /O, to display terms, to flush terminal output buffers... ; all of these predicates provide a more convenient environment and improved interaction means for the user.

ii) arithmetic :
Better suited arithmetic functors, comprising bitwise operations, are defined ; extensions to real or complex numbers are easy to define (e.g see (23)).

iii) internal data base.

Several predicates allow to search the data base of clauses and to modify it.

iv) meta-logical :

Predicates allow to test whether X is a variable (free), or not, whether it is an atom, an integer, whether its value is formally identical to another variable value ; more complicated predicates give ways to examine clause or literal bodies (recall UNIV seen previously) ; such predicates are useful for doing symbolic analysis of the program itself.

An important extension is the Setof construct :

Setof (X, P, S) which instantiates S to be the "set" of (i.e the ordered list) all instances of X such that P is provable ; S is represented without duplicates ; variables appearing in X should not appear anywhere else in the clause except within P ; it is possible for the provable instance of P to contain variables, but in this case S is only a finite approximation for a potentially infinite set (24).

It is possible to backtrack on Setof, generating alternative values for S corresponding to alternate instantiations of free variables in P. S cannot be empty (it fails otherwise)

Example

Setof (X, Likes (X, Y), S)

could produce for a given data base

Y = beer   S = {dick, harry, tom}

and on backtracking

Y = wine , S = {bill, john, tom}

Setof predicates can, as all predicates, be used as literals too, thus allowing such constructs as :

Setof ((Y,S), Setof (X, Likes (X,Y), S), SS)

which would produce for instance

SS = {(beer,{dick, harry, tom }), (wine {bill, john, tom})}

Existential quantifiers can also be used in such constructs

← Setof (X, Y^ (Likes (X,Y)), S) would yield

S = {bill, dick, harry, john, tom} if Y^ stands for ∃Y.

v) Debugging facilities

Byrd (24) has developped a model for control flow in which he views the program control flow in terms of movement about the the program text, palliating the confusion stemming from backtracking. A description of these facilities is out of order ; let us stress this attempt to adapt a debugging procedure to a given language.

2. COMPILERS AND NEW IMPLEMENTATIONS

The Edinburgh implementation (5), -6), (24) includes a compiler as well as an interpreter ; its performance is increased by an order of magnitude and has been compared to that of LISP (25). More recently LISP implementations (27), (28) of PROLOG have been seen ; the real interest of such implementations is that they provide a common environment to LISP and to PROLOG, allowing one to mix PROLOG predicates to LISP programs. Further, in (28) the actual language implemented is not PROLOG but it is a more general Horn clause interpreter using LUSH resolution.

Also interpreters on smaller machines,even microcomputers, are being built (22), (50), (51), (52) and even marketted (53). Alternative memory management schemes are being tried (50), (51) avoiding the classical structure-sharing invented by Boyer and Moore. PROLOG implementation has been tuned due to the possibility of dealing easily with tail recursion and some other features. It seems as if the memory management problems were easier to solve than in LISP implementation.

3. FURTHER ENHANCEMENTS

Applications already developed have shown some of the deficiencies of logic languages in general and of PROLOG in particular. Most important among them is the problem of controlling the LUSH interpreter (I.2.2) in two ways :

- controlling the forward process i.e the choice of literal and of matching clause at each interpreter step
- controlling the backtracking process, i.e making it more intelligent.

Many proposals have been made in order to cater with the forward control problems, ranging from ad-hoc solutions, to new built-in predicates allowing coroutining control, to meta predicates expressing control outside of the Prolog programs themselves. Most of these proposals have been presented in (3) as for instance (29), (30), (31) ; other proposals were presented earlier in (9), (10) ; others have been discussed at (4). Whether such constructs have a future or not, which ones will prevail is yet to be seen. Implementation such as (28) offer means to try some of these constructs. It is clear that both general strategies (problem solving strategies) and specific ones (program or even literal specific ones) should be expressible in order to truly improve the interpretation process.

One can see the intelligent backtracking process as an alternative, or as a complement to controlling the forward process : finding out what are the actual reasons for a failure and backtracking to a meaningful action is equivalent in many cases to finding out, during the forward process how to better organize a node expansion ; for instance implementing a data flow forward control avoids many useless backtracking steps. Work on this topic is also reported on in (13) (32), (33) (34) and (35).

In both cases it remains to experiment these constructs on a large basis. Doubtless some of the proposed constructs will have to be used in large applications. One would hope that they do not alter too much PROLOG (or for this matter logic programming) spirit, i.e that of separating, "as much as possible" logic and control (9), (10), (36), (37).

Chapter 3

APPLICATIONS TO PROGRAM MANIPULATIONS

Logic programming has been used in many different ways to deal with problems of program specification, program transformations, program properties proofs. These matters are very much related and we shall leave it to real experts of this field to discuss these relations in other chapters of the book. Let us just point out the following :

i) because logic in general, and logic programming to a less extent, aim at separating logic from control, it is tempting to take logic as a specification language (expressing "what" as opposed to "how") ; certainly natural, or standard logic appears more qualified in this respect than is Horn clause logic, or than is PROLOG itself ; such a view has been widely advocated by Bibel (40), Manna and Waldinger (41).

ii) granted that specifications ought to be written in natural logic, one can try to prove properties of specifications (such as termination, invariants,...) ; procedures for proving properties have been devised by several authors, as in (41),(54)

iii) such specifications cannot be run as few systems allow to run natural logic directly (see (56), (58)). Thus transformations from natural logic to Horn-clause logic have been studied (40) (42), (43), (54), (60), which must preserve properties of specifications or allow to prove properties about them ; such transformations have obviously been influenced by the work of Burstall and Darlington (45).

iv) programming from examples, i.e program synthesis which had been mainly developed for LISP programs (57), (46). Recent work by Shapiro (44) deals with similar problems for Horn-clause logic.

v) program specifications through algebraic data types have been said to be easily "translatable" to Horn-clause programs (47), provided one adopts a common view , i.e that of rewriting rules ; this work has been illustrated by R. Davis (47), and although it seems to need further developments, such as (59), this approach should yield important results in the future.

CONCLUSION

This paper had limited ambition, that of introducing to the basic and important features of PROLOG ; this could not be attained without investigationg logic programming itself because it gives a better understanding as well as a better foundation to PROLOG itself. Programming examples have been given, but clearly it is expected that the interested reader will go to more complete sets (e.g (4)) and will actually practice. We have indicated in what directions PROLOG was being improved, both for its implementation and for its features. We believe PROLOG, or logic languages will take a larger share of the programming community, and influence programming activities very deeply.

ACKNOWLEDGMENTS

Although I assume full responsability for all misconceptions about PROLOG, misunderstandings or lack of giving it the right emphasis, I greatly acknowledge the influence of people who struggled to let PROLOG be known, A. Colmerauer and his team, R. Kowalski and D. Warren ; the various logic programming meetings which I attended gave me the taste of and for logic programming and PROLOG while I was at CERT-DERI working with my colleagues on relational data bases and on deductive systems of data bases ; I owe a debt to colleagues at CERT-DERI with whom ideas on PROLOG, its nice and not so nice features, its possible improvements, were discussed : C. Lasserre, A. Fahmi, M. Dincbas.

REFERENCES

(1) COLMERAUER, A., KANOUI, H., PASERO R. and P. ROUSSEL : Un système de communication homme-machine en Français Rapport, Groupe d'Intelligence Artificielle, Université d'Aix-Marseille, Luminy-(1972)

(2) BATTANI, G. and H. MELONI : Interpréteur du langage de programmation PROLOG - Rapport, Groupe d'Intelligence Artificielle, Université d'Aix-Marseille, Luminy - (1973)

(3) ROUSSEL, Ph. : PROLOG, Manuel de référence et d'utilisation Rapport, Groupe d'Intelligence Artificielle, Université d'Aix-Marseille, Luminy - 1975

(4) COELHO, H., COTTA J.C. and L.M. PEREIRA : How to solve it with PROLOG - 2nd Ed - Laboratorio National do Engenharia Civil - Lisboa - 1980

(5) WARREN D.H. : Implementing PROLOG - Compiling predicate logic programs - Vol. 1 & 2 -D.A.I. Research Report 39 - University of Edinburgh - 1977

(6) PEREIRA L.M., PEREIRA F. and D.H. WARREN : User's guide to DEC system 10 PROLOG - Laboratorio Nacional do Engenharia Civil - Lisboa - 1978

(7) SZEREDI, P. : PROLOG, a very high level language based on Predicate logic - Preprints of second Hungarian Computer Science Conference - Budapest - 1977

(8) ROBERTS, G.M. : An implementation of PROLOG - DCS,University of Waterloo, Canada

(9) KOWALSKI, R.A. : Logic for problem solving - DCL Memo 75, Dept of AI - University of Edinburgh - 1974

(10) KOWALSKI, R.A. : Logic for problem solving - The Computer Science library - Elsevier - 1979

(11) COLMERAUER, A. : Les grammaires de métamorphose - Rapport Groupe d'Intelligence Artificielle - Université d'Aix-Marseille Luminy - 1975

(12) VAN EMDEN, H. : Programming with resolution logic - Report C.S-75-30 - Dept of Computer Science - University of Waterloo - Canada - 1975

(13) TARNLUND, S.A : Logic Programming Workshop - Proceedings Debrecen - Hungary 1980 (Ed. Address : UPMAIL - C.S. Dept Uppsala University S-752 23 UPPSALA - SWEDEN) : A revised version of papers is included in a book to appear (Academic Press)

(14) Workshop on Logic Programming - University of Syracuse N.Y. - April 1981. No proceedings available

(15) Mc DERMOTT, D. : The PROLOG Phenomenon - SIGART newsletter 72 - 1980

(16) NILSON, N.J. : Principles of Artificial Intelligence Tioga Publishing Co.1980

(17) CHANG, C.L. and R.C.T. LEE : Symbolic logic and mechanical theorem proving Academic Press 1973

(18) NICOLAS, J.M. and H. GALLAIRE : Data base : theory vs. interpretation. Logic and data bases, (GALLAIRE H. and J. MINKER (eds) - Plenum Press (1978)

(19) CLARK, K : Negation as failure. Logic and Data Bases (H. GALLAIRE and J. MINKER, eds) Plenum Press, New-York (1978)

(20) REITER, R : On closed world data bases. Logic and Data Bases (H. GALLAIRE and J. MINKER (eds) - Plenum Press, New-York (1978)

(21) WARREN, D : Higher order extensions to PROLOG. Are they needed?D.A.I. Research Paper 154 - University of Edinburgh - 1980

(22) COLMERAUER, A., KANOUI H. and M. VAN CANEGHEN : Etude et réalisation d'un système PROLOG - Rapport, Groupe d'Intelligence Artificielle - Université d'Aix-Marseille - 1979

(23) DINCBAS, M. : A knowledge-based expert system for automatic analysis and synthesis in CAD - Proc. IFIP 1980, Tokyo - 1980

(24) BYRD L., PEREIRA, F. and W. WARREN : A guide to version 3 of DEC-10 PROLOG - D.A.I. occasional paper n° 19 - University of Edinburgh - 1980

(25) WARREN, D., PEREIRA L.M., and F. PEREIRA : PROLOG, the language and its implementation compared with LISP - SIGART/SIGPLAN Symposium, 1977

(26) COLMERAUER A. and J.P. PIQUE : About natural logic - In Advances in Data base theory, (H. GALLAIRE, J. MINKER and J.M. NICOLAS, eds) Plenum Press, New-York, 1981

(27) KOMOROWSKI, I. : QLOG - The software for PROLOG and logic programming In (13)

(28) ROBINSON, J.A. and E. SIBERT : LOGLISP - An alternative to PROLOG. Report School of Computer and Information Science - University of Syracuse - New-York

(29) DINCBAS, M. : The METALOG problem solving system - an informal presentation - In (13)

(30) GALLAIRE H. and C. LASSERRE : A control metalanguage for logic programming. In (13)

(31) CLARK, K. and F. Mc CABE : IC-PROLOG - Language features. In (13)

(32) PEREIRA, L. : Intelligent backtracking and sidetracking in Horn-clause programs - implementation - In (13)

(33) LASSERRE, C. and H. GALLAIRE : Controlling backtrack in Horn-clause programming - In (13)

(34) BRUYNOOGHE, M. : Intelligent backtracking for an interpreter of Horn-clause logic programs. Colloquium on Mathematical Logic in Programming, Salgotarjan, Hungary 1978

(35) BRUYNOOGHE, M. and L. M. PEREIRA : Revision of top-down logical reasoning through intelligent backtracking - This book.

(36) KOWALSKI, R. : Algorithm = logic + control.CACM, 8-1979

(37) HAYES, P.J. : Computation and deduction. Proc 2 nd MFCS Symp. Czechoslovak Academy of Sciences - 1973

(38) PEREIRA, F.C. and D.H. WARREN : Definite clause grammars for language analysis - a survey of the formalism and a comparison with augmented transition networks - Artificial Intelligence 13 - 1980

(39) PEREIRA, F.C. : Extraposition grammars. In (13)

(40) BIBEL, W. : On strategies for the synthesis of algorithms. Proc. AISB/GI Conf. on A.I., Hamburg-1978

(41) MANNA, Z. and R.J. WALDINGER : A framework for deductive programming - Computer science Dept - Stanford University and SRI international - 1978

(42) HOGGER, C.J. : Program synthesis in predicate logic - Proc AISB/GI Conf. on A.I., Hamburg - 1978

(43) CLARK, K and J. DARLINGTON : Algorithm classification through synthesis - Computer Journal

(44) SHAPİRO, E.Y. : Inductive inferences of theories from facts. Research report 192, Department of Computer Science - Yale University - 1981

(45) BURSTALL, R.M. and J. DARLINGTON : Transformation for developing recursive programs, JACM 24-1 1977

(46) PAPON E. and Y. KODRATOFF : A system for program synthesis and program optimization, Proc AISB, Amsterdam, pp. Kodratoff 1-10, 1980

(47) DAVIS, R. : Runnable specifications as a design tool - In (13)

(48) HILL, R. : Lush resolution and its completeness - DCL Memo n° 78 - University of Edinburgh - 1974

(49) BRUYNOOGHE, M : An interpreter for predicate logic programs Part 1 : basic principles - Report CW10 - Applied Mathematics and Prog. Dept. Katholicke Univ. LEUVEN (Belgium)

(50) MELLISH, C.S. : An alternative to structure-sharing in the implementation of a PROLOG interpreter. In (13)

(51) BRUYNOOGHE M. : The memory management of PROLOG implementation - In (13)

(52) MELLISH, C.S. and M. CROSS: The UNIX PROLOG System - Software Report 5, Dept of A.I., University of Edinburgh

(53) Logic Programming Associates, Ltd : MICRO PROLOG 2.0

(54) CLARK, K. and S.A. TARNLUND : A first order theory of data and programs. IFIP 77 - North Holland - 1977

(55) GALLAIRE, H. and J. MINKER : Logic and Data bases - Plenum Press - 1978

(56) WEYHRAUCH, R.W. : Prolegomena to a mechanized theory of formal reasoning.Artificial Intelligence Journal-1980

(57) BIERMAN, A.W. : The inference of regular LISP programs from examples IEEE Transactions on Systems, Man and Cybernetics 8- 1978

(58) MUSSER, D.R. : Abstract data type specification in the AFFIRM system. IEEE Tr. on Soft. Eng. SE-6,1 - 1980

(59) FOISSEAU J. et al. : Recherches sur la conception de programmes assistée par ordinateur. Rapports CERT-DERI 1980

(60) CLARK K. and S. SICKEL : Predicate logic, a calculus for the formal derivation of programs. Proc. IJCAI 77 - 1977.

PROGRAM SYNTHESIS FROM EXAMPLES OF BEHAVIOR.

Jean-Pierre JOUANNAUD* and Yves KODRATOFF**

*Centre de Recherche en Informatique de Nancy
Université NANCY I - 54037 NANCY CEDEX (FRANCE)

**GR 22 du C.N.R.S. - Université Paris VI
Institut de Programmation
75230 PARIS CEDEX 05 (FRANCE)

After a fast survey of the field, we describe Summers methodology and its extensions, by the mean of an example. The theoretical foundations of the method are fully detailed, together with an essential remaining open problem. A particular application is pre sented : the derecursivation of program schemes.

1. INTRODUCTION

Automatic synthesis of programs is a field which has been extensively studied during the ten past years. In the same time, many people are working on program transformation methods. It is well accepted now that these two fields are strongly connected. We emphasize here this important link and show how it is possible to perform program transformation using synthesis techniques.
The synthesis problem can be approached from two points of view, which differ in the specification mode of the program to be synthesized.
The synthesis from a "full description" was first studied by Waldinger (25), more recently by Manna and Waldinger (20), Bibel (1), Guiho and Gresse (7). The program to be synthesized is described in a formal way, using a specification language (generally closed to the first order predicate logic). In this approach, the first problem is the specification of the program itself and much progress was done with the use of algebraic specification of abstract data types. The second problem is to obtain the desired program from the given specification, i.e. to transform the given specification into the desired program using deduction rules. But

A. W. Biermann and G. Guiho (eds.), Computer Program Synthesis Methodologies, 213–250.

this transformation process requires generally some invention. The synthesis of functions from a finite set $\{x_i \rightarrow F(x_i)\}$ of input-output examples (or computational traces was first studied by Bierman (2-4), then Shaw (24), Hardy (8), Summers (25), Kodratoff (17) and Papon (22). The program to be synthesized is partially described by a finite set of examples (or computational traces). The first problem is to infer the infinite set of examples (or computational traces), the second one is to obtain the desired program. The interest of the method lies in five primal points :

- some programs defined in an abstract data type (for example LISP programs) are easier to describe by a finite set of examples than by a formal complete description.
- There is a strong connection between examples of a recursive program and the approximations of the functionnal which defines it.
- The method is supported by a theoretical framework which enables us to consider synthesis from examples as a kind of interpolation between terms of an abstract data type.
- It is possible to precise the kind of inference which is needed in order to synthesize programs from examples. More precisely, it is possible to study the class of functions which can be synthesized from a finite number of examples.
- The synthesis method provides a very simple but powerfull transformation method.

Let now give some information about the history of the theory. The first research in this domain dates from 1971 (2). In this work, Bierman tries to synthesize a program (actually, an automaton as pointed at by Blum (5)). Hedrick (9), and Shaw (24) tried to synthesize programs specified by examples, using ad'hoc techniques. A crucial result was obtained by Summers (25)(26) who had the idea of computing computational traces from a set of examples of a LISP program. He also set the basis of the theory. This work was also developped by other people (19), (14), (15), (17). Similar ideas may be found in (3), (4), (11), (12).
From this material, two approaches are coming out :
first : a heuristic approach (8), (9), (22) using backtracking when it fails. It was not carried on.
Second : an analytic approach by the others : given a set $\{x_i \rightarrow y_i\}$ of input-output examples of a function F, the goal is to look for a recursive expression of F, which extends some properties of the set of input-output examples. The approaches differ by the properties which are extended :
In (3), (4), (11), (12), the output y_i is expressed as a composition of parts of the input x_i. This composition property is extended to any i.
In (11), the output y_{i+1} is expressed as a composition of parts of the output y_i and of the input x_i. This composition property is extended to any i.

In (15), (17), (19), (22), (26), the output y_i is expressed as a composition of parts of the input x_i and recurrence relations are then computed between y_{i+1} and y_i. These recurrence relations are extended to any i.
All these approaches have provided LISP systems, Bierman's system (3), the system SISP (11), the system THESYS (25), Papon's system (19), (22).
To be complete, we have to mention the recent work of Shapiro (23) about inductive inference, using a general frame including both the heuristic approaches mentionned above and some of the analytic ones. His system is more general than the previous ones but (up to now) not so powerfull.

2. METHODOLOGY OF SYNTHESIS FROM EXAMPLES

We will be concerned with LISP unary functions defined on the data type "List" and this could be easily extended to functions of greater arity (18), or to functions of an other data type.
Before the synthesis process takes place, two transformations of the input-output values are necessary. The first transforms these values into a set of computational traces expressed in terms of the constructors and selectors of the data type. The second transforms the computational traces into partial functions, defined on increasing subdomains of the function F to be synthesized.
The transformation of the input-output values into computational traces and the synthesis process itself will be presented by the use of an example of a LISP function, the parenthesis function which transforms a right-parenthesized list into a left-parenthesized one :

$$x_0 = (A),\ Par(x_0) = (A)$$
$$x_1 = (A(B)),\ Par(x_1) = ((A)B)$$
$$x_2 = (A(B(C))),\ Par(x_2) = (((A)B)C)$$
$$x_3 = (A(B(C(D)))),\ Par(x_3) = ((((A)B)C)D).$$

Before going on, we give some

2.1. Notations and definitions

Let us first give the algebraic specification of the abstract data type LIST :

<u>Type</u> LIST (external types : ATOM,BOOL)
<u>Constructors</u>
nil : $\rightarrow$ LIST
cons : (ATOM $\cup$ LIST) $\times$ LIST $\rightarrow$ LIST
<u>Selectors</u>
head : LIST\nil $\rightarrow$ ATOM $\cup$ LIST
tail : LIST\nil $\rightarrow$ LIST
atom : ATOM $\cup$ LIST $\rightarrow$ BOOL

Axioms

head (cons(x,y)) → x
tail (cons(x,y)) → y
atom (cons(x,y)) → False
a ∈ ATOM ⇒ atom(a) → True
atom(nil) → True

From now up, the head and tail selectors will be abreviated in h and t respectively.
Notice that we do not care about the problem of undefinition. This forces us to consider partial functions head and tail which will lead to no contradiction in the following.
Terms may be viewed as labelled trees (6). We give now notations about two particular classes of terms.

A cons-tree is a tree each node of which is labeled by a "cons". For instance, the tree

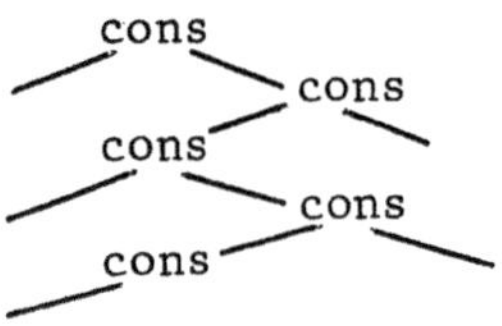

or simply

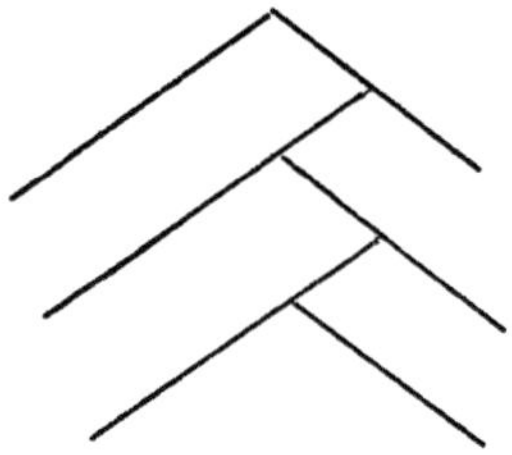

is a cons-tree.
Let L be the set of LISP lists. Each such list can be viewed as a cons-tree, whose leaves are atoms. For instance,

x_o = (A(B)C) =

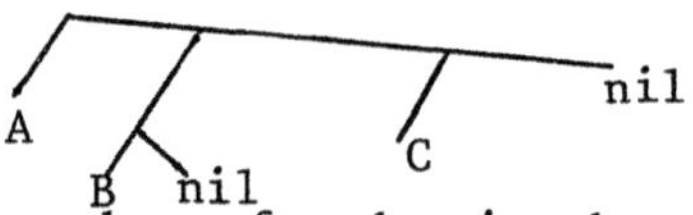

Let |x| be the number of nodes in the associated cons-tree (the "complexity" of the list x).
Let x[* ← y] denote the list obtained by substitution of y to all the *-labelled atoms in x. For instance

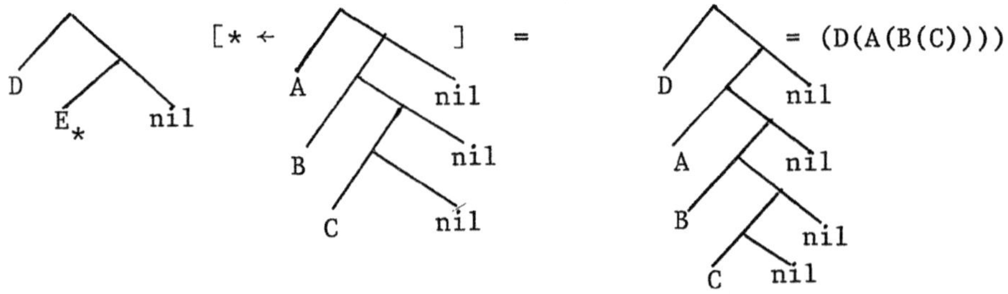

It is well known that the order associated with the substitution relation is a partial ordering of L.
Definition : A reduction function is a finite composition of selectors (h and t here). Let R be the set of reduction functions. The identity function I is assumed to belong to R, as the empty

composition of selectors.

Definition : Let A_i be the ith occurence of the atom A in the list x and u be the unique reduction function such that $u(x)=A_i$. Then u is said to be the functional name of A_i in x.

For instance, in the above example, the functional name of E_* is h t .

Definition : Given a set of terminal functions (for example : $R \cup \{nil\}$), a cons-expression is a cons-tree, each leaf of which is a terminal function.

For instance, f =

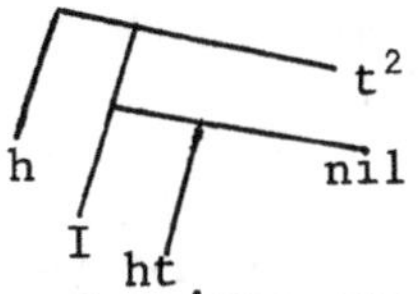

Notice that cons-expressions are terms in normal form of the term algebra.
Conversely, given a cons-expression f, the leaves of the cons-trees are called the terminal functions of f.
Let ω be the undefined atom, D(f) the domain of the function f, i.e. the set $\{x \mid x \in L,\ f(x) \neq \omega\}$.

2.2. Computational traces

A main step in Summers's method is to transform the sequence $\{x_i \to F(x_i)\}$ of input-output examples into a sequence $\{p_i(x) \to f_i(x)\}$ of "computational traces", where p_i is a predicate characterizing the input x_i and f_i a cons-expression characterizing the output $F(x_i)$.

2.2.1. Ascending linear domains. The further step of synthesis requires the existence of recurrence relations between the predicates p_i. It is thus natural to study input domains that have this property.

Definition : Given $x_o \in L$, $\alpha \in L$, α having a single $*$-labelled atom, the set $\{x_o, x_{i+1} = \alpha[* \leftarrow x_i]\}$ is called ascending linear domain $A(x_o,\alpha)$.

In fact, we change the names of the atoms in order to avoid several occurences of the same atom in x_i. For instance, with x_o =

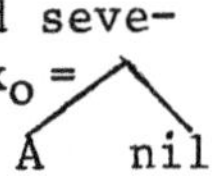

and α =

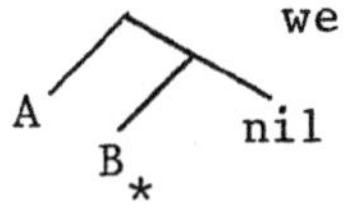

we obtain : x_1 =

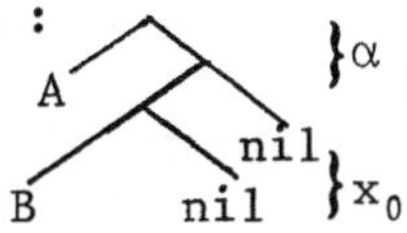

x_2 =

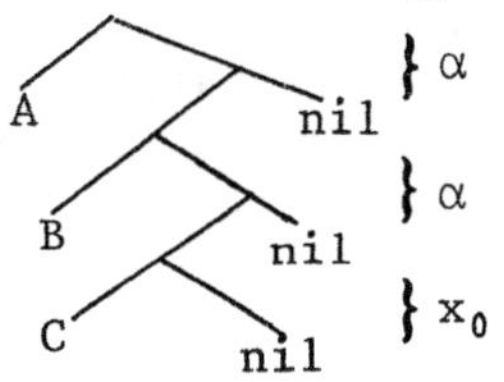

, . . .

and $A(x_o,\alpha)$ is the set {(A), (A(B)), (A(B(C))), (A(B(C(D)))),...}
Notice that the partial ordering on L is a total ordering on $A(x_o,\alpha)$.
We now have to exhibit a sequence p_i of predicates which verify (see 2.3) : $p_i(x_i)$ = True, $p_i(x_j)$ = False if $j > i$.

Definition : An atom occurence in a list x is said to be in the direction of $f \in R - \{I\}$ iff its functional name has the form $f'.f^p$ where f' is a proper right factor of f, i.e. $f = f''.f'$, $f'' \neq I$.

For instance, nil_* in x = 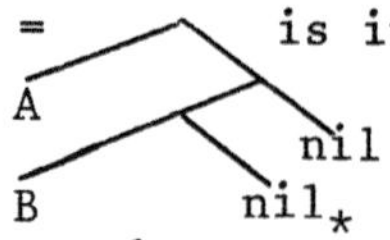 is in the direction of h t : its functional name is t h t = t.h t .

The intuitive meaning of this concept is the following : if we compute h t(x) , we obtain 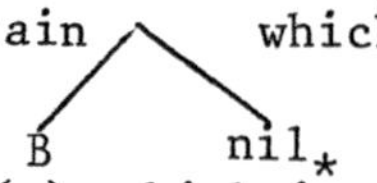 which is not yet an atom. We cannot compute h t .h t(x) which is not defined, but we can start the computation and keep computing as long as an atom is not reached : t.h t(x) = nil_* is an atom, we therefore stop at this point.

Property of the ascending linear domains : given an ascending linear domain $A(x_o,\alpha)$ and the name d of the *-labelled atom of α, let π be the functional name of the atom of x_o which is in the direction of d. Then p_i = atom.π.d^i is a sequence of predicates for the domain.
Proof in (14)

Example 1 : Let us consider the previous set $A(x_o,\alpha)$. We obtain d = h t and π = t. Thus p_o = atom.t, p_1 = atom. t h t, p_2 = atom.t h t h t, ... The match between p_i and p_{i+1} succeeds and we obtain : $p_{i+1}(x) = p_i(h\,t(x))$

Remark : It is possible to study different input domains that have similar properties (22). It is important in practice, but has no strong influence on the theory.

2.2.2. Fragments. The output $F(x_i)$ is taken in its cons-tree form. Each atom, except nil, of this cons-tree is replaced by its functional name in x_i. The process is ambiguous iff there are not several atoms (except nil) bearing the same name in any x_i. Each obtained cons-expression $f(x_i)$ is then turned into a partial function $f_i(x)$ by the substitution of x to each x_i of $f(x_i)$. The function f_i is called the ith fragment of F. Notice that f_i is defined for any $x \geq x_i$ and is equal to $F(x_i)$ when $x = x_i$.
In the following, we detail the construction of the fragments for the function Par :

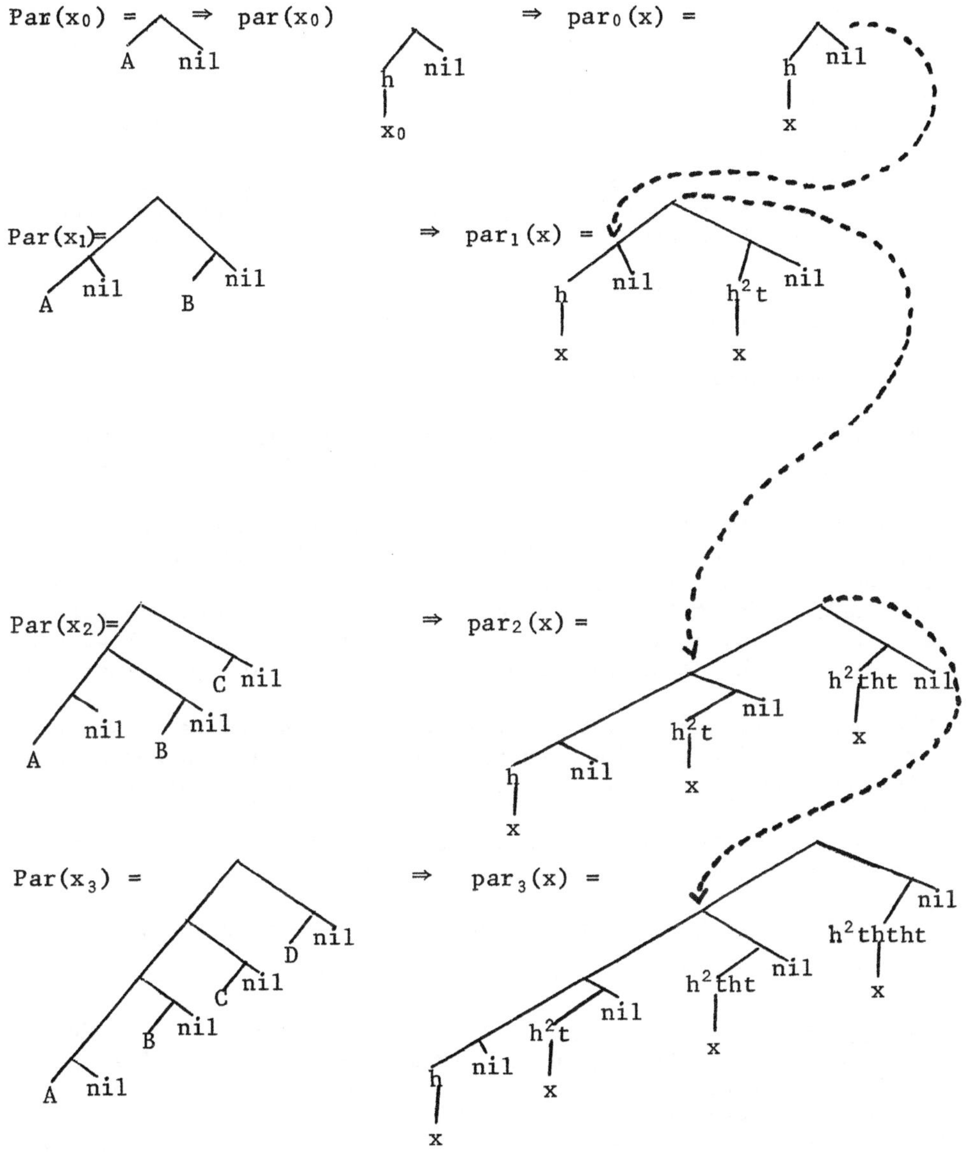

2.2.3. Approximations. The input-output sequence $\{x_i \rightarrow F(x_i)\}$ has been changed into a sequence of predicates and fragments $\{p_i(x) \rightarrow f_i(x)\}$ which is embedded into an if ... then ... else ... function in order to obtain the following n-approximation F_n of F

<u>Case</u> $p_0(x)$ <u>then</u> $f_0(x)$
$p_1(x)$ <u>then</u> $f_1(x)$
...
$p_n(x)$ <u>then</u> $f_n(x)$ <u>else</u> $\perp$ ($\perp$ is the undefined value).

The computation of $F_n(x_j)$ runs as follows : $p_o(x_j)$ is computed and takes the value False if $j \neq o$ by the definition of the predicates p_i. Then $p_1(x_j)$ is computed and so on as long as $p_j(x_j)$ is not reached. If $j > n$, $F_n(x_j)$ takes the value $\perp$ and if $j \leq n$, then $F_n(x_j)$ takes the value $f_j(x_j)$ equal to $F(x_j)$ by the fragment construction technique.
Notice that the names of the atoms are lost in F_n : we therefore cannot synthesize functions depending on these names, like the function which is true iff all the names of the atoms are different.
Our problem now is to get a finite expression of the ∞-approximation, which is an infinite expression of F.

2.3 The synthesis technique

The principle of the method consists in detecting a repetitive computation on the successive traces, using first order pattern matching techniques.
When these recursion relations are found, we make an induction hypothesis which is (and will stay) the really unprovable part of any work on inductive inference : we suppose that the function we are looking for fits always these recursion relations. An alternate way of saying is that we synthesize a function which extends to infinity the recursion properties of the first few values given as examples.
When this hypothesis is taken for granted, then it is possible to prove the equivalence between the recursion relations and a recursive program, using Summers Main Synthesis theorem and its extensions (15), (17), (22).
We first show the technique using the example of the function Par and detail the theoretical material in a further section.

Predicates : It has still be done for the Par function :

$$p_o(x) = atom(t(x)), \quad p_{i+1}(x) = p_i(h\,t(x)) \qquad (1)$$

Fragments : The classical pattern matching techniques (10) do not succeed to match the sequence of fragments of F. In such a case, a first order pattern matching attemps to use generalization techniques together with matches between subtrees :
The match succeeds easily between Par_i and the first left subtree of Par_{i+1} as it is shown by the dotted lines which link the successive fragments of the example 1.
We obtain the following recurrence relation between Par_i and Par_{i+1} :

$par_o(x) =$ 

$$par_{i+1}(x) = g_{i+1}(x, par_i(x)) \qquad (2)$$

with : $g_1(x,y) =$ 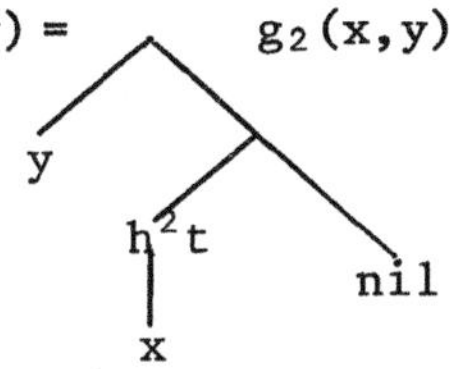

$g_2(x,y) =$ 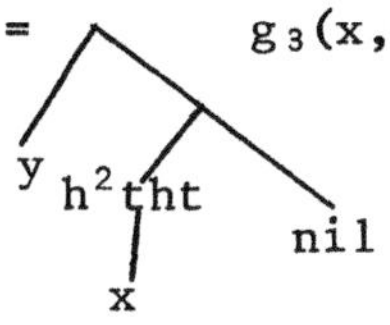

$g_3(x,y) =$ 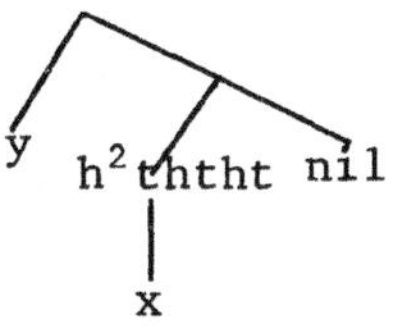

In order to achieve our goal, we now have to solve the subproblem appearing during this process ; more precisely, we try to match the new set of fragments g_i ; we find the substitution $x \setminus h\,t(x)$, which provides the following recurrence relation :

$g_1(x,y) =$ 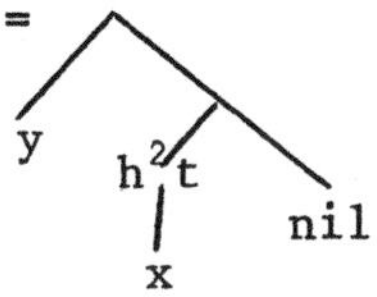

$$g_{i+1}(x,y) = g_i(h\,t(x),y) \qquad (3)$$

The matching process is thus successfull and from the relations (1), (2) and (3), we obtain the following program :

Par(x) ← Par'(x,x)

Par'[x,xx] ← <u>If</u> atom(t(xx)) <u>then</u> cons(h(x),nil)
<u>else</u> G(x,xx,Par'[x,h t(xx)])

G[x,xx,y] ← <u>If</u> atom(t h t(xx)) <u>then</u> cons(y,cons(h^2t(x),nil))
<u>else</u> G[h t(x), h t(xx), y]

Notice the extra-variable xx, defined as the <u>domain variable</u>, which appears in the definition of P'_i. It is introduced in order to take into account the relation on $p_i(x)$.
Notice also that the first fragment of G, namely g_1, is associated with the predicate p_1.

<u>Remarks</u> : Since we use matching between subtrees, there might be several different matchings, therefore several different (equivalent ?) programs. This problem will be emphasized in a next section.

Since recurrence relations are found by the mean of a matching technique, the size of the fragment f_i is obviously an increasing function of i. However, non-increasing functions may be synthesized by solving equations in a well suited equational theory : any equation has solutions in this theory (13), (16).

2.4. The synthesis theorems

Our goal in this section is to prove that the synthesized function F is the upper bound of the sequence $\{F_n\}$ of approximations defined in section 2.3.
The synthesis theorem we shall prove follows from the fixed point

theorem. Its goal is to construct the functional τ, whenever it exists, the approximation set of which $\{\tau^i[\Omega]\}$, satisfies given recurrence relations ($\Omega(x)=\bot$ for any x). Before stating and proving the general theorem, we first prove Summers main synthesis theorem :

Theorem 1 : Let
$$F_n(x) \leftarrow \underline{\text{Case}}\ p_o(x)\ \underline{\text{then}}\ f_o(x)$$
$$p_1(x)\ \underline{\text{then}}\ f_1(x)$$
$$\dots$$
$$p_n(x)\ \underline{\text{then}}\ f_n(x)$$
$$\underline{\text{else}}\ \bot$$

with
$$p_{i+1}(x) = p_i(b(x))$$
$$f_i(x) = g_i(\vec{a}(x))$$
$$g_{i+1}(\vec{y}) = u[\vec{y}, g_i(\vec{c}(\vec{y}))]$$

with $p_o, b, u, \vec{c}, g_o$ and $\vec{a}$ monotonous functions and u strictly defined (i.e. $u(\bot) = \bot$ where $\bot$ is the undefined value).
Then the upper bound F_∞ of the chain $\{F_n\}$ is the function F :

$$F(x) \leftarrow g_\tau(x, \vec{a}(x))$$
$$\tau(G)(xx,y) \leftarrow \underline{\text{If}}\ p_o(xx)\ \text{then}\ g_o(\vec{y})$$
$$\underline{\text{else}}\ u\ [\vec{y}, G(b(xx), \vec{c}(\vec{y}))]$$

where g_τ denotes the least fixed point of the functional τ.

Remarks : 1) The theorem says that F and $\lim_{n\to\infty}\{F_n\}$ are the same function. In particular, they are undefined for the same input values. This is the case when the vector function $\vec{c}$ leads to forbidden computations.

2) In order to give more readability to the paper, the functions which begin with particular cases are ruled out. The recurrence relations begin therefore with the first example of the function.

Proof : We apply the fixed point theorem to the functional τ ; we thus exhibit the generalized approximation G_n :

$$G_n(xx, \vec{y}) \leftarrow \underline{\text{Case}}\ p_o(xx)\ \underline{\text{then}}\ g_o(\vec{y})$$
$$\dots$$
$$p_n(xx)\ \underline{\text{then}}\ g_n(\vec{y})$$
$$\underline{\text{else}}\ \bot$$

We now prove by induction on n that the chains $\{G_{n-1}\}$ and $\{\tau^n[\Omega]\}$ are identical, which implies that $\lim_{n\to\infty} G_n$ is the least fixed point of the continuous functional τ.

Basic step : $\tau[\Omega](xx,\vec{y})$ = If $p_o(xx)$ then $g_o(\vec{y})$ else $\bot = G_o(xx,\vec{y})$.

Induction step : by definition of τ^{n+1}

$$\tau^{n+1}[\Omega](xx,\vec{y}) = \tau[\tau^{n}[\Omega](xx,\vec{y})$$
$$= \tau[G_{n-1}](xx,\vec{y}), \text{ by induction hypothesis}$$
$$= \underline{\text{If}}\ p_o(xx)\ \underline{\text{then}}\ g'_o(y)$$
$$\underline{\text{else}}\ u[\vec{y},G_{n-1}(b(xx),\vec{c}(\vec{y}))], \text{ by definition of } \tau$$
$$= \underline{\text{If}}\ p_o(xx)\ \underline{\text{then}}\ g_o(\vec{y})$$
$$\underline{\text{else}}\ \underline{\text{case}}\ p_o(b(xx))\ \underline{\text{then}}\ u[\vec{y},g_o(\vec{c}(\vec{y}))]$$
$$\ldots$$
$$p_{n-1}(b(xx))\ \underline{\text{then}}\ u[\vec{y},g_{n-1}(\vec{c}(y))]$$
$$\underline{\text{else}}\ u(\vec{y},\bot), \text{ by definition of } G_{n-1}$$

Notice we have used the distributivity of u relative to the conditional, because u is a strict function (21). From the same reason, we get $u(y,\bot) = \bot$ and using then the recurrence relations we prove that $\tau^{n+1}[\Omega](xx,\vec{y}) = G_n$, which ends this first part.

Getting F from g_τ is now easy by a simple use of the following instantiation lemma. □

Instantiation lemma : Let $\{G_n\}$ be a chain with least upperbound G and $\vec{a}$ a vector of monotonous functions. Then the chain $\{F_n(x) = G_n(\vec{a}(x))\}$ has $G(\vec{a}(x))$ as least upper bound.

Proof : It follows from the continuity of the functional ρ : $\rho[H](x) = H(a(x))$ and from the chain property of F_n and G_n :

$$\lim F_n(x) = \lim \rho\,[G_n](x)) = \rho\,[\lim(G_n)](x)$$
$$= \rho\,[G](x) = G(\vec{a}(x)).$$

This ends theorem 1.

As seen above, the recursion relations that may be found are more complicated then those allowed by theorem 1 : we need to generalize them to any level of embedded recursive calls.

Definition. A sequence f_i is defined by embedded recursive calls and will be further named a recurrent sequence, if it verifies the recursion property :

either (1)
$$f_i(x) = g_i(\vec{a}(x))$$
$$g_{i+1}(\vec{x}) = u(\vec{x},g_i(\vec{c}(\vec{x})))$$

or

(2)
$$f_i(x) = g_i(\vec{a}_i(x))$$
$$g_{i+1}(\vec{x}) = u_i(\vec{x},g_i(\vec{c}_{i+1}(\vec{x})))$$

where the sequences $\vec{a}_i$, $\vec{u}_i$, c_i are themselves recurrent sequences. A chain of approximations is said to be recurrent when $p_{i+1}(x) = p_i(b(x))$ and the sequence of fragments is recurrent.

Example : The sequence par_i of fragments of the function Par is a recurrent sequence of type (2), while u_i is a recurrent sequence of type (1).

Summers synthesis theorem extends to recurrent computational trace in the following way :

Theorem 2 : The least upper bound F of a recurrent chain of approximations $\{F_n\}$ is given

- in case (1), by theorem 1,
- in case (2), by the following. Let

$$F_n(x) \leftarrow \underline{\text{Case}}\ p_o(xx)\ \underline{\text{then}}\ f_o(x)$$
$$p_n(xx)\ \underline{\text{then}}\ f_n(x)\ \underline{\text{else}}\ \perp,$$ with the relations :

$$p_{i+1}(x) = p_i(b(x)),\ f_i(x) = g_i(\vec{a}_i(x)),\ g_{i+1}(\vec{x}) = u_{i+1}(\vec{x}, g_i(\vec{c}_{i+1}(\vec{x})))$$

$F(x) = \lim F_n(x) = g_\tau(x, \vec{A}(x,x))$ where g_τ is the least fixed point of the functional τ :

$$\tau[G](xx,\vec{x}) = \underline{\text{If}}\ p_o(x)\ \underline{\text{then}}\ g_o(\vec{x})$$
$$\underline{\text{else}}\ U(xx,\vec{x},G(b(xx),\vec{C}(xx,\vec{x})))$$

where $\vec{A}$, $\vec{C}$, H are the limits of the recurrent traces :

$$\vec{A}_n(xx,x) = \underline{\text{Case}}\ p_o(xx)\ \underline{\text{then}}\ \vec{a}_o(x)\ \ldots$$
$$p_n(xx)\ \text{then}\ \vec{a}_n(x)\ \underline{\text{else}}\ \perp,$$
$$\vec{C}_n(xx,\vec{x}) = \underline{\text{Case}}\ p_1(xx)\ \underline{\text{then}}\ \vec{c}_1(\vec{x})\ \ldots$$
$$p_n(xx)\ \underline{\text{then}}\ \vec{c}_n(\vec{x})\ \text{else}\ \perp,$$

$$U_n(xx,\vec{x},y) = \underline{\text{Case}}\ p\ (xx)\ \underline{\text{then}}\ u_1(x,y)\ \ldots$$
$$p_n(xx)\ \underline{\text{then}}\ u_n(\vec{x},y)\ \underline{\text{else}}\ \perp.$$

When one starts with n fragments, the functions $\vec{C}$ and H are defined from n-1 fragments generated as sub-problems of the original problem. This explains why c_o and h_o do not exist.

Proof : It uses a double recurrence, first on the recurrent property of the computational trace F_n, second on n as in theorem 1. It uses also the fact that U is strict if u_o is strict, which we suppose to be true.

- if F_n is of type (1) one then comes back to theorem 1
- if F_n is of type (2) then one proves first

that $\tau^{n+1}[\Omega] = G_n$, then that $G_n(x,\vec{A}(x,x)) = F_n(x)$.

Prove that $\tau^{n+1}[\Omega] = G_n$

The basic step is identical to the initial step in theorem 1.
Induction step : (the induction hypothesis is $\tau^n[\Omega] = G_{n-1}$)

$$\tau^{n+1}[\Omega](xx,\vec{x}) = \tau[\tau^n[\Omega]](xx,\vec{x})$$

$$= \tau[G_{n-1}(xx,\vec{x})](xx,\vec{x}) \quad \text{(induction hypothesis)}$$

$$= \underline{\text{If}}\ p_o(xx)\ \underline{\text{then}}\ g_o(\vec{x})\ \underline{\text{else}}\ U(xx,\vec{x},G_{n-1}(b(xx),\vec{C}(xx,\vec{x}))) \quad \text{(def. of } \tau)$$

$$= \underline{\text{Case}}\ p_o(xx)\ \underline{\text{then}}\ g_o(\vec{x})$$
$$p_o(b(xx))\ \underline{\text{then}}\ U(xx,\vec{x},g_o(\vec{C}(xx,\vec{x})))\ \ldots$$
$$p_{n-1}(b(xx))\ \underline{\text{then}}\ U(xx,\vec{x},g_{n-1}(\vec{C}(xx,\vec{x})))$$
$$\underline{\text{else}}\ U(xx,\vec{x},\bot)$$

(definition of G_{n-1} and distributivity of U relative to the conditional)

$$= \underline{\text{Case}}\ p_o(xx)\ \underline{\text{then}}\ g_o(\vec{x})$$
$$p_1(xx)\ \underline{\text{then}}\ U(xx,\vec{x},g_o(\vec{C}(xx,\vec{x})))\ \ldots$$
$$p_n(xx)\ \underline{\text{then}}\ U(xx,\vec{x},g_{n-1}(\vec{C}(xx,\vec{x})))\ \underline{\text{else}}\ \bot$$

(recursion property of p_i, U is strict).

If $p_i(xx) = \text{True}$, we have for all $n \geq i$:

$$U_n(xx,\vec{x},y) = u_i(\vec{x},y) \quad \text{and} \quad \vec{C}_n(xx,\vec{x}) = \vec{c}_i(\vec{x}).$$

By the first induction hypothesis, U is the least upper bound of U_n and C of C_n. Thus, if $p_i(xx)$ = True, we get $U(xx,\vec{x},y) = u_i(\vec{x},y)$ and $\vec{C}(xx,\vec{x}) = \vec{C}_i(\vec{x})$.
We can write :

$$\tau^{n+1}[\Omega](xx,x) = \underline{\text{Case}}\ p_o(xx)\ \underline{\text{then}}\ g_o(\vec{x})$$
$$p_1(xx)\ \underline{\text{then}}\ u_1(\vec{x},g_o(\vec{c}_1(\vec{x})))\ \ldots$$
$$p_n(xx)\ \underline{\text{then}}\ u_n(\vec{x},g_{n-1}(\vec{c}_n(\vec{x})))\ \underline{\text{else}}\ \bot$$

Using the recursion relations of the g_i, it follows

$$\tau^{n+1}[\Omega](xx,\vec{x}) = G_n(xx,\vec{x}).$$

Prove that $G_n(x,\vec{A}(x,x)) = F_n(x)$

By definition :

$$F_n(x) = \underline{\text{Case}}\ p_o(x)\ \underline{\text{then}}\ f_o(x)\ \cdots$$
$$p_n(x)\ \underline{\text{then}}\ f_n(x)\ \underline{\text{else}}\ \bot$$
$$= \underline{\text{Case}}\ p_o(x)\ \underline{\text{then}}\ g_o(a_o(x))\ \ldots$$
$$p_n(x)\ \underline{\text{then}}\ g_n(a_n(x))\ \text{else}\ \bot$$

(since $f_i(x) = g_i(a_i(x))$)

Using the first induction hypothesis applied to $\vec{A}$, we get in the same way as before : $p_i(x) = \text{True} \Rightarrow \vec{A}(x,x) = \vec{a}_i(x)$. Thus :

$$F_n(x) = \underline{\text{Case}}\ p_o(x)\ \underline{\text{then}}\ g_o(\vec{A}(x,x)) \ldots$$
$$p_n(x)\ \underline{\text{then}}\ g_n(\vec{A}(x,x))\ \underline{\text{else}}\ \bot$$
$$(\text{since } p_i(x) = \text{True implies } \vec{A}(x,x) = \vec{a}_i(x)$$
$$= G_n(x,\vec{A}(x,x)).$$

The instantiation lemma implies that their limits are equal, which achieves the proof.

Remarks : The above scheme does not forbid cross-recursive call. If want to forbid them, this will be achieved by defining a <u>level of recursive embedding</u> as follows :

Definition : A program is of <u>recursive embedding</u> o if it appears in no recursive definition or only once in its own recursive definition. For instance, $F(x), g_\tau(xx,x)$, are of recursive embedding o in the above scheme.
A program is of <u>recursive embedding n+1</u> if it appears in the recursive definition of a program of recursive enbedding n.

Property of our scheme : The above definition is consistant with our scheme ; i.e. a program of recursive embedding level m can call an other of recursive embedding level n, if and only if $m < n$.

Consequence : The programs of recursive embedding level o are defined by the original sequence of input-output examples, their halting condition is thus $p_o(xx) = \text{True}$. The programs of recursive embedding 1 are defined by a sequence of examples which starts with the second example, their halting condition is thus $p_1(xx) = \text{True}$. For instance, the fragments u_i are defined by $g_i(\vec{x}) = u_i(\vec{x}, g_{i-1}(\vec{c}_i(\vec{x})))$, it follows that $g_1(\vec{x}) = u_1(\vec{x}, g_o(\vec{c}_1(\vec{x})))$ and $u_1(x,z)$ is computed when $g_1(x)$ is computed, i.e. when $p_1(xx) = \text{True}$.

2.5. The synthesis algorithm

The synthesis algorithm we have used is made up of five steps : Given the ascending linear domain $A(x_o,\alpha)$ and the sequence of input-output examples of the function F to be synthesized :

1. Compute the functional name d of the unique $\star$-labeled atom of α
2. Compute the functional name π of the unique atom of x_o which is in the direction of d.
3. Compute the fragments f_i.
4. Look for recurrence relations between consecutive fragments using the DETREC Algorithm (see next section).
5. If recurrence relations are found, then transform them into program using the Synthesis Theorems.

In practice, x_o and α can be computed from the set of inputs.

3. DETECTION OF RECURRENCE RELATIONS : the BMWk algorithm.

3.1. The basic algorithm

Given a finite sequence of fragments, we shall now describe an algorithm which tries to decide whether the sequence is recurrent or not. This algorithm uses two basic tools, the sequence matching and the sequence generalization ; moreover, it applies recursively on new sequences of fragments which are generated. Notice that this algorithm is not a decision procedure for recurrent sequences. However, it seems to be powerful enough in practice.

Definitions : Let T be the term algebra over a set x of variables associated to the abstract data type in use (without axioms). Let $D(\sigma) = \{x | \sigma(x) \neq \emptyset\}$ be the domain of a substitution σ. The sequence $\{f_i | i=o,\ldots,n\}$ of fragments matches (or is a matching sequence) iff there exists a sequence $\{\sigma_i | i=1,\ldots,n\}$ of substitutions such that $f_i=\sigma_i\, f_{i-1}$, $\forall 1 \leq i \leq n$. For each $x \in D(\sigma_i)$, $\{\sigma_i x | i=1,\ldots,n\}$ is said to be a generated sequence of fragments.

Notice that we use only head to head matchings in this definition. It can be easily extended to matchings with subtrees (in this case, a generated sequence $\{u_i\}$ of fragments appears which comes from the head part of f_i which is not matched with f_{i-1} -see more details in section 3.5), we are in fact specially interested with head to head matchings which lead to tail recursive synthesized definitions.

The algorithms which matches a sequence $\{f_i | i=o,\ldots,n\}$ is easy to design : it uses n times a classical matching algorithm. When a sequence does not match, we generalize it.

Definition : The sequence $\{g_i | i=o,\ldots,n-1\}$ generalizes the sequence $\{f_i | i=o,\ldots,n\}$ iff g_i generalizes the two terms f_i and f_{i+1} for any $o \leq i \leq n-1$. Let $G(\{f_i\})$ the set of sequences which generalizes the sequence $\{f_i\}$.

Let us now define in T^n the following partial ordering :

$$\{g_i\} \leq \{g_i'\} \text{ iff} \exists\ \sigma \text{ s.t. } g_i' = \sigma g_i, \quad \forall o \leq i \leq n-1.$$

It is easy to verify that the set $G(\{f_i\})$ is a complete lattice for this ordering (modulo the equivalence generated by the partial ordering). This lattice has a bottom (modulo the equivalence) that we call least generalized sequence of $\{f_i\}$.

Proposition : Given any one to one mapping θ between T^2 and X, the least generalized sequence $\{g_i | i=o,\ldots,n-1\}$ of a given sequence $\{f_i | i=o,\ldots,n\}$ is given by $g_i = f_i \hat{\theta} f_{i-1}$, where $f_i \hat{\theta} f_{i-1}$ is the least generalized terms of f_i and f_{i-1} relative to θ (see 10).

Proof : Left to the reader.

Examples :

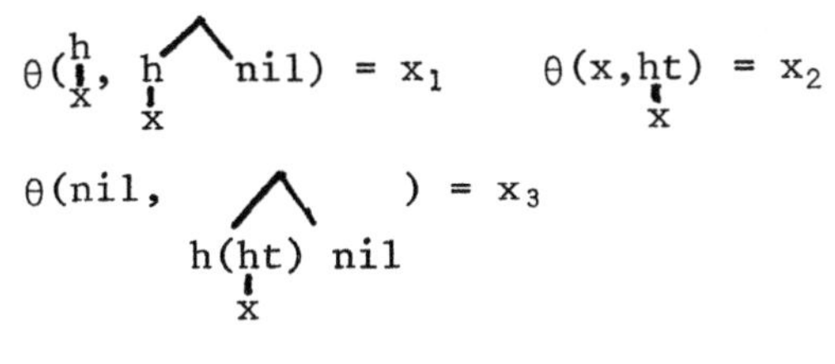

par_0 =

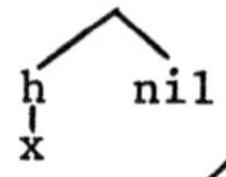

g_0 =

par_1 =

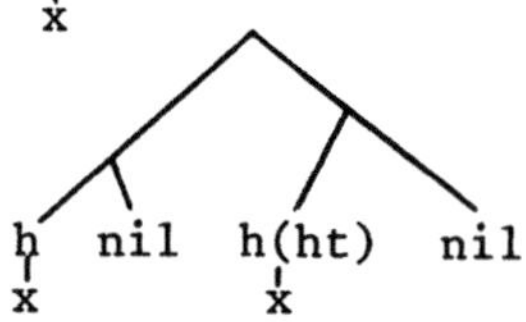

g_1 =

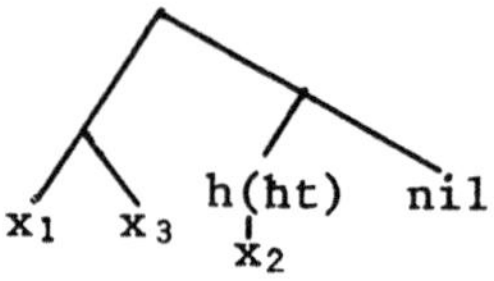

par_2 =

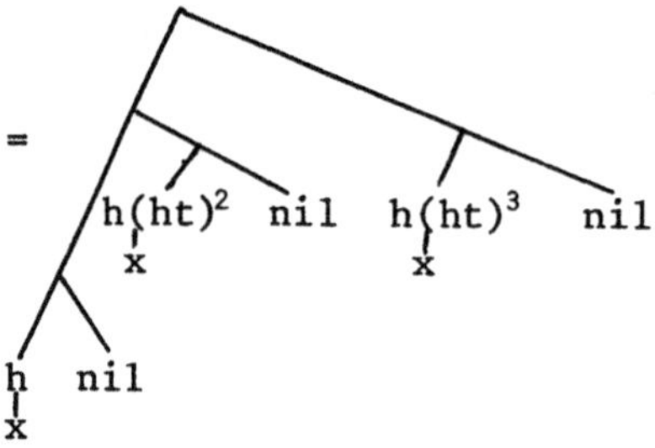

g_2 =

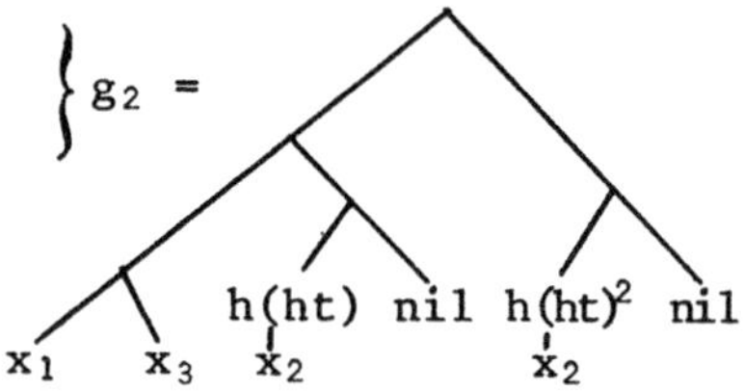

par_3 =

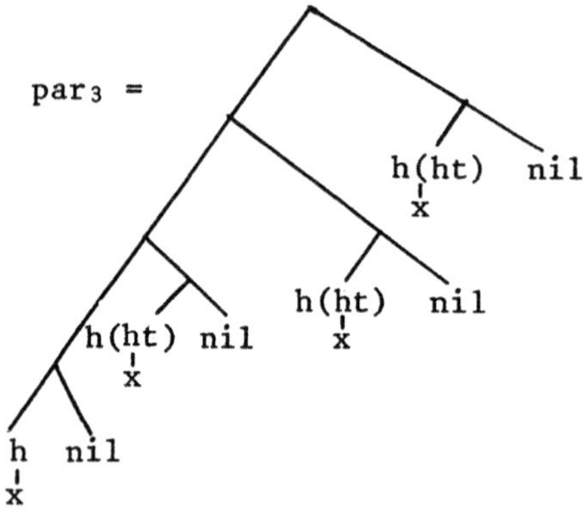

g_3 =

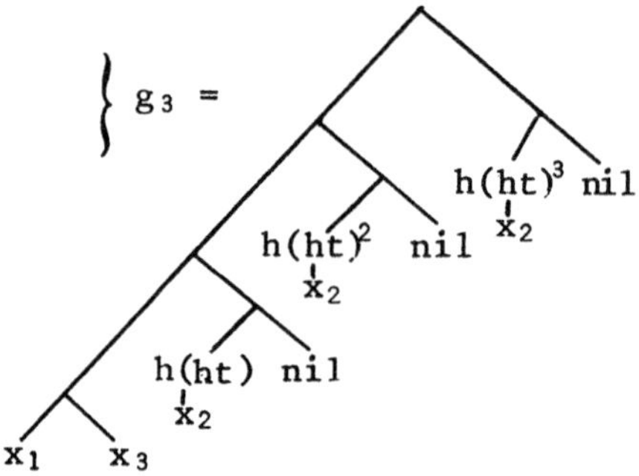

par_4 =

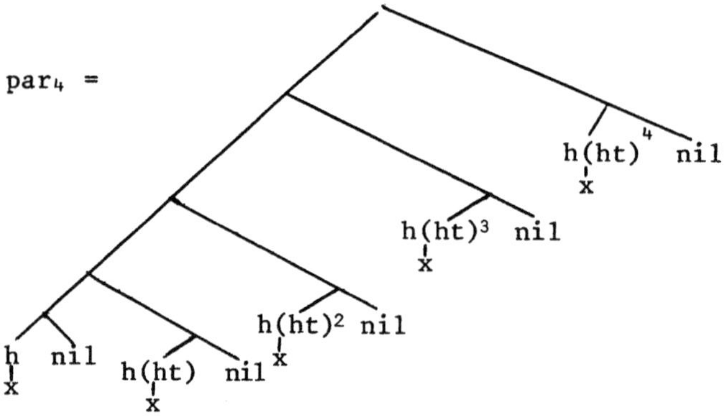

Notice that the aim of the mapping θ is to memorize the name of the variables (here x_1,x_2,x_3) introduced all along the generalization process.
It is now easy to match the obtained sequence $\{g_i\}$. If we consider g_i as a fragment, i.e. a function of the variables x_1,x_2,x_3, we get the following recurrence relations :

$$g_0 = \underset{x_1 \quad x_3}{\wedge} \qquad g_{i+1}(x_1,x_2,x_3) = g_i\Big(\underset{x_1 \quad x_3}{\wedge}\ ,\ \underset{x_2}{\overset{ht}{|}}\ ,\ \underset{h(\underset{x_2}{ht}) \qquad nil}{\wedge}\Big)$$

Notice that par_i is obtained from g_i by applying a substitution σ which is the left projection of the mapping θ. For the preceeding example :

$$par_i = g_i\Big(\underset{x}{\overset{h}{|}}\ ,\ x,\ nil\Big)$$

Using now the synthesis theorem, we get a new definition of the par function :

$$NPar(x) \leftarrow G\Big(\underset{x}{\overset{h}{|}}\ ,\ x,\ nil\Big)$$

$$G(x_1,x_2,x_3) \leftarrow \underline{If}\ atom\ (ht(x_2))\ \underline{then}\ \underset{x_1 \quad x_3}{\wedge}$$
$$\underline{else}\ G\Big(\underset{x_1 \quad x_3}{\wedge}\ ,\ \underset{x_2}{\overset{ht}{|}}\ ,\ \underset{h(\underset{x_2}{ht}) \qquad nil}{\wedge}\Big)$$

It is easy to prove that Par and NPar are equivalent functions
We shall come back to this problem in the next section.

An other remark is that we did not introduce the domain variable xx : since xx has the same recurrence relation as x_2, this was not necessary.
Before describing a difficult problem raised by the generalization process, let us write the basic algorithm which detects recurrence relations :

```
Proc DETREC({f_i})
     If {f_i} is a matching sequence
          then For each generated sequence of fragment {u_i}
          Do DETREC ({u_i}) ;
     else compute {g_i} the least generalized sequence of {f_i}
          DETREC  {g_i} ;
     Endif ;
End Proc ;
```

Notice that this procedure ends when there is no longer any generated sequence of fragments.
We shall now give and prove an extension of this basic algorithm for detecting recurrence relations. This extension allows to solve a wide class of examples. Let us first show how it works.

3.2. An example of the difficulties raised by the generalization

We consider the input-output sequence given by :

(A) → (A), (A B) → (B A), (A B C) → (C A B), (A B C D) → (D A B C),

(A B C D E) → (E A B C D) and consider the sequence $\{rot_i\}$ of the fragments of the function Rotate(x) defined by the above input-outputs :

rot_0 = 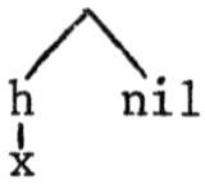
rot_1 = 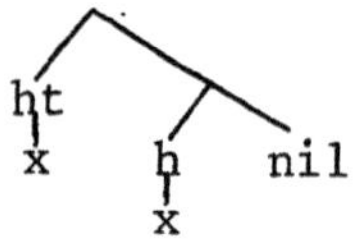
rot_2 = 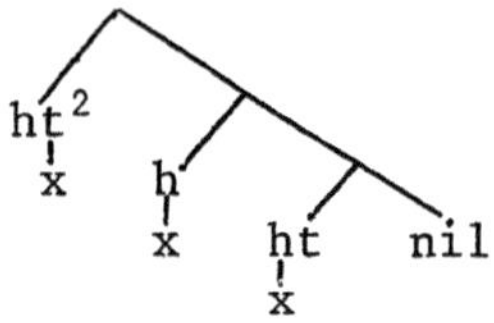

rot_3 = 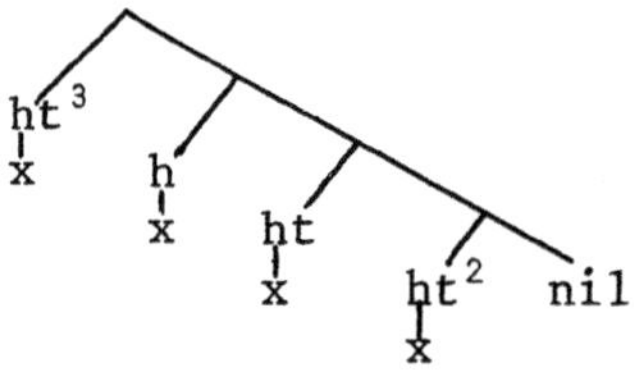
rot_4 = 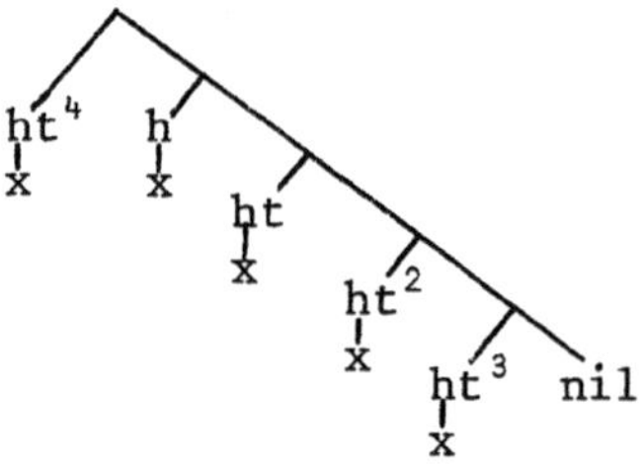

The matching of rot_i and rot_{i+1} fails because one obtains the contradictory substitutions x\cdr x and x\x. It fails also because the constant nil cannot be substituted by 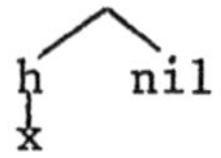 . According to the generalization process, BMWk will compute :

g_0 = 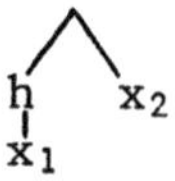 with : x_1\tx, x_2\ 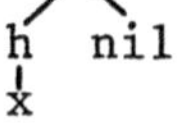

g_1 = 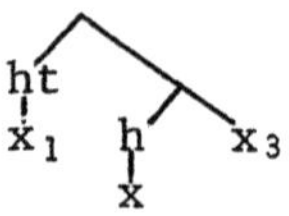 with : x_1\tx, x_3\ 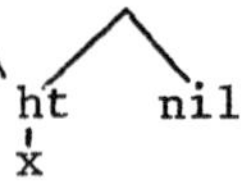

g_2 = 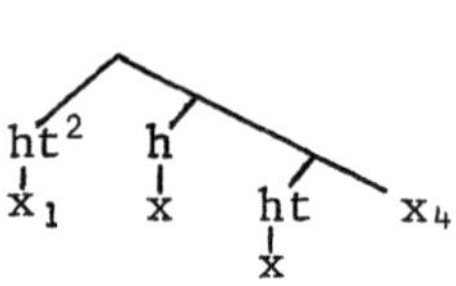 with x_1\tx, x_4\ 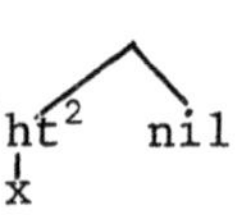

g_3 = 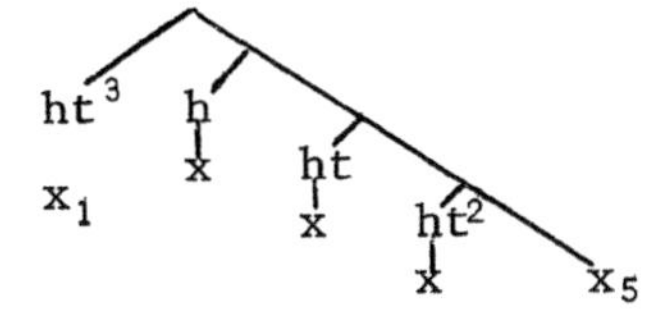 with $x_1\backslash tx$, $x_5\backslash$ 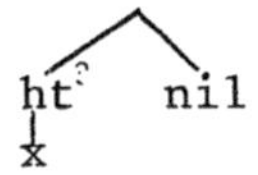

The matching between g_i and g_{i+1} succeeds with the substitutions :

g_0 matches g_1 with $\sigma_1 = (x_1\backslash t(x_1), x_2\backslash$ 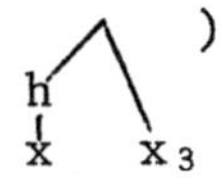 $)$

g_1 matches g_2 with $\sigma_2 = (x_1\backslash t(x_1),\ x\backslash x,\ x_3\backslash$ 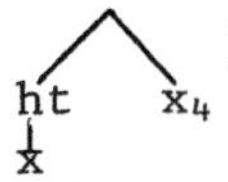 $)$

g_2 matches g_3 with $\sigma_3 = (x_1\backslash t(x_1), x\backslash x,\ x_4\backslash$ 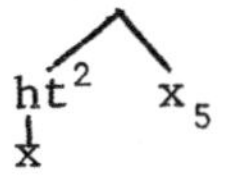 $)$

from which we are unable to deduce recursion relations between g_i and g_{i+1} since they would need and infinite set of variables. The reader's intuition must tell him how one has to proceed on : remark that the sequence $\{x_2, x_3, x_4, x_5, \ldots\}$ may reduce to one unique variable z and iterate the above process on the generated sequence of substitutions on z :

$z\backslash$ 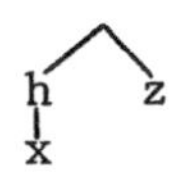 , $z\backslash$ 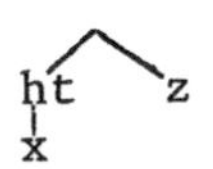 , $z\backslash$ 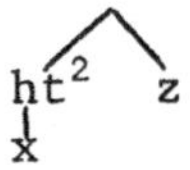

Since during this process one retrieves links between apparently independent variables $x_2, \ldots, x_5, \ldots$ we have called this process : retrieving of variables and given an algorithm that automatically retrieves the variables.
We shall describe it and its conditions of use in section 3.3., but let us first suppose that we have such an algorithm and show how it allows us to consider non-constant substitutions as new subproblems and achieve the synthesis of Rotate.

Since the variables $x_2, x_3, x_4, \ldots$ are supposed to be a same variable z, we have the following generalized fragments :

g_0 = 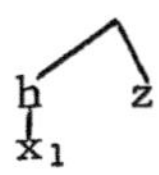 , g_1 = 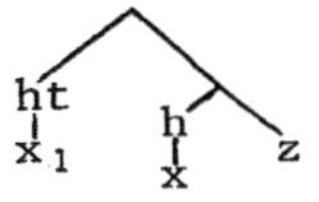 g_2 = 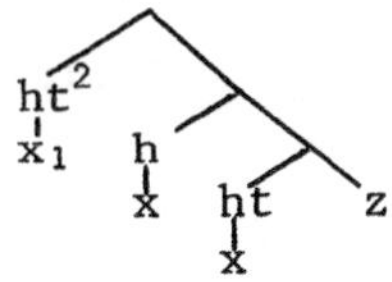

g_3 = 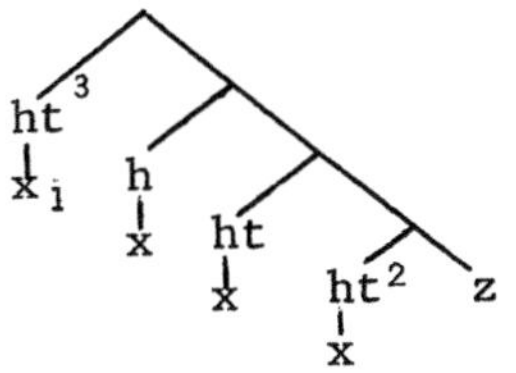

The matching process provides a constant substitution $x_1\backslash \begin{array}{c} t \\ | \\ x_1 \end{array}$

and a non constant one : $z\backslash a_1(x,z) = \begin{array}{ccc} & \wedge & \\ \begin{array}{c} h \\ | \\ x \end{array} & & z \end{array}$, $z\backslash a_2(x,z) = \begin{array}{ccc} & \wedge & \\ \begin{array}{c} ht \\ | \\ x \end{array} & & z \end{array}$

$z\backslash a_3(x,z) = \begin{array}{ccc} & \wedge & \\ \begin{array}{c} ht^2 \\ | \\ x \end{array} & & z \end{array}$

As indicated before, we now consider the values of the z-substitution as a new set of fragments a_i to be matched. These fragments belong to a function A which has the following approximation A_i :

<u>Case</u> $p_1(x)$ <u>then</u> $a_1(x,z)$

....

$p_i(x)$ <u>then</u> $a_i(x,z)$

<u>else</u> $\perp$.

The synthesis of the function A is viewed as a subproblem generated during the matching process. Notice that an other kind of subproblem appeared during the first attempt of synthesizing the Par function (section 2.3).
The matching between a_i and a_{i+1} succeeds with the substitution $x\backslash \begin{array}{c} t \\ | \\ x \end{array}$. It follows that we have found the recursion relations that characterize our starting sequence $\{f_i\}$:

$f_i(x) = g_i(x,x,nil)$

$g_0(x_1,x,z) = \begin{array}{ccc} & \wedge & \\ \begin{array}{c} h \\ | \\ x_1 \end{array} & & z \end{array}$, $g_{i+1}(x_1,x,z) = g_i(t(x_1),x,a_{i+1}(x,z))$

$a_1(x,z) = \begin{array}{ccc} & \wedge & \\ \begin{array}{c} h \\ | \\ x \end{array} & & z \end{array}$, $a_{i+1}(x,z) = a_i(t(x),z)$

Using the synthesis theorems, we obtain the terminal recursive program Rotate :

Rotate(x) $\leftarrow$ G(x,x,nil)

$G(x_1,x,z) \leftarrow$ <u>If</u> atom $(h(x_1))$ <u>then</u> $\begin{array}{ccc} & \wedge & \\ \begin{array}{c} h \\ | \\ x_1 \end{array} & & z \end{array}$

<u>else</u> $G(t(x_1),x,A(x_1,x,z))$

$A(x_1,x,z)$ <u>If</u> atom $(h^2(x_1))$ <u>then</u> $\begin{array}{ccc} & \wedge & \\ \begin{array}{c} h \\ | \\ x \end{array} & & z \end{array}$

<u>else</u> $A(t(x_1),t(x),z)$

3.3. Retrieving links between the variables issued from a generalization

In this section, we assume that the non-matching starting sequence of fragment has been generalized in a new sequence $\{f_i(x_1,\ldots,x_p)\}$, and we have to retrieve links between the introduced variables $x_1,\ldots,x_p$. If the matching of the sequence $\{f_i(x_1,x_2,\ldots,x_p\}$ of generalized fragments succeeds, we obtain a substitution

$$\sigma = (x_1\backslash a_1(x_1,\ldots,x_p),\ldots,x_p\backslash a_p(x\ ,\ldots,x_p))$$

which can be looked at as a variable rewriting system :

$$x_1 \to a\ (x_1,\ldots,x_p)$$
$$x_p \to a_p(x_1,\ldots,x_p)$$

<u>Notations</u> :

- Let $\#(x,t)$ the number of occurrences of the variable x in the term t. If X is a set of variables, $\#(X,t) = \sum_{x\in X} \#(x,t)$.

- Let $\sigma = (x_1\backslash a_1(x_1,\ldots,x_p),\ldots,x_p\backslash a_p(x_1,\ldots,x_p))$ a substitution of domain $X_p = (x_1,\ldots,x_p)$. For each $1\le i\le p$, $a_i(x_1,\ldots,x_p)$ is a term of the algebra over X_p.

- Let $\vec{\sigma} = \{x_i \to a_i(x_1,\ldots,x_p)\,|\,1\le i\le p\}$ denote the variable rewriting system associated to σ.

- Let $n_i = \#(X_p,a_i(x_1,\ldots,x_p))$ and $(n_i^k = \#(X_p,\sigma^k(x_i))$.

Notice that n_i is n_i^1.

<u>Definition</u> : The <u>speed</u> n_i^k of a variable x_i of the variable rewriting system $\vec{\sigma}$ is said to be polynomial if n_i^k is a polynomial in the variable k. We denote by $d^0(n_i^k)$ the degree of this polynomial in k.
The speed of the variable rewriting system $\vec{\sigma}$ is said to be polynomial if the speed of each variable x_i, $1\le i\le p$, is polynomial.

<u>Example</u> :

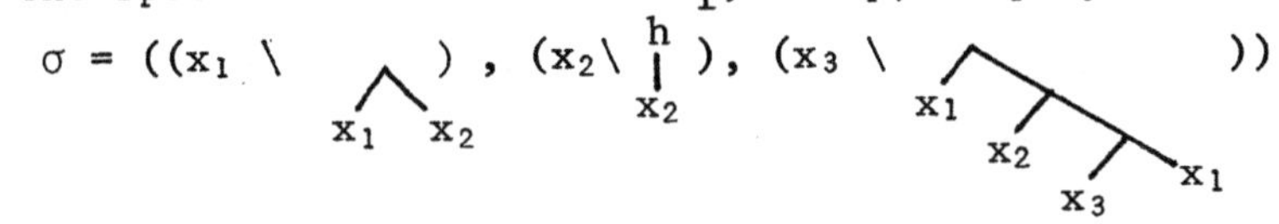

$n_1 = 2$, $n_2 = 1$, $n_3 = 4$
$n_1^k = (k+1)$, $n_2^k = 1$, $n_3^k = (k+1)^2$

<u>Remark</u> : There are two reasons to be interested in this polynomial property :
- first, we will show that it is decidable. This is a generaliza-

tion of results from (22) ;

- second, it is related to the growth of the fragments. More precisely, the fragments are growing polynomially iff any variable rewriting system $\vec{\sigma}$ computed by the DETREC algorithm has a polynomial speed. We thus restrict ourselves to polynomially growing functions. But this is still a restriction in our method (see the discussion at the end : we cannot synthesize exponentially growing functions using an algorithm looking for finite differences between fragments).

The problem now is to decide whether a variable rewriting system has a polynomial speed or not. We prove technical lemma before the main result.

Let $G = \{X_p, \Gamma\}$ be the rewriting multigraph defined by :

$$\Gamma(x_i, x_j) = \#(x_j, a_i(x_1, \ldots, x_p))$$

Example : For the above example

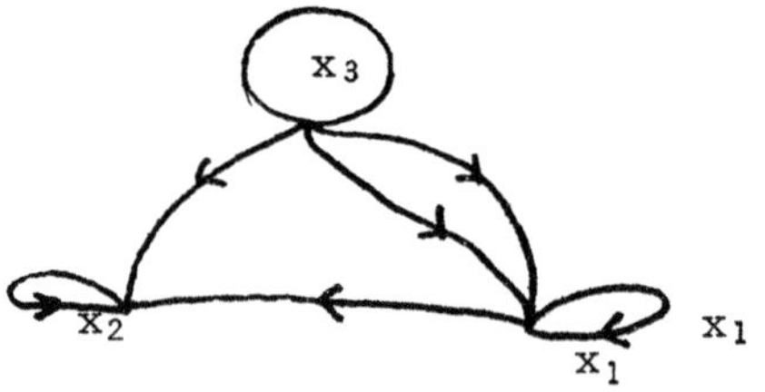

Lemma 1: If a node x_i of G belongs to two circuits at least in the multigraph, then n_i^k increases at least as an exponential in k.

Proof : Let n and m be the lengths of two such circuits. Then $\#(x_i, \sigma^{n+m}(x_i)) \geq 2$ and thus $\#(x_i, \sigma^{k(n+m)}(x_i)) \geq 2^k$. ■

Example 1 : $x_1 \backslash$ 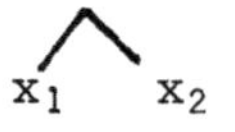 , $x_2 \backslash$ 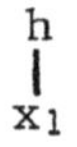

$G = x_1$  and $\sigma^2(x_1) =$ 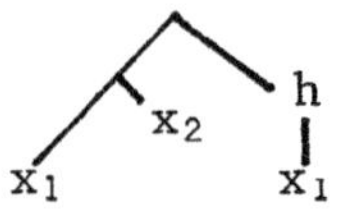

Example 2 : $x_1 \backslash$ 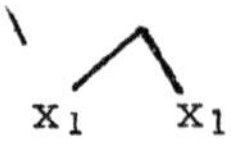 $G = x_1$ 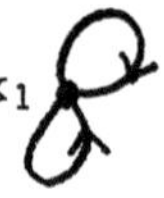 and $n_1^k = 2^k$

Lemma 2: If the nodes x_i and x_j of G, such that $\Gamma(x_i, x_j) \geq 1$, have a polynomial speed, then $d^0(n_i^k) \geq d^0(n_j^k)$.

Proof : Obvious.

Theorem : A variable rewriting system $\{x_i \to a_i(x_1,\ldots,x_p) \mid i \le p\}$ has a polynomial rewriting speed iff there does not exist a node x_i belonging to more than one elementary circuit in the associated multigraph G.
(This condition is obviously decidable).

Proof : $\Rightarrow$ by lemma 1

$\Leftarrow$ by induction on the number of nodes in G.

- basic case : $G = x\circlearrowleft$ $\sigma = \left(x \setminus \begin{array}{c} a \\ | \\ x \end{array}\right)$ then $\sigma^n(x) = \begin{array}{c} a^n \\ | \\ x \end{array}$

- general case : Let us first remark that the result is true (by induction) if G is divided into disconnected parts. We thus assume now that G is connected. Two cases may occur :

Case 1 : $\exists\, x_i$ which verifies (a) $\nexists\, x_j \neq x_i,\ \Gamma(x_j,x_i) \ge 1$

(b) $\exists\, x_j \neq x_i,\ \Gamma(x_i,x_j) \ge 1.$

By induction hypothesis, the multigraph $G_i = \{X_p - \{x_i\}, \Gamma\}$ has a polynomial rewriting speed. Using (a), all nodes $x_j \neq x_i$ have the same rewriting speed in G and G_i.
Let us now compute the rewriting speed of x_i in G :

$$n_i^k = \#(X_p, \sigma^k(x_i)) = \sum_{j \le p} |\Gamma(x_i,x_j)| \times n_j^{k-1} .$$

Two sub-cases occur in case 1.

- If $\Gamma(x_i,x_i) = 0$, then $n_i^k = \sum_{\substack{j \le p \\ j \neq i}} |\Gamma(x_i,x_j)| \times n_j^{k-1}$

  Thus x_i has a polynomial speed.

- If $\Gamma(x_i,x_i) = 1$, then $n_i^k = \sum_{\substack{j \le p \\ j \neq i}} |\Gamma(x_i,x_j)| \times n_j^{k-1} + n_i^{k-1}$

$$= \sum_{\substack{j \le p \\ j \neq i}} |\Gamma(x_i,x_j)| \sum_{1 \le h < k} n_j^h + n_i^1 .$$

But $\sum_{1 \le h < k} n_j^h$ is an polynomial of degree $1 + d^\circ(n_j^k)$, which proves that x_i has a polynomial rewriting speed.

Case 2 : $\forall x_i,\quad x_j \neq x_i,\ \Gamma(x_j,x_i) \ge 1.$

Therefore, there exists an elementary circuit in the graph G (the nodes of which are assumed to be $x_1,\ldots,x_q$ without lost of generality, with $\Gamma(x_i,x_{i+1}) = 1$, $1 \le i \le q-1$) such that $\nexists\, x_h$, $h > q$

with $\Gamma(x_h, x_i) \geq 1$, $1 \leq i \leq q-1$. In other words, there does not exist any incidental arc in this circuit.

Let n be the length of the circuit and consider the substitution $x_i \backslash \sigma^n(x_i)$, with $1 \leq i \leq q-1$. Two sub-cases occur in case 2 :

- $\#(X_p, \sigma^n(x_i)) = 1 = \#(x_i, \sigma^n(x_i))$. Then G is the circuit $(x_1, \ldots, x_q)$ and the result is true.

- $\#(X_p, \sigma^n(x_i)) > \#(x_i, \sigma^n(x_i)) = 1$. Thus $q<p$. Let us consider the graph $G' = \{x_{q+1}, \ldots, x_p\}$. By induction hypothesis, G' has a polynomial speed and by the property of the circuit, a node x_h of G' has the same rewriting speed in G and G'. The same reasonning as in case 1 allows to obtain the result for x_i. Applying now lemma 2 along the circuit, we notice that $\forall 1 \leq i, j \leq q-1$, $d^0(n_i^k) \geq d^0(n_j^k) \geq d^0(n_i^k)$. Thus, all variable on the circuit have the same speed (End of proof).

We now show how the polynomial condition allows the retrieving of variables. The problem is the following : given a sequence σ_i of variable rewriting systems, with $\bigcup_i V(\sigma_i)$ increasing, find a new sequence σ'_i of variable rewriting systems that verify a finite condition, that is :

$$\text{for some } j, \qquad \bigcup_i V(\sigma'_i) = \bigcup_{i=1}^{j} V(\sigma'_i).$$

This new sequence of variable rewriting systems is obtained by merging as many variables as possible, with respect to the polynomial property of each σ_i. More precisely :

Definition : Let $\leq$ be the following ordering on sequences of variable rewriting systems :

$$\{\sigma'_i\} \leq \{\sigma_i\} \text{ iff } \exists\, \theta : V(\sigma_i) \to V(\sigma'_i) \quad \text{such that :}$$

$$x_j \backslash a_j(x_1, \ldots, x_n) \in \sigma_i \quad \text{iff} \quad \theta(x_j) \backslash a_j(\theta(x_1), \ldots, \theta(x_n)) \in \sigma'_i.$$

Our problem is thus to compute a minimum sequence of variable rewriting systems less than the sequence $\{\sigma_i\}$ and verifying the finite condition and the polynomial properties. The algorithm is thus very simple : As long as the conditions are verified, choose two new variables and merge them !

Let us now come back to the example and show at work the algorithm. We obtained the following sequence of substitutions :

σ_1 : $x_1 \backslash t(x_1)$ $\qquad$ G_1 :

$x_2 \backslash$ 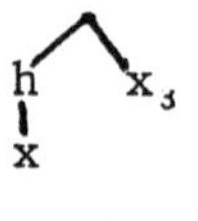

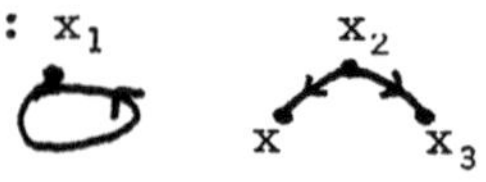

σ_2 : $x \backslash x$

$x_1 \backslash t(x_1)$

$x_3 \backslash$

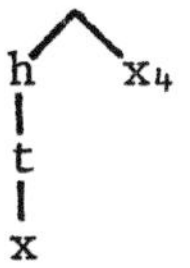

G_2 :

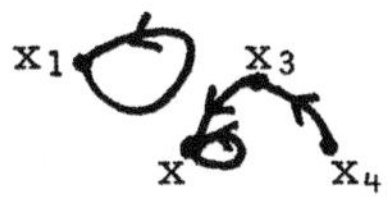

σ_3 : $x \backslash x$

$x_1 \backslash t(x_1)$

$x_4 \backslash$

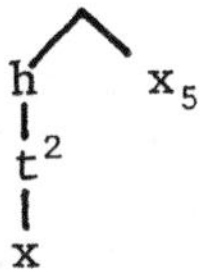

G_3 :

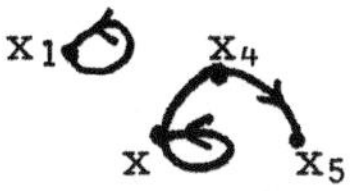

It is clearly possible to merge x_2 and x_3 in G_1, x_3 and x_4 in G_2, x_4 and x_5 in G_3. Let us call x_2 the name of this new variable. We have the multigraphs

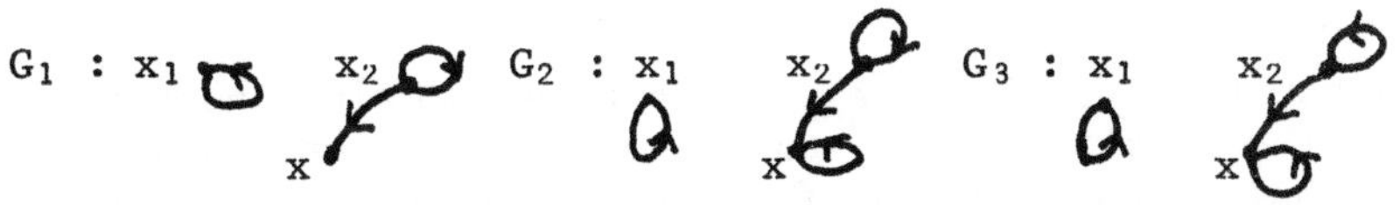

It is clearly no longer possible to merge now nodes without losing the polynomial property. We stop and the variables retrieving is achieved.
Notice that in practice, it is possible to choose the nodes to be merged in a standard way : we choose the nodes associated to the variables which have been introduced during the generalization process. Using such a heuristic, the process is not combinatorial and can be performed during the generalization.

3.4. Heuristic part of BMWk : resolution of the "lethal successes".

The scheme studied in the preceding section describes a rather restricted class of programs and we should like to enlarge this class even by introducing the possibility that BMWk does not end its computations for peculiar examples.

It happens that the obtained substitutions do not lead to recursion relations. This event is called : <u>lethal success</u>.

3.4.1 Lethal successes due to a domain undefinition.

Let a_o be the recursion relation of the domain, $p_{i+1}(x)=p_i(a_o(x))$. There exists an integer n such that, for any given x, $p_o(a_o{}^n(x))=$ True and $p_o(a_o^{n+p}(x))$ = Undefined, $p > o$.
This condition follows directly from the assumption that one generates the terms by a function which can be described by :
if $p_o(x)$ then $f_o(x)$ else ... if $p_i(x)$ then $f_i(x)$ else ...

Let us consider the substitutions of speed one : $x_k \backslash a_k(x_k)$, $k > o$. Suppose that one of the a_k is such that $a_o^n(x)$ = True implies $a_k^m(x)$ = True for any x and $m < n$. Since the computation of f_n is done up to the time where $a_o^n(x)$ = True, the computation of $a_k^{m+p}(x)$, $o < p < n-m$, will give an undefined value.

Example : Let the sequence of terms be obtained from the input-output sequence (A B C D) → A, (A B C D E) → C, (A B C D E F) → E,... One has $p_{i+1}(x) = p_i(t(x))$ and the sequence of terms is $f_o = h(x)$, $f_1 = ht^2(x)$, $f_2 = ht^4(x)$,... therefore $f_{i+1}(x) = f_i(t^2(x))$ and $f_o = h(x)$ describe this sequence of terms. (see also section 2.4)

The term $f_i(x) = ht^{2i}x$ is computed for variables x such that atom $t^{3+i}x$ = True. When $2i > 3+i$, i.e. $i > 3$, the computation of $f_i(x)$ leads to an undefined value.

Resolution of the lethal success. On a given domain, one knows the nature of the possible a_i and they are ordered (for linear ascending domains) by their speed of "destruction" of x as shown above. Each time a new substitution $x_k \backslash a_k(x_q)$ is found (i.e. before the variable retrieving process), BMWk verifies that a_k destructs x as or less rapidly than a_o.

3.4.2. Matching failures

Let us consider an extanded substitution of the type $\sigma'_i = (\ldots, f'^{\circ}_{ki} \backslash f^{\circ}_{ki}, \ldots)$ where the r.h.s.* of the substitution is itself a term. We shall replace f'_{ki} by a variable the initial value of which is computed by a function equivalent to $\{f'^{\circ}_{ki}\}$. This is possible only when the "size" of each f'_{ki} is less than the size of f°_{ki}, because the size of terms cannot decrease by substitutions.

*r.h.s. : *right hand side*.

We made more precise the notion of size by the following definition. Let t_1 and t_2 be two terms. On will say that t_1 is smaller than t_2 if

(i) the number of functions of arity greater than 1 is less in t_1 than in t_2

(ii) if these numbers are equal, the number of functions of arity 1 is less in t_1 than in t_2 .

Examples : 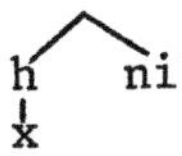 is smaller than 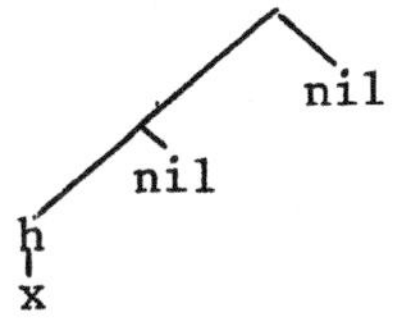 because of (i) and than 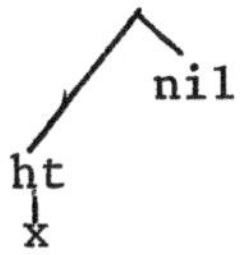 because of (ii). 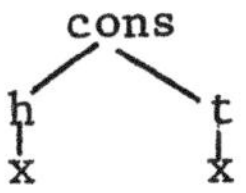 has the same size as 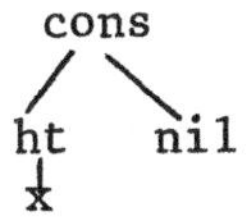

Detection of the lethal success. When substitutions of the type $\sigma'_i = (\ldots, f'^{\circ}_{ki}\backslash f^{\circ}_{ki}, \ldots)$ are found, the relative size of f'°_{ki} and f°_{ki} is smaller than f'°_{ki}, a lethal success is detected.

3.4.3. The resolution of the lethal successes.

Suppose that a lethal success between $f_i(x)$ and $f_{i+1}(x)$ is detected. The variable or sub-term of $f_i(x)$ undergoing the lethal substitution gives the place where $f_i(x)$ is cut into two pieces following the rule : consider the nearest father of arity ≥ 2 of this variable or sub-term. Call $g_i(x,z)$ the new term obtained by replacing in $f_i(x)$ by a variable z the son (of the above father) which contains the variable or sub-term undergoing the lethal success. Call $u_i(x)$ this son. One has $f_i(x) = g_i(x,u_i(x))$. Repeat this operation for each i (if one always has a lethal success). The whole is a transformation of $\{f_i(x)\}$ into $\{g_i(x,u_i(x))\}$. Apply BMWk recursively to $\{g_i(x,z)\}$ and $\{u_i(x)\}$.

Examples : $f_0(x) =$ 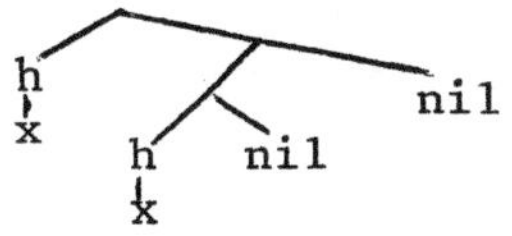

$f_1(x) =$ 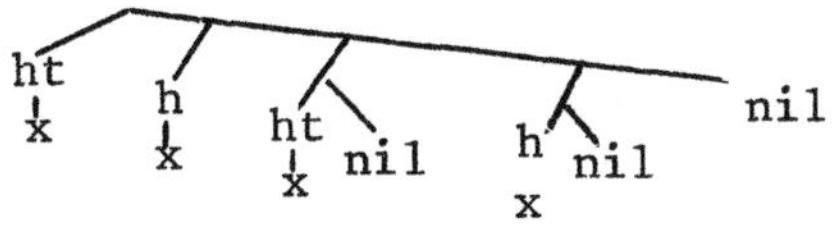

$f_2(x) =$ 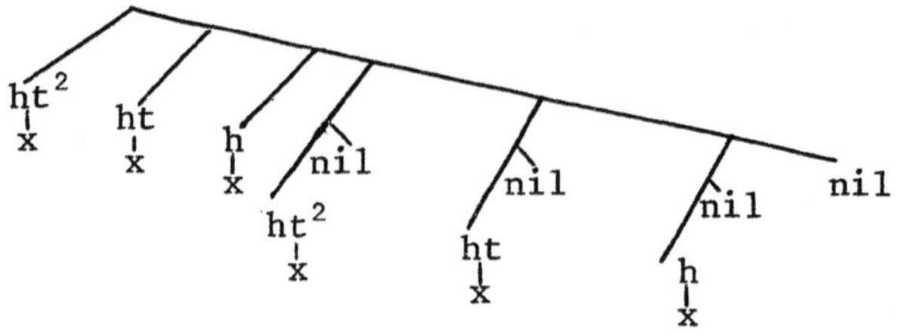

$f_3(x) =$ 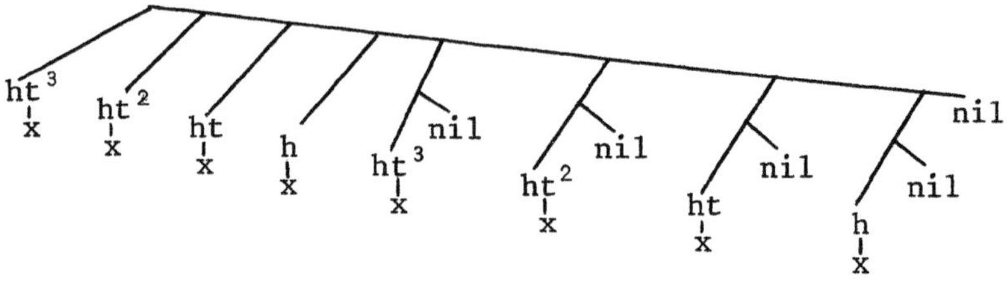

The matching of f_0 anf f_1 leads to a substitution 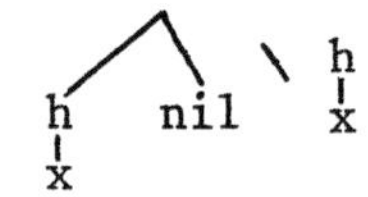

which is a lethal success . It follows that
BMWk cuts f_0 into two parts $g_0(x,z) =$ 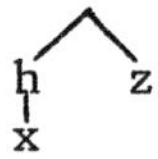 and $u_0(x) =$ 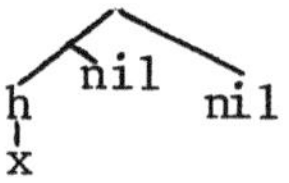

so that $f_0(x) = g_0(x,u_0(x))$.
The matching of f_1 and f_2 would lead to 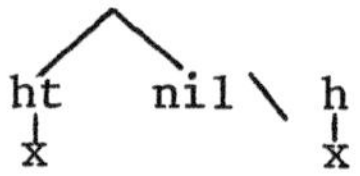 which again gives $f_1(x) = g_1(x,u_1(x))$ with

$g_1(x,z) =$ 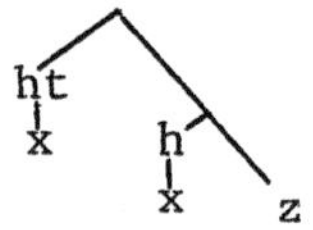 and $u_1(x) =$ 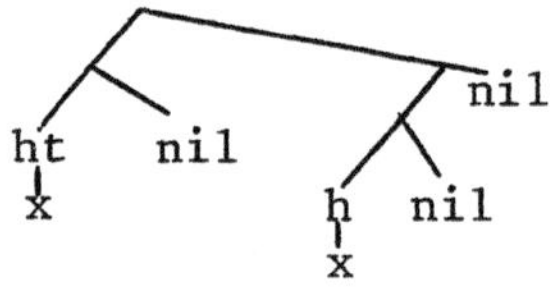

Similarly, the matching of f_2 and f_3 gives :

$f_2(x) = g_2(x,u_2(x))$, $g_2(x,z) =$ 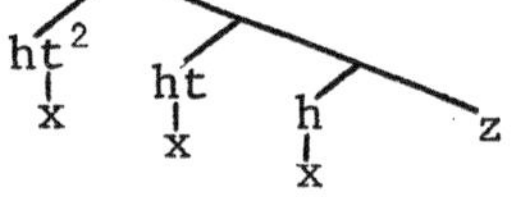

$u_2(x) =$ 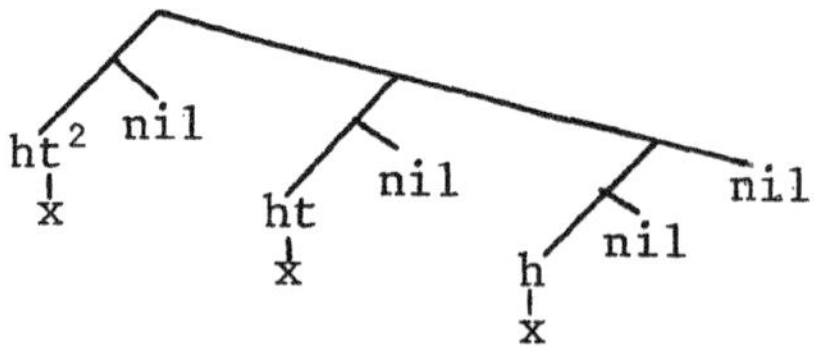

Now, BMWk applies to the two sequences $\{g_i\}$ and $\{f_i\}$, finds no new lethal successes and ends with the recursion relations :

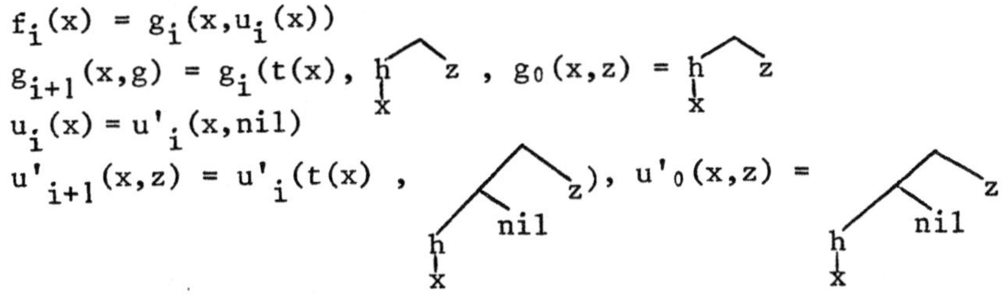

Discussion

One can be led to infinite computation loops when there are substitutions $f'^{\circ}_{ki} \backslash f^{\circ}_{ki}$.

One then cannot know which f'° belong to a given sequence $\{f'^{\circ}_{ki}\}$, i variable. Our heuristic is clearly ad'hoc since to might well be that one has to "zig-zag" in the sequence of terms $\{f_i\}$ in order to construct $\{f'^{\circ}_{ki}\}$.

Concerning 3.4., the lethal successes are clearly heuristics by themselves. It is quite easy to see that 3.4.1. strongly depends on the domain, for instance it might not be valid when a_0 has an inverse. Similarly, 3.4.2. could be formulated in many different ways, for instance by looking first at the number of functions the arity of which is greater than 2...

Cutting the terms of two pieces :

- may drive to failures since nothing insures that the two generated sequences will have recursion relations,
- may also drive to infinite computation loops when the lethal successes recur in the generated sequences for a greater index. For instance, the lethal success that lead to $f_i(x) = g_i(x,u_i(x))$ may well recur during the matching of $u_i(x)$ and $u_{i+1}(x)$.

  One would then obtain a non constant number of compositions.

4. APPLICATIONS OF BMWk

4.1. LISP Program synthesis from input-output pairs

The synthesis process has already been described in previous sections. Let us give a last example of such an application, using classical LISP primitives.

Consider the function (F X) where X contains k atoms at its top-level. F concatenates to the first atom of X, the (i+1)th and the ith for i = 1 to k-1. An input-output sequence is therefore : {(A)->(A), (A B)->(A B A), (A B C)->(A B A C B), (A B C D)-> (A B A C B D C),...}. It is easy to show that $(P_i X)$ = (ATOM(CDRI X)) where CDRI is the ith power of CDR. It follows that $(P_{i+1} X) = (P_i (CDR X))$.
The functional name of A in X_2 = (A B) is (CAR X_2) and B in X_2 is (CADR X_2), since Y_2 = (CONS A(CONS B(CONS A NIL))) we have :

(F X_2)=(CONS(CAR X_2)(CONS(CADR X_2)(CONS(CAR X_2)NIL))) and

(F_2 X)=(CONS(CAR X)(CONS(CADR X)(CONS(CAR X)NIL))).

One finds in the same way that :

```
(F1 X)= (CONS(CAR X)NIL),

(F3 X)= (CONS(CAR X)(CONS(CADR X)(CONS(CAR X)
                 (CONS(CADDR X)(CONS(CADR X)NIL)))))

(F4 X)= (CONS(CAR X)(CONS(CADR X)(CONS(CAR X)(CONS(CADDR X)
        (CONS(CADR X)(CONS(CADDDR X)(CONS(CADDR X)NIL))))))).
```

This procedure has been described several times (15), (17), (19).

-Transforming a computation trace into a program.
BMWk detects recursion relations in the trace and by a straightforward use of the theorems of section 2. These recursion relations are transformed in a program.
In this way, one obtains the following function from the above computation trace :

```
(F X)=(FF X X NIL)
(FF X Y Z)=IF (ATOM(CDR Y))
      THEN (CONS(CAR X)Z)
      ELSE (FF X(CDR Y)(H X Y Z))
(H X Y Z)=IF(ATOM(CDDR Y))
      THEN(CONS(CADR X)(CONS(CAR X)Z))
      ELSE(H(CDR X)(CDR Y)Z).
```

4.2. Program transformation

This section shows the use of a synthesis technique in program transformation. The idea is quite simple : starting from a recursive definition, we compute its predicates and fragments. The following is identical to synthesis : compute recurrence relations using BMWk and get the associated program.
The only difference between transformation and synthesis is in the way the predicates and fragments are computed. The recurrence relation on the predicates is directly obtained from the recursive definition. The obtention of the fragments from the recursive definition is explained in the following.

4.2.1. Obtention of the computational trace. In a formal way, the nth trace of a recursive function F can be defined as the nth iterate of Tarski's sequence, i.e. $\tau_F^n[\Omega](x)$, where τ_F is the functional which recursively defines F. We now give a practical way to compute it.
A recursive definition contains a recursive symbol (let us call it F) and symbols of auxiliary functions. For instance, a classical program scheme reads : (F X) = IF (P X) THEN (H_1 X) ELSE (H_2(G_1 X)(F(E X))) where it is supposed that the auxiliary predicate P takes the value True when E has been applied n times to X and where H_1,H_2,G and E are the auxiliary functions.
We define the nth trace of (F X) by the result obtained when

1- one suppose that (P(E power i)X) = False if i < n-1 and (P(E power n) = True.
2- The following computation is carried on :
   - the computation of the auxiliary functions is suspended
   - the computation of the branches that contain the recursive symbol is carried on as far as ((E power n)X) is not reached.

For example, if X is such that (P X) = True then the trace of the above scheme is (H_1 X). If X is such that (P(E(E X))) = True then its trace is (H_2(G_1 X)(H_2(G_1(E X))(H_1(E(E X))))).
In practice, one obtains easily the function FP associated to F and which writes down the trace of F :

```
(FP X) = (FP1 X 'X)
(FP1 X Z) = IF (P X) THEN (CONS 'H1(CONS Z NIL)) ELSE
(CONS 'H2(CONS(CONS 'G1(CONS Z NIL))(CONS(FP1(E X)(CONS 'E(CONS
Z NIL)))NIL))))
```
where P and E are now actually evaluated.

4.2.2. Recursive to tail-recursive transforms. When we choose a version of BMWk which tries to find first tail-recursive recursion relations, i.e. we attempt matchings and compositions first we shall obtain tail-recursive programs.
The differences between the synthesis method and the transformation method are simply :

1- the domain is not explicitly given. One has to check carefully the substitutions on the x type variables do not introduce infinite computation loops.

2- A lethal success introduces a composition but one cannot come back to the father CONS since we no longer have the CONS function. We have seen that a good rule is to come back to the first father which is a function of arity greater than 1. It might be possible also to give to the system the name of the wished function.

<u>Example</u> : We start from the function : (DE F(X)(FP X X X X))

```
(DE FP(X1 X2 X3 Y)(CON
((P Y)(H X (G X2 (L X3))))
(T(H X  (G X  (FP (M X1)(N X2)(O X3)(Q Y))))))
```

where P is a predicate and H,G,L,M,N,O,Q are supposed to be known functions with no special properties.
By the method of section 3 so get the traces :

```
IF(P X) THEN (H X(G X(L X))) ELSE
IF(P(Q(X)) THEN (H X(G X(H5M X)(G(N X)(L (O Y)))))) ELSE
IF(P(Q(Q X))) THEN (H X(G X(H(M X)(G(N X)(H(M(M X))
                          (G(N(N X))(L(O(O X)))))))))...
```

The traces lead to a lethal success which cuts the problem into two parts :

```
Problem 1 : F1=(H X(G X Z))
            F2=(H X(G X(H(M X(G(N X)Z))))
            F3=(H X(G X(H(M X(G(N X)(H(M(M X))(G(N(N X))Z)))))))
Problem 2 : FP1=(L X) FP2=(L(O X)) FP3=(L(O(O X)))
```

Problem 2 is readily solved by the substitution (X\(O X)). Problem 1 leads to the substitutions :

```
s1=(X\X),(Z\(H(M X)(G(N X)Z))))

s2=(X\X),(Z\(H(M X))(G(N(N X))Z))))
```

The substitution on Z introduces a new subproblem to problem 1 which is easily solved by a generalization of X into two variables X_1 and X_2, and the sunstitutions $(X_1\backslash(M\ X_2))$, $(X_2\backslash(N\ X_2))$, (Z\Z).
It follows that we obtain a new program F :

```
(DE F(X)(FP X X(FPP X X)))
(DE FPP(X Y)(COND
    ((P Y)(L X))
    (T(FPP(O X)(Q Y)))))
(DE FP(X Y Z)(COND
    ((P Y)(H X(G X Z)))
    (T(FP X(Q Y)(FPPP X Y Z)))))
(DE FPPP(X Y Z)(FPPPP X X Y Z))
(DE FPPPP(X1 X2 Y Z)(COND
    ((P(Q Y))(H(M X1)(G(N X2)Z)))
    (T(FPPPP(M X1)(N X2)(Q Y)Z))))
```

This example brings two comments.
First by following the computation, if H and G are strict and if the existence of an X such that (P X) = True implies the existence of an X such that (P(Q X)) = True then the two (F X) are strongly equivalent.
Second it illustrates well the time-place dilemma since the original F needs a stack but has a linear complexity while the second F does not need a stack but has a complexity in X power 2.

4.2.3. Iterative to recursive transforms. It may be interesting in the context of theorem proving to have a recursive program rather than a iterative one (see (28) for the description of the difficulties raised by proving properties of iterative definitions).

It is clear that favoring subtree matchings will lead to recursive programs.

4.3. Equivalence proofs

Specially relative to section 4.2, it would be nice to be able to prove the equivalence between a program and its transform, obtained by our methodology. The principle of the proof is the following.

From a program P, we construct traces $\{t_i\}_{o \leq i \leq q_o}$, we detect recursion relations valid for the q_o+1 traces, we hypothesize that these relations are valid for any $q > q_o$ and deduce a program P' from these relations. We have shown that P and P' are equivalent on their q_o+1 first traces. It is therefore enough to prove that : if P and P' are equivalent on their q first traces then they are equivalent on their q+1 first traces (induction on the number of terms). This is simply done once the general form of the q-th trace is known: let f_q be this trace. We compute f_{q+1} by applying P of f_q, we compute f'_{q+1} by applying P' to f_q and verify that $f_{q+1} = f'_{q+1}$.

In the polynomial case the expression of f_q is trivial to find and our methodology always proves easily the equivalence of two programs which have traces whose size increases polynomially. (see more details, and exponential cases in (22)).

Example : Let

P be : f(x,z) ⇐ <u>if</u> p(x) <u>then</u> z
<u>else</u> h(f(a(x),b(x,z)),x)

The methodology of section 4.2. leads to the tail-recursive transform :

```
P' : F(x,z) ⇐ G(x,x,K(x,z))
     G(x,y,z) ⇐ if p(y) then z
                        else G(x,a(y),H(x,y,z))

     H(x,y,z) ⇐ if p(a(y)) then h(z,x)
                           else H(a(x),a(y),z)

     K(x,z) ⇐ if p(x) then z
                      else K(a(x),b(x,z))
```

Let f_n be a term of the trace of f, F_n, G_n, H_n, K_n be terms of the traces of F, G, H, K respectively.

From the definition of f, one sees at once that

$$f_{n+1}(x,z) = h(f_n(a(x),b(x,z)),x)$$

and we want to prove that

$$F_{n+1}(x,z) = h(F_n(a(x),b(x,z)),x).$$

The proof proceeds as follows

• $\underline{K}$. $K_o(x,z) = z$

If $K_n(x,z) = b(a^{n-1}(x), b(a^{n-2}(x), \ldots, b(x,z))\ldots)$

then $K_{n+1}(x,z) = K_n(a(x), b(x,z))$

$= b(a^n(x), b(a^{n-1}(x), \ldots, b(a(x), b(x,z))\ldots)$

which proves by induction our hypothesis on the form of K_n.

• H, G. similarly, it is proven that

$H_n(x,z) = h(z, a^n(x))$,

$G_n(x,z) = h(h(\ldots(f(z, a^{n-1}(x)), \ldots, a(x)), x)$

• $\underline{F}$.$F_o(x,z) = z$

$F_n(x,z) = G_n(x, K_n(x,z))$

$= h(h(\ldots(b(a^{n-1}(x), \ldots, b(x,z)\ldots), a^{n-1}(x)), \ldots, x)$

It follows that

$h(F_n(a(x), b(x,z)), x) = h(h(\ldots(h(b(a^n(x), \ldots, b(x,z), \ldots),$
$a^n(x))), \ldots, x)$

$= F_{n+1}(x,z)$.

It follows that $F_{n+1}(x,z) = h(F_n(a(x), b(x,z)), x)$ which completes the proof of the equivalence between P and P'.

5. POLYNOMIAL SCHEME

The critical problem with such an approach using inductive inference, is the knowledge of the class of functions which can be theoretically synthesized from a finite set of examples. In other words, which is the class of functions fully described by a finite behavior, i.e. by a finite set of examples or by a finite set of computational traces ? Let us call <u>polynomials</u> these functions. Surely, the Par function is a polynomial. Many other ones are also polynomials, for instance the LISP functions REVERSE, ROTATE, HALF, DOUBLE, ... ; more generally, any function which can be synthesized from a finite set of examples (using our method) belongs to the set of polynomials.

Let us assume we know a decidable subset of polynomials (perhaps the whole set itself is decidable ?). Let now choose one polynomial in this set, for instance the Par function. By the above assumption, we know that it is fully characterized by a finite set of examples, or computational traces. It follows that all the possible synthesized expressions of the Par function are equivalent. The problem of correctness of the transformation method described here is thus reducible to the preceding problem.

We have therefore risen two open problems of main interest. First the characterization of the classes of functions which have the property of being fully characterized by a finite set of examples (or computational traces), second the characterization of decidable subclasses (perhaps the whole class itself).

Let us now describe a class of functions which are surely polynomials.

Definition : The polynomial scheme is inductively defined as follows :

Polynomial scheme PS_o :

PS_o is the set of cons-trees whose terminal functions belong to $R \cup \{nil\}$.

Polynomial scheme PS_{i+1} :

$$F(x) \leftarrow G(x,\vec{A}(x))$$

$$G[xx,\vec{y}] \leftarrow \underline{\text{If}}\ (\text{atom}.\pi(xx))\ \underline{\text{then}}\ f'_o(\vec{y})\ \text{else}\ U(xx,\vec{y},G\ d(xx),\vec{C}(xx,\vec{y})])$$

where :

1- $\pi = d'.d^p$, $p \geq o$, $d = d''.d'$, $d'' \neq I$ (ascending domain property)

2- f'_o belongs to PS_o

3- $A,G,U,\vec{C}$ belongs to $PS_o \cup PS_1 \cup \ldots \cup PS_i$.

Some more restrictions on the terminal functions of the cons-expressions a'_o, f'_o, u'_1, g'_1, c'_1, ... should also be added (14). As they are rather cumbersome to describe and not used in the following, we omit them.

The polynomial scheme PS is $\bigcup_{i=o}^{\infty} PS_i$.

A program instance of PS is called a list polynomial.

Conjecture : PS is a decidable subset of polynomials. Any function in PS may be synthesized from a finite set of examples using our methodology.
It seems to be a hard conjecture. We have proven this conjecture for a very small subclass of PS : $PS_o \cup PS_1$, and the proof method does not extend to the general case (14).

The conjecture was also proven for another small subclass by D. Angluin (29). Her proof uses the concept of "convergence in the limit" (30), also used by Shapiro (23).

An important property of PS is that any program instance of PS computes a function F such that $|F(x)|$ is an increasing polynomial of $|x|$. This result is easy to prove by induction on PS_i. Notice therefore that the non-increasing list polynomials do not belong to the scheme PS : may be, an extension of the method allows to synthesize them (13).

The similarities between list polynomials and the polynomials on the integers may be illustrated as follows :

Let P_n be a polynomial of degree n. Then $P_n(x+1) - P_n(x)$ is a polynomial of degree n-1.

Let now F be a list polynomial belonging to PS_n. By matching $F(x_{i+1})$ with $F(x_i)$ we obtain list polynomials in PS_{n-1}.

The matching is thus an operation which works on list as the difference operation works on integers. It is therefore very natural that a technique which obtains recursive definitions of polynomials on integers using the difference operation (17), obtains in the same way recursive definitions of list polynomials using the matching operation.
We may now use this deep analogy between polynomials on integers and list polynomials to infer answers to many various questions. For instance, it is very natural to define the degree of a list polynomial F as the smallest n such that $F \in PS_n$.
The key question is then : How many list polynomials of degree n accept a sequence of n+1 examples, the input of which are successive points of an ascending linear domain ? The answer should be only one and would imply the equivalence between two progams synthesized from the same set of examples.

Before closing this question, let us come back to our conjecture. It is not too hard to prove that $F \in PS$ iff it verifies some iteration lemma : more precisely, if after some matching steps we get constant substitutions, then it is not possible to get new recurrence relation by using new examples. And it is well known that iteration lemmas about algebraic trees constitute up to now a difficult open problem (perhaps easier in our particular case ?)

REFERENCES

(1) Bibel, W. : "On strategies for the synthesis of algorithms" Proc. AISB Meeting, Hamburg (1978), pp. 22-27.
(2) Bierman, A.W. : "On the inference of Turing machines from sample computations", Artificial Intelligence 3,(1972) 181-198.
(3) Bierman, A.W. : "The inference of regular LISP programs from examples", I.E.E.E. Trans. on Systems, Man and Cybernetics, vol. SMC-8, (1978), pp.585-600.
(4) Bierman, A.W., Smith, D.R. : "The hierarchical synthesis of LISP scanning programs", Information Processing 77, B. Gilchrist ed., North Holland (1977), pp. 41-45.
(5) Blum, L., Blum, M. : "Inductive inference : a recursion theoretic approach", Recursion Function Theory Newsletter, n° 6, (1973), pp. 15-29.
(6) Doner : "Tree acceptors and some of their applications", J.C.S.S. 4 (1970).

(7) Guiho, G., Gresse, C. : "Program synthesis from incomplete specifications", Proc. 5th Conf. on Automated Deduction, Les Arcs (1980), pp. 53-62.

(8) Hardy, J. : "Synthesis of LISP functions from examples", Proc. 4th I.J.C.A.I. (1975), pp. 268-273.

(9) Hedrick, C.L. : "Learning production systemsfrom examples", Proc. 4th I.J.C.A.I. (1975).

(10) Huet, G. : "Unification dans les théories d'ordre 1,2,...,ω" Thèse de Doctorat, Paris VII (1976).

(11) Jouannaud, J.P., Guiho, G. : "Inference of functions with an interactive system", Machine Intelligence 9, (1979), pp. 227-250.

(12) Jouannaud, J.P., Cuiho, G., Treuil, J.P. : "SISP/1, an interactive system able to synthesize functions from examples", Proc. 5th I.J.C.A.I. (1977), pp. 412-418.

(13) Jouannaud, J.P., Kirchner, C., Kirchner H., Picard, M. : "Signed trees : an algebraic frame for solving equations on binary trees . Application to LISP program synthesis and derecursivation", Pub. CRIN, Univ. Nancy, (1980).

(14) Jouannaud, J.P., Kodratoff, Y. : "Characterization of a class of functions synthesized from examples by a Summers like method using a BMW matching technique", Proc. 6th I.J.C.A.I. (1979), pp. 440-447, Proofs in Pub. CRIN, Univ. Nancy (1979).

(15) Jouannaud, J.P., Kodratoff, Y. : "An automatic construction of LISP programs by transformations of functions synthesized from their input-output behavior", Intern. J. of Policy Analysis and Information Systems 4, (1980), pp. 331-358.

(16) Kirchner, C., Kirchner, H., Jouannaud, J.P. : "Algebraic manipulations as a unification and matching strategy for linear equations in signed binary trees", Pub. CRIN, Univ. Nancy (1981) and Proc. I.J.C.A.I. 1981.

(17) Kodratoff, Y. : "A class of functions synthesized from a finite number of examples and a LISP program scheme", Interntl. J. of Comp. and Information Sciences 8, (1979), pp. 489-521.

(18) Kodratoff, Y., Jouannaud, J.P. : "A methodology for two variables functions synthesis from examples", Pub. CRIN, Univ. Nancy (1979).

(19) Kodratoff, Y., Papon, E. : "A system for program synthesis and program optimization", Proc. A.I.S.B. meeting, Amsterdam (1980), pp. 1-10.

(20) Manna, Z., Waldinger, R. : "A deductive approach to program synthesis", A.C.M. ToP LaS 2, (1980), pp. 90-121.

(21) Manna, Z., Ness, S., Vuillemin, J. : "Inductive methods for proving properties of programs", Comm. A.C.M. 16, (1973), pp. 491-502.

(22) Papon, E. : "Algorithmes de détection de relations de récurrence - Application à la synthèse et à la transformation de programmes", Thèse de 3ème cycle, Univ. Orsay (1981).

(23) Shapiro, E.Y. : "Inductive inference of theories from facts" Yale University, Dept. Comp. Sci., Rep. 192 (Feb. 1981).
(24) Shaw, D., Swartout W., Green C. : "Inferring LISP programs from examples", Proc. 4th I.J.C.A.I., Tbilisi, pp. 268-273.
(25) Summers, P.D. : "Programs construction from examples", Ph. D. Thesis, Yale, (1975).
(26) Summers, P.D. : "A methodology for LISP program construction from examples", J. A.C.M. 24, (1978), pp. 161-175.
(27) Waldinger, R., Lee, R.C.T. : "PROW : A step toward automatic program writing", Proc. I.J.C.A.I., (1969), pp. 241-252.
(28) Moore, J S. : "Introducing iteration into the pure LISP theorem prover", I.E.E.E., SE-1, (1975), pp. 328-338.
(29) Angluin, D. : "Summers revisited", Internal Report, Yale University, (1981).
(30) Gold, E.M. : "Language identification in the limit", Information and Control 28, (1975).

DEDUCTIVE SYNTHESIS OF THE UNIFICATION ALGORITHM

Zohar Manna
Computer Science Department
Stanford University
and
Applied Mathematics Department
Weizmann Institute

Richard Waldinger
Artificial Intelligence Center
SRI International

ABSTRACT

The *deductive approach* is a formal program-construction method in which the derivation of a program from a given specification is regarded as a theorem-proving task. To construct a program whose output satisfies the conditions of the specification, we prove a theorem stating the existence of such an output. The proof is restricted to be sufficiently constructive so that a program computing the desired output can be extracted directly from the proof. The program we obtain is applicative and may consist of several mutually recursive procedures. The proof constitutes a demonstration of the correctness of this program.

To exhibit the full power of the deductive approach, we apply it to a nontrivial example — the synthesis of a *unification algorithm*. Unification is the process of finding a common instance of two expressions. Algorithms to perform unification have been central to many theorem-proving systems and some programming-language processors.

The task of deriving a unification algorithm automatically is beyond the power of existing program-synthesis systems. In this paper, we use the deductive approach to derive an algorithm from a simple, high-level specification of the unification task. We will identify some of the capabilities required of a theorem-proving system to perform this derivation automatically.

This research was supported in part by the National Science Foundation under Grants MCS-78-02591 and MCS-79-09495, in part by the Office of Naval Research under Contracts N00014-75-C-0816 and N00014-76-C-0687, and in part by the Air Force Office of Scientific Research under Contract AFOSR-81-0014.

A. W. Biermann and G. Guiho (eds.), Computer Program Synthesis Methodologies, 251–307.

INTRODUCTION

In an earlier paper (Manna and Waldinger [1980]) we describe a deductive approach to program synthesis. In this approach, program synthesis is regarded as a theorem-proving task: Given a high-level specification of the purpose of the program, we prove a theorem that establishes the existence of an output satisfying this specification. The proof is restricted to be sufficiently constructive so that the desired program can be extracted directly. This approach is the direct descendant of the technique applied, *e.g.*, by Green [1969] and by Waldinger and Lee [1969].

In the earlier paper, we only applied the technique to very simple examples. In this paper, we consider a somewhat more difficult task: the synthesis of a unification algorithm.

Unification is the process of finding a common instance of two expressions. If such an instance exists, the algorithm is to produce a substitution that will yield that instance when applied to either of the expressions. If no common instance exists, the algorithm is to produce a special symbol indicating this situation. The first unification algorithm appeared in Herbrand's [1930] thesis, but the procedure did not come to widespread attention until it was rediscovered by Prawitz [1960] and employed by Robinson [1965] in his resolution principle for automatic theorem proving. Since then, the algorithm has been used not only for resolution theorem proving but also in many nonresolution theorem provers (see Bledsoe [1977]) and some programming-language processors (*e.g.*, PLANNER, see Hewitt [1971] or PROLOG, see Warren *et al.* [1977] or Colmerauer *et al.* [1979]).

Because of its importance in theorem proving and other applications, some effort has gone into the design of efficient unification algorithms (e.g., Martelli and Montanari [1976]; Paterson and Wegman [1978]) and the extension of the algorithm to more complex logical theories (*e.g.*, higher-order logic, Huet [1975]; associative and commutative theories, Stickel [1975], Livesay *et al.* [1979]).

The unification algorithm was the subject of partial verification efforts (*e.g.*, Waldinger and Levitt [1974]) and an example of automatic debugging (von Henke and Luckham [1974]). An early attempt to synthesize such a program appeared in Manna and Waldinger [1975]. Nevertheless, no complete automatic synthesis, or even verification, of the algorithm has been completed by any system.

The derivation presented in this paper depends on the formulation of a theory of expressions and substitutions. Intuitive observations about these objects can then be expressed and proved within the theory. In this paper, we set down, without proof, those results necessary for the derivation. A full presentation of the theory of expressions and substitutions is included in our forthcoming book (Manna and Waldinger [1982]).

The proof on which the derivation is based is presented in full. A summary of those aspects of the deductive approach necessary to understand the derivation is included. Although this proof can be expressed in the deductive tableau formalism of our [1980] paper, it is given here informally. We do not attempt to describe strategies under which the proof could be generated automatically. However, afterwards we consider what capabilities would be required of a theorem-proving system to discover such a proof.

THE DEDUCTIVE APPROACH

The specification of a program allows us to express the purpose of the desired program without indicating an algorithm by which that purpose is to be achieved. In general, we are considering the synthesis of programs whose specifications have the form

$$\begin{aligned} f(a) \ &\Leftarrow \ \textit{find } z \textit{ such that } R(a,z) \\ &\textit{where } P(a). \end{aligned}$$

Here, a denotes the input of the desired program and z denotes its output. The *input condition* $P(a)$ expresses the class of legal inputs to which the program is expected to apply. The *output condition* $R(a,z)$ describes the relation the output z is intended to satisfy.

For example, to specify a program to compute the integer square-root of a nonnegative integer n, we would write

$$\begin{aligned} sqrt(n) \ \Leftarrow \ &\textit{find } z \textit{ such that} \\ &\quad integer(z) \textit{ and } z^2 \leq n < (z+1)^2 \\ &\textit{where } integer(n) \textit{ and } 0 \leq n. \end{aligned}$$

A specification of the above form describes an applicative program, one that yields an output but produces no side effects. To derive a program from such a specification, we attempt to prove a theorem of the form

$$(\forall a)(\exists z)[\textit{if } P(a) \textit{ then } R(a,z)].$$

This theorem states that, for every input a, there exists an output z satisfying the output condition, provided that the input satisfies the input condition. The proof of this theorem must be constructive, in the sense that, in proving the existence of a satisfactory output z, it must tell us how to find such an output. From this proof, a program to compute z can be extracted.

WELL-FOUNDED INDUCTION

The formation of repetitive program constructs in the deductive approach depends on the application of the principle of mathematical induction. The induction principle we use is the principle of "well-founded induction," which applies to a wide variety of mathematical structures and results in the formation of a recursive procedure in the program being constructed. Before we can present the induction principle, we must introduce the notion of a "well-founded ordering."

Definition: If $\succ$ is a relation over a set S, we will say that $\succ$ satisfies the *decreasing sequence* condition if there are no infinite decreasing sequences $x_1, x_2, x_3, \ldots$ of elements of S; *i.e.*, there are no sequences such that

$$x_1 \succ x_2 \succ x_3 \succ \ldots$$

For example, the ordinary "greater-than" relation $\succ$ over the nonnegative integers satisfies the decreasing sequence condition. The same relation over all the integers does not.

Definition: If $\succ$ is a relation over a set S, we will say that $\succ$ is a *well-founded ordering* and S is a well-founded set under this ordering if

- $\succ$ is transitive and
- $\succ$ satisfies the decreasing sequence condition.

We will regard $y \prec x$ as synonymous with $x \succ y$.

For example, the following are all well-founded orderings:

- The $>$ relation over the nonnegative integers
- The subset relation over the finite sets
- The subtree relation over the finite trees.

The *principle of well-founded induction* may be expressed as follows:

> Let $\succ$ be a well-founded ordering over a set S. Then
> to prove that a statement $Q(a)$ is true for all elements a of S,
> consider an arbitrary element a of S,
> assume the induction hypothesis
> $(\forall x)[if\ x \prec a\ then\ Q(x)]$
> and prove that the conclusion
> $Q(a)$
> then follows.

When we are applying the well-founded induction principle in a program-synthesis context, we may use the following special form:

Let $\succ$ be a well-founded ordering over a set S. To construct a program f satisfying the specification

$$\begin{array}{ll} f(a) \Leftarrow & \text{find } z \text{ such that } R(a,z) \\ \text{where} & P(a), \end{array}$$

where the inputs a belong to a well-founded set, consider an arbitrary input a, assume the induction hypothesis

$$(\forall x)\begin{bmatrix} if\ x \prec a \\ then\ if\ P(x) \\ \qquad then\ R(x, f(x)) \end{bmatrix}$$

and then prove the conclusion

$$(\exists z)R(a,z).$$

In other words, we consider an arbitrary input a, and find an output z satisfying the given specification, under the following induction hypothesis: the program $f(x)$ we are trying to construct will satisfy the specification for all inputs x that are less than a in the well-founded ordering.

Application of the induction hypothesis during a proof will cause a recursive call $f(x)$ to appear in the program being constructed. The condition $P(x)$ will ensure that the input x of the recursive call will be a legal input; *i.e.*, it will satisfy the given input condition. The condition $x \prec a$ will ensure that the new recursive call cannot result in an infinite computation.

ℓ-EXPRESSIONS

In this section, we define a class of ℓ-expressions that will contain not only the expressions but also nested lists of expressions formed from a given alphabet.

• *the alphabet*

Suppose that S is an alphabet of symbols, consisting of three disjoint sets:

C: the *constants*

X: the *variables*, and

F: the *function symbols*.

Together, the constants and variables will be referred to as the *atoms* of S. With each function symbol of F is associated a unique positive integer, called its *arity*, indicating how many arguments the function takes.

• *generation rules*

The *expressions* of S are constructed by repeated application of the following *generation rules*:

Any constant of C is an expression;

Any variable of X is an expression;

If f is a function symbol (of arity n)
  and ℓ is a list of expressions (of length n)
then the result of applying f to the expressions in ℓ,
  denoted by $f \bullet \ell$, is an expression.

Note that if $\ell = [\ell_1, \ell_2, \ldots, \ell_n]$ then $f \bullet \ell$ is the expression informally denoted by $f(\ell_1, \ell_2, \ldots, \ell_n)$.

The ℓ-expressions of S are constructed by repeated application of the following generation rules:

The empty list [ ] is an ℓ-expression;

Any expression is an ℓ-expression;

If s is an ℓ-expression
  and m is a list of ℓ-expressions

then the result of inserting s before the first element of m, denoted by $s \circ m$, is an ℓ-expression.

Note that if $m = [m_1, m_2, \ldots, m_n]$ then $s \circ m$ is the list of ℓ-expressions informally denoted by $[s, m_1, m_2, \ldots, m_n]$.

• *uniqueness properties*

We assume that each ℓ-expression can only be produced in a unique way from the above rules. This assumption is expressed by the following properties:

$$c \neq x$$

$$c \neq f \bullet \ell$$

$$x \neq f \bullet \ell$$

if $f \bullet \ell = f' \bullet \ell'$
then $f = f'$ *and* $\ell = \ell'$

$$c \neq [\,]$$

$$c \neq s \circ m$$

$$x \neq [\,]$$

$$x \neq s \circ m$$

$$f \bullet \ell \neq [\,]$$

$$f \bullet \ell \neq s \circ m$$

$$[\,] \neq s \circ m$$

if $s \circ m = s' \circ m'$
then $s = s'$ *and* $m = m'$

for all constants c, variables x, function symbols f and f' (of arity n and n', respectively), lists of expressions ℓ and ℓ' (of length n and n', respectively), ℓ-expressions s and s', and lists of ℓ-expressions m and m'.

• *the occurs-in relation*

We will say that an ℓ-expression s *occurs in* an ℓ-expression s', denoted by

$s \preceq s'$ or $s' \succeq s$,

if s is a subexpression of s'; we will say that s *occurs properly in* s', denoted by

$$s \prec s' \quad \text{or} \quad s' \succ s$$

if s occurs in s' but is distinct from s'. We will abbreviate

$$\textit{not } s \prec s' \quad \text{as} \quad s \not\prec s',$$

$$\textit{not } s \preceq s' \quad \text{as} \quad s \not\preceq s',$$

and so forth. Formally, these relations are defined by the following properties:

$$s \preceq s' \quad \textit{if and only if} \quad s \prec s' \;\; \textit{or} \;\; s = s' \qquad (\textit{partiality})$$

$$s \not\prec a \qquad (\textit{atom})$$

$$s \not\prec [\,] \qquad (\textit{empty})$$

$$s \prec (f \bullet \ell) \quad \textit{if and only if} \quad s \preceq \ell \qquad (\textit{application})$$

$$s \prec (s' \circ m) \quad \textit{if and only if} \quad s \preceq s' \;\; \textit{or} \;\; s \preceq m \qquad (\textit{insertion})$$

for all atoms a, function symbols f (of arity n), lists of expressions ℓ (of length n), ℓ-expressions s and s', and lists of ℓ-expressions m.

We assume as part of its definition that the occurs-in relation is well-founded. This is a way of expressing formally that all the ℓ-expressions are finite. It follows that the relation is irreflexive, *i.e.*,

$$s \not\prec s \qquad (\textit{irreflexivity})$$

for all ℓ-expressions s.

The definition implies the following *component* properties of $\prec$:

$$\ell \prec f \bullet \ell$$

$$s \prec s \circ m$$

$$m \prec s \circ m$$

for every function symbol f (of arity n), list of expressions ℓ (of length n), ℓ-expression s, and list of ℓ-expressions m.

- *the vars function*

The value of $vars(s)$ is the set of variables that occur in the ℓ-expression s. Formally, we define

$$vars(c) = \{\} \qquad (constant)$$

$$vars(x) = \{x\} \qquad (variable)$$

$$vars([\,]) = \{\} \qquad (empty)$$

$$vars(f \bullet \ell) = vars(\ell) \qquad (application)$$

$$vars(s \circ m) = vars(s) \cup vars(m) \qquad (insertion)$$

for all constants c, variables x, function symbols f (of arity n), lists of expressions ℓ (of length n), ℓ-expressions s, and lists of ℓ-expressions m.

Now let us state a proposition relating the occurs-in relation with the *vars* function.

Proposition (variables): For every variable y and ℓ-expression s,

$$y \in vars(s) \quad \textit{if and only if} \quad y \preceq s.$$

In other words, the elements of $vars(s)$ are indeed those variables that occur in s. The proof is by well-founded induction over the occurs-in relation itself.

SUBSTITUTIONS

Substitution is the operation of replacing certain variables of an ℓ-expression by other expressions. We begin by giving an informal exposition of substitutions; subsequently, we give a formal treatment of the same notion.

Informally, we will represent a substitution θ as a set of *replacements*

$$\theta = \{x_1 \leftarrow e_1,\ x_2 \leftarrow e_2,\ \ldots,\ x_n \leftarrow e_n\},$$

where $x_1, x_2, \ldots, x_n$ are distinct variables in X and $e_1, e_2, \ldots, e_n$ are expressions such that $e_i \neq x_i$. Thus, replacements of the form $x \leftarrow x$ are excluded from substitutions, and substitutions of the form $\{x \leftarrow e,\ x \leftarrow e'\}$ are not allowed.

If $x_i \leftarrow e_i$ is a replacement in the substitution θ, we will refer to x_i as the *variable* and e_i as the *expression* of the replacement. We will denote by

$dom(\theta)$: the set of variables $\{x_1, x_2, \ldots, x_n\}$

affected by the substitution θ,

called the *domain* of θ, and

$$range(\theta): \quad \text{the set of variables that occur in } e_1, e_2, \ldots, \text{ or } e_n,$$

called the *range* of θ.

The result $s \blacktriangleleft \theta$ of applying such a substitution θ to an ℓ-expression s is obtained by simultaneously replacing every instance of the variables $x_1, x_2, \ldots,$ and x_n by the corresponding expressions $e_1, e_2, \ldots,$ and e_n.

For example, if θ is the substitution

$$\{x \leftarrow f(y),\ y \leftarrow g(a,z)\},$$

then

$$dom(\theta) = \{x, y\},$$

$$range(\theta) = \{y, z\};$$

furthermore, if s is the ℓ-expression

$$[\,x,\ [\,g(a,x),\ y\,]\,],$$

then the result $s \blacktriangleleft \theta$ of applying θ to s is

$$[\,f(y),\ [\,g(a,f(y)),\ g(a,z)\,]\,].$$

Note that the replacements are performed simultaneously: thus, the variable y in $f(y)$ above was *not* replaced by $g(a,z)$ even though θ contains a replacement $y \leftarrow g(a,z)$.

Let us be more precise: suppose that S is an alphabet consisting of the constants C, the variables X, and the function symbols F, as before.

• *generation rules*

The *substitutions* are constructed by repeated application of the following *generation rules.* We define the domain and range sets for each substitution at the same time.

- The empty substitution $\{\}$ is a substitution,

- $dom(\{\ \}) = \{\ \}$,
- $range(\{\ \}) = \{\ \}$;

If θ is a substitution,
 x is a variable not in $dom(\theta)$,
 and e is an expression distinct from x,
then
 - the result of adding the replacement $x \leftarrow e$ to the substitution θ, denoted by $(x \leftarrow e) \circ \theta$, is also a substitution,
 - $dom((x \leftarrow e) \circ \theta) = \{x\} \cup dom(\theta)$,
 - $range((x \leftarrow e) \circ \theta) = vars(e) \cup range(\theta)$.

Note that if $\theta = \{x_1 \leftarrow e_1, x_2 \leftarrow e_2, \ldots, x_n \leftarrow e_n\}$ then $(x \leftarrow e) \circ \theta$ is the substitution informally denoted by $\{x \leftarrow e, x_1 \leftarrow e_1, \ldots, x_n \leftarrow e_n\}$. Furthermore,

$$dom((x \leftarrow e) \circ \theta) = \{x, x_1, \ldots, x_n\}$$

$$range((x \leftarrow e) \circ \theta) = vars(e) \cup vars(e_1) \cup \ldots \cup vars(e_n).$$

We call $\circ$ the *addition* function for substitutions.

Substitutions do not have uniqueness properties: we may regard two substitutions as equal even though they have been constructed in different ways. We will say that two substitutions are equal if they have the same effect when applied to an arbitrary ℓ-expression. But first let us define more precisely what we mean by applying a substitution to an ℓ-expression.

- *the apply function*

If θ is a substitution, and s is an ℓ-expression, then the *apply* function $s \blacktriangleleft \theta$ is defined to satisfy the following properties:

$$s \blacktriangleleft \{\ \} = s \qquad \textit{(empty substitution)}$$

$$c \blacktriangleleft \theta = c \qquad \textit{(constant)}$$

$$x \blacktriangleleft ((x \leftarrow e) \circ \theta) = e \qquad \textit{(same variable)}$$

$$y \blacktriangleleft ((x \leftarrow e) \circ \theta) = y \blacktriangleleft \theta \quad \text{if } x \neq y \qquad \textit{(distinct variable)}$$

$$[\,] \blacktriangleleft \theta = [\,] \qquad \textit{(empty list)}$$

$$(f \bullet \ell) \blacktriangleleft \theta = f \bullet (\ell \blacktriangleleft \theta) \qquad \textit{(application)}$$

$$(s \circ m) \blacktriangleleft \theta = (s \blacktriangleleft \theta) \circ (m \blacktriangleleft \theta) \qquad \textit{(insertion)}$$

for all substitutions θ, constants c, variables x and y, expressions e, function symbols f (of arity n), lists of expressions ℓ (of length n), ℓ-expressions s, and lists of ℓ-expressions m.

Note that, in the *same variable* and *distinct variable* properties, we do not require that x and e be distinct or that $x \not\in dom(\theta)$ even though these are conditions on the generation rules for substitutions. This means that $(x \leftarrow e) \circ \theta$ is defined in these cases.

Let us now introduce a simple property relating ℓ-expressions and substitutions.

Proposition (monotonicity): For all ℓ-expressions s and s' and substitutions θ,

(a) if $s \prec s'$ then $s \blacktriangleleft \theta \prec s' \blacktriangleleft \theta$,

(b) if $s \preceq s'$ then $s \blacktriangleleft \theta \preceq s' \blacktriangleleft \theta$.

In other words, the subexpression and proper subexpression relations are maintained after the application of a substitution.

- *agreement and equality*

We will say that two substitutions θ_1 and θ_2 *agree* on an ℓ-expression s if

$$s \blacktriangleleft \theta_1 \;=\; s \blacktriangleleft \theta_2.$$

Now we can define the notion of equality for substitutions. We will say that two substitutions θ_1 and θ_2 are *equal*, denoted by $\theta_1 = \theta_2$, if they agree on all ℓ-expressions; *i.e.*,

$$\theta_1 = \theta_2$$

if and only if

$$s \blacktriangleleft \theta_1 = s \blacktriangleleft \theta_2 \quad \text{for all } \ell\text{-expressions } s.$$

We assume that the only substitutions are those that have been constructed by a finite number of applications of the generation rules; it follows that any substitution θ is a finite set of replacements and that the sets $dom(\theta)$ and $range(\theta)$ are finite sets of variables. This finiteness can be expressed formally by an appropriate induction principle over the substitutions.

• *characterization of domain and range*

Let us state two propositions that characterize the domain and range of a substitution.

Proposition (domain): For every substitution θ and variable x,

$$x \in dom(\theta) \quad \text{if and only if} \quad x \blacktriangleleft \theta \neq x.$$

That is, the domain is the set of all variables affected by the substitution.

Proposition (range): For every substitution θ and variable y,

$$y \in range(\theta) \quad \text{if and only if} \quad \begin{bmatrix} \text{there exists a variable } x \text{ such that} \\ x \in dom(\theta) \text{ and } y \in vars(x \blacktriangleleft \theta) \end{bmatrix}.$$

That is, the range is the set of all variables that may be introduced by a substitution.

Let us now introduce the notion of the composition of two substitutions.

COMPOSITION OF SUBSTITUTION

We define the composition $\theta \diamond \theta'$ of two substitutions θ and θ' to be the substitution satisfying the following property:

$$s \blacktriangleleft (\theta \diamond \theta') = (s \blacktriangleleft \theta) \blacktriangleleft \theta'$$

for all ℓ-expressions s. In other words, applying the composition $\theta \diamond \theta'$ to an ℓ-expression is the same as applying θ first, and then applying θ' to the result.

For example, if

$$\theta = \{x \leftarrow f(y)\}$$

$$\theta' = \{y \leftarrow g(a, x),\ x \leftarrow b\}$$

and

$$s = h(x, y, z),$$

then

$$s \blacktriangleleft \theta = h(f(y),\ y,\ z)$$

and

$$s \blacktriangleleft (\theta \diamond \theta') = (s \blacktriangleleft \theta) \blacktriangleleft \theta' = h(f(g(a,x)),\ g(a,x),\ z).$$

It follows from the definition that composition has the following properties:

$$\theta \diamond \{\} = \theta \qquad (right\ empty)$$

$$\{\} \diamond \theta = \theta \qquad (left\ empty)$$

$$(\theta_1 \diamond \theta_2) \diamond \theta_3 = \theta_1 \diamond (\theta_2 \diamond \theta_3) \qquad (associativity)$$

for any substitutions θ, θ_1, θ_2, and θ_3. Because $\diamond$ is associative, we may write expressions of form $(\theta_1 \diamond \theta_2) \diamond \theta_3$ and $\theta_1 \diamond (\theta_2 \diamond \theta_3)$ as $\theta_1 \diamond \theta_2 \diamond \theta_3$ without fear of ambiguity.

SYNTACTIC CATEGORIES

We will regard constants, variables, expressions, and lists as being of distinct "syntactic categories."

Definition: Let S be an arbitrary alphabet. Then the *syntactic categories* of the ℓ-expressions of S are the following five sets:

- the set C of constants,
- the set X of variables,
- the set consisting of the empty list $[\,]$,
- the set of functional expressions of form $f \bullet \ell$,
  where f is a function symbol (of arity n)
  and ℓ is a list of expressions (of length n).
- the set of nonempty lists of form $t \circ m$,
  where t is an ℓ-expression and m is a list of ℓ-expressions.

By the uniqueness properties of ℓ-expressions, every ℓ-expression belongs to precisely one syntactic category. The power of a substitution to change the syntactic category of an ℓ-expression is severely limited by the following observation:

Proposition (syntactic category): For any substitution θ and ℓ-expression s,

(a) if s is not a variable
then s and $s \blacktriangleleft \theta$ are in the same syntactic category.

(b) if e is an expression
then $e \blacktriangleleft \theta$ is an expression.

Note that it is not necessarily true that, if s is a variable, then $s \blacktriangleleft \theta$ is also a variable; it may be a constant or a functional expression. However, the converse is true:

Corollary: For any substitution θ and ℓ-expression s,

if $s \blacktriangleleft \theta$ is a variable
then s is also a variable.

THE AGREEMENT PROPOSITION AND ITS CONSEQUENCES

The following proposition has many useful consequences:

Proposition (agreement): For all substitutions θ and θ' and ℓ-expressions s, we have

$s \blacktriangleleft \theta = s \blacktriangleleft \theta'$.

if and only if

for every x, if $x \in vars(s)$ then $x \blacktriangleleft \theta = x \blacktriangleleft \theta'$.

In other words, two substitutions agree on an ℓ-expression precisely when they agree on all the variables of the ℓ-expression.

An immediate consequence of this proposition tells what happens when all the variables of an ℓ-expression are unchanged by a substitution:

Corollary (invariance): For every substitution θ and ℓ-expression s,

$s \blacktriangleleft \theta = s$

if and only if

$$vars(s) \cap dom(\theta) = \{\}.$$

In other words, a substitution has no effect on an ℓ-expression precisely when no variable in the domain of the substitution actually occurs in the ℓ-expression. The proof depends on taking θ' to be the empty substitution $\{\}$ in the Agreement Proposition.

Corollary (replacement invariance): For every ℓ-expression s, expression e, and variable x,

$$\begin{aligned}&\text{if } x \notin vars(s)\\ &\text{then } s \blacktriangleleft \{x \leftarrow e\} = s.\end{aligned}$$

In other words, applying a single replacement to an ℓ-expression has no effect if the variable of the replacement does not occur in the ℓ-expression.

Another consequence of the Agreement Proposition gives a useful characterization of the equality between substitutions. We have defined two substitutions to be equal if they agree on all ℓ-expressions. In fact it suffices to show that they agree on all variables:

Proposition (equality): For all substitutions θ and θ', we have

$$\theta = \theta'$$

if and only if

$$\text{for every variable } x,\ x \blacktriangleleft \theta = x \blacktriangleleft \theta'.$$

According to this proposition, to prove equality between substitutions it suffices to show that they agree on all variables. In fact, it suffices to consider only variables in their domains:

Corollary (equality): For all substitutions θ and θ', we have

$$\begin{bmatrix}\text{for every variable } x, & \\ \quad \text{if } x \in dom(\theta) \cup dom(\theta') & \\ \quad \text{then } x \blacktriangleleft \theta = x \blacktriangleleft \theta' & \end{bmatrix} \quad \text{if and only if} \quad \theta = \theta'.$$

The following proposition relates the addition and composition functions:

Proposition (addition-composition): For every substitution θ, variable x, and expression e,

$$(x \leftarrow e \blacktriangleleft \theta) \circ \theta = (x \leftarrow e) \diamond \theta.$$

The proof relies on the Equality Proposition.

- *the subtraction function*

We denote by $\theta - x$ the substitution that has no effect on the variable x, but that agrees with θ on all other variables. Formally, we define $\theta - x$ by the following properties:

$$x \blacktriangleleft (\theta - x) = x$$

$$y \blacktriangleleft (\theta - x) = y \blacktriangleleft \theta \quad \text{if} \quad x \neq y$$

for all substitutions θ and variables x and y. It follows (by the Domain Proposition) that

$$x \not\in dom(\theta - x)$$

for all substitutions θ and variables x and y.

Proposition (subtraction): For any variable x, expression e, and substitution θ, where $x \not\prec e$,*i.e.*, $x \not\in vars(e)$, we have

$$e \blacktriangleleft (\theta - x) = e \blacktriangleleft \theta.$$

The following proposition enables us to break down a substitution into its component replacements:

Proposition (decomposition): For every substitution θ and variable x,

$$\theta = (x \leftarrow x \blacktriangleleft \theta) \circ (\theta - x).$$

SET THEORETIC PROPERTIES OF SUBSTITUTIONS

In this section, we give some properties that relate the domain and range of a substitution with the variables of the ℓ-expressions.

Proposition (variable elimination): For any ℓ-expression s, substitution θ, and variable y,

$$\text{if } y \in dom(\theta) \text{ and } y \not\in range(\theta)$$
$$\text{then } y \not\in vars(s \blacktriangleleft \theta).$$

Thus, if a variable occurs in the domain of a substitution but not in its range, then the substitution will remove that variable from any ℓ-expression in which it occurs.

Proposition (variable introduction): For any ℓ-expression s, substitution θ, and variable y,

$$\text{if } y \in vars(s \blacktriangleleft \theta)$$
$$\text{then } y \in range(\theta) \text{ or } y \in vars(s),$$

i.e.,

$$vars(s \blacktriangleleft \theta) \subseteq range(\theta) \cup vars(s).$$

In other words, if a variable occurs in an ℓ-expression after a substitution, then it was introduced by the substitution or it occurred in the ℓ-expression originally.

UNIFIERS

Suppose that s and s' are ℓ-expressions and θ is a substitution of some alphabet S of constants, variables, and function symbols. We will say that θ is a *unifier* of s and s' if

$$s \blacktriangleleft \theta = s' \blacktriangleleft \theta.$$

Example: If

$$s = g(x, z)$$

and

$$s' = g(y, f(y))$$

then the substitution

$$\theta = \{x \leftarrow y,\ z \leftarrow f(y)\}$$

is a unifier, because

$$s \blacktriangleleft \theta = s' \blacktriangleleft \theta = g(y, f(y)).$$

Note that θ is not the only unifier of s and s'; *e.g.*,

$$\lambda = \{x \leftarrow b,\ y \leftarrow b,\ z \leftarrow f(b)\}$$

is a unifier, because

$$s \blacktriangleleft \lambda = s' \blacktriangleleft \lambda = g(b, f(b));$$

also,

$$\rho = \{y \leftarrow x,\ z \leftarrow f(x)\}$$

is a unifier, because,

$$s \blacktriangleleft \rho = s' \blacktriangleleft \rho = g(x, f(x)). \quad \blacksquare$$

Not all pairs of ℓ-expressions have a unifier. For example, there is no unifier for

$$s = g(a, b)$$

and

$$s' = g(x, x).$$

For, the result of applying any substitution θ to $g(a, b)$ will be the expression

$$g(a, b)$$

itself, in which the arguments are distinct. On the other hand, the result of applying θ to $g(x, x)$ will be an expression of form

$$g(e, e),$$

in which the arguments are identical. These two results can never be the same.

Two ℓ-expressions will be said to be *unifiable* if they have (at least one) unifier. Thus, $g(x, z)$ and $g(y, f(y))$ are unifiable, but $g(a, b)$ and $g(x, x)$ are not.

The following propositions characterize the unifiers for different categories of ℓ-expressions.

Proposition (application unifier): For any function symbol f (of arity n) and lists ℓ and ℓ' (of length n), we have

λ is a unifier of $f \bullet \ell$ and $f \bullet \ell'$

if and only if

λ is a unifier of ℓ and ℓ'.

Proposition (insertion unifier): For any ℓ-expressions t and t' and lists of expressions m and m',

λ is a unifier of $t \circ m$ and $t' \circ m'$

if and only if

λ is a unifier of t and t'
and
λ is a unifier of m and m'.

GENERALITY

We will say that a substitution θ is *more general than* a substitution λ, denoted by $\theta \succeq_{gen} \lambda$ or $\lambda \preceq_{gen} \theta$, if there exists some substitution ψ such that

$$\lambda = \theta \diamond \psi.$$

In this case, we will also say that λ is an *instance* of θ.

Example:

The substitution

$$\theta = \{y \leftarrow b\}$$

is more general than the substitution

$$\lambda = \{x \leftarrow a,\ y \leftarrow b\},$$

i.e., $\theta \succeq_{gen} \lambda$, because (taking ψ to be $\{x \leftarrow a\}$ in the definition)

$$\lambda = \theta \diamond \{x \leftarrow a\}.$$

In other words, λ is an instance of θ. ∎

Note that, according to our terminology, a substitution is always more general than itself, *i.e.*,

$$\theta \succeq_{gen} \theta \qquad (\textit{reflexivity})$$

for every substitution θ. Also, the empty substitution is more general than any substitution, *i.e.*,

$$\{\} \succeq_{gen} \theta \qquad (\textit{empty})$$

for every substitution θ.

MOST-GENERAL UNIFIERS

We have observed that there may be several distinct unifiers for a given pair of ℓ-expressions. In fact, any instance of a unifier is also a unifier.

Proposition (instance of a unifier): For all substitutions θ and λ and ℓ-expressions s and s',

if θ is a unifier of s and s'
and $\theta \succeq_{gen} \lambda$
then λ is a unifier of s and s'.

Definition: A substitution θ is a *most-general unifier* of two ℓ-expressions s and s' if

- θ is a unifier of s and s'; *i.e.*,
  $s \blacktriangleleft \theta = s' \blacktriangleleft \theta$,

and

- θ is more general than any other unifier of s and s'; *i.e.*,
  if $s \blacktriangleleft \lambda = s' \blacktriangleleft \lambda$
  then $\theta \succeq_{gen} \lambda$

for any substitution λ.

Combining the above proposition and definition, we see that the unifiers of two ℓ-expressions are precisely the instances of a most-general unifier.

Corollary (most-general unifier): For all substitutions θ and λ and ℓ-expressions s and s',

$$\theta \text{ is a most general unifier of } s \text{ and } s'$$

$$\text{if and only if}$$

$$\begin{bmatrix} \lambda \text{ is a unifier of } s \text{ and } s' \\ \text{if and only if} \\ \theta \succeq_{gen} \lambda \end{bmatrix}$$

Example:

We have seen that the two ℓ-expressions

$$s = g(x, z)$$

and

$$s' = g(y, f(y))$$

have many unifiers; *e.g.*,

$$\theta = \{x \leftarrow y,\ z \leftarrow f(y)\},$$

$$\lambda = \{x \leftarrow b,\ y \leftarrow b,\ z \leftarrow f(b)\}, \text{ and}$$

$$\rho = \{y \leftarrow x,\ z \leftarrow f(x)\}.$$

It turns out that θ is a most-general unifier of s and s'. In particular, θ is more general than λ and ρ, because

$$\lambda = \theta \diamond \{y \leftarrow b\}$$

and

$$\rho = \theta \diamond \{y \leftarrow x\}.$$

The unifier λ is not most-general. In particular, λ is not more general than θ; *i.e.*,

$$\theta \neq \lambda \diamond \psi$$

for any substitution ψ. For instance,

$$x \blacktriangleleft \theta = y$$

but (by the definition of composition and the *constant* property of the apply function)

$$x \blacktriangleleft (\lambda \diamond \psi) = (x \blacktriangleleft \lambda) \blacktriangleleft \psi = b \blacktriangleleft \psi = b \neq y.$$

Most-general unifiers are not unique; for example, the above substitution ρ is also a most-general unifier of s and s'. In particular, ρ is more general than θ and λ, because

$$\theta = \rho \diamond \{x \leftarrow y\}$$

and

$$\lambda = \rho \diamond \{x \leftarrow b\}.$$

The following proposition concerns the unifiers of a variable and an expression:

Proposition (variable unifier): For any variable x and expression e such that $x \not\prec e$, we have

$$\{x \leftarrow e\} \text{ is a most-general unifier of } x \text{ and } e.$$

We include the proof of this proposition, because it is not straightforward, and because it may be regarded as an integral part of the synthesis of the unification algorithm.

Proof:

It suffices to show (by the Most-General Unifier Corollary) that, for an arbitrary substitution λ,

$$\lambda \text{ is a unifier of } x \text{ and } e$$

if and only if

$$\{x \leftarrow e\} \succeq_{gen} \lambda,$$

i.e. (by the definition of the generality relation $\succeq_{gen}$),

$$x \blacktriangleleft \lambda = e \blacktriangleleft \lambda$$

if and only if

$$\lambda = \{x \leftarrow e\} \diamond \lambda^* \text{ for some substitution } \lambda^*$$

On the one hand, if

$$\lambda = \{x \leftarrow e\} \diamond \lambda^*$$

for some substitution λ^*, then

$$\begin{aligned} x \blacktriangleleft \lambda &= x \blacktriangleleft (\{x \leftarrow e\} \diamond \lambda^*) \\ &= (x \blacktriangleleft \{x \leftarrow e\}) \blacktriangleleft \lambda^* \end{aligned}$$

by the definition of composition

$$= e \blacktriangleleft \lambda^*$$

by the *same-variable* property of the apply function,

and

$$\begin{aligned} e \blacktriangleleft \lambda &= e \blacktriangleleft (\{x \leftarrow e\} \blacktriangleleft \lambda^*) \\ &= (e \blacktriangleleft \{x \leftarrow e\}) \blacktriangleleft \lambda^* \end{aligned}$$

by the definition of composition

$$= e \blacktriangleleft \lambda^*$$

by the Replacement Invariance Corollary, because $x \not\in vars(e)$.

In short,

$$x \blacktriangleleft \lambda = e \blacktriangleleft \lambda.$$

On the other hand, suppose

$$x \blacktriangleleft \lambda = e \blacktriangleleft \lambda.$$

Then

$$\lambda = (x \leftarrow x \blacktriangleleft \lambda) \circ (\lambda - x)$$
by the Decomposition Proposition

$$= (x \leftarrow e \blacktriangleleft \lambda) \circ (\lambda - x)$$
by our supposition that $x \blacktriangleleft \lambda = e \blacktriangleleft \lambda$

$$= (x \leftarrow e \blacktriangleleft (\lambda - x)) \circ (\lambda - x)$$
by the Subtraction Proposition, because $x \not\prec e$

$$= \{x \leftarrow e\} \diamond (\lambda - x)$$
by the Addition-Composition Proposition.

Therefore, if λ^* is taken to be $\lambda - x$, we have

$$\lambda = \{x \leftarrow e\} \diamond \lambda^*$$

as we had intended.

This concludes the proof. ∎

IDEMPOTENT SUBSTITUTIONS

We will say that a substitution θ is *idempotent* if it has the special property that

$$\theta = \theta \diamond \theta.$$

Example:

The substitution

$$\theta = \{x \leftarrow f(y)\}$$

is idempotent, because

$$\theta \diamond \theta = \{x \leftarrow f(y)\} \diamond \{x \leftarrow f(y)\}$$
$$= \{x \leftarrow f(y)\}$$

$$= \theta.$$

On the other hand,

$$\phi = \{x \leftarrow f(x)\}$$

is not idempotent, because

$$\begin{aligned}\phi \diamond \phi &= \{x \leftarrow f(x)\} \diamond \{x \leftarrow f(x)\} \\ &= \{x \leftarrow f(f(x))\} \\ &\neq \phi. \quad \blacksquare\end{aligned}$$

The property of idempotence is characterized by the following proposition.

Proposition (idempotence): A substitution is idempotent if and only if its domain and range are disjoint; *i.e.*,

$$\theta = \theta \diamond \theta \quad \text{if and only if} \quad dom(\theta) \cap range(\theta) = \{\},$$

for all substitutions θ.

MOST-GENERAL IDEMPOTENT UNIFIERS

Let us return to the example from the beginning of this section.

Example:

We have seen that the two ℓ-expressions

$$s = g(x, z)$$

and

$$s' = g(y, f(y))$$

have among their most-general unifiers the substitutions

$$\theta = \{x \leftarrow y,\ z \leftarrow f(y)\}$$

and

$$\rho = \{y \leftarrow x,\ z \leftarrow f(x)\}.$$

Both of these substitutions happen to be idempotent, *i.e.*,

$$\theta = \theta \diamond \theta \quad \text{and} \quad \rho = \rho \diamond \rho.$$

However, not all most-general unifiers are idempotent.

For instance, the substitution

$$\phi = \{x \leftarrow z,\ z \leftarrow f(z),\ y \leftarrow z\}$$

also turns out to be a most-general unifier of s and s'. It is a unifier, because

$$s \blacktriangleleft \phi \;=\; g(z, f(z)) \;=\; s' \blacktriangleleft \phi.$$

It is more general than the most-general unifier θ, because

$$\theta = \phi \diamond \{z \leftarrow y\}.$$

But ϕ is not idempotent, because

$$z \in \{x, y, z\} = dom(\phi)$$

and

$$z \in \{z\} = range(\phi);$$

therefore

$$dom(\phi) \cap range(\phi) \;=\; \{z\} \;\neq\; \{\},$$

and hence (by the Idempotence Proposition) ϕ is not idempotent. ∎

Most-general, idempotent unifiers have some properties we will find useful.

Proposition (most-general, idempotent unifier): If θ is a unifier of two ℓ-expressions s and s', then

θ is most-general and idempotent

if and only if

for every unifier λ of s and s', $\quad \lambda = \theta \diamond \lambda$.

Proposition (domain and range): If θ is a most-general, idempotent unifier of two ℓ-expressions s and s', then

(a) $dom(\theta) \subseteq vars(s) \cup vars(s')$

(b) $range(\theta) \subseteq vars(s) \cup vars(s')$.

In other words, the only variables that may appear in θ are those that occur in s or s'.

THE UNIFICATION ALGORITHM

A *unification algorithm* is a procedure for finding a most-general, idempotent unifier for two ℓ-expressions, if any unifiers exist at all. Otherwise, it produces a special symbol *nil*, which is assumd to be distinct from any substitution.

The specification for the unification algorithm may be expressed as follows:

$$\begin{array}{|l|}\hline
\mathit{unify}(s, s') \Leftarrow \text{find } \theta \text{ such that} \\
\qquad\qquad \theta \text{ is a most-general, idempotent unifier of } s \text{ and } s' \text{ and} \\
\qquad\qquad \theta \neq \mathit{nil} \\
\qquad\qquad\quad \text{or} \\
\qquad\qquad s \text{ and } s' \text{ are not unifiable and} \\
\qquad\qquad \theta = \mathit{nil} \\
\text{for all } \ell\text{-expressions } s \text{ and } s'. \\ \hline
\end{array}$$

According to the deductive approach, then, we must prove the existence of an output θ satisfying this specification; *i.e.*, we prove the following theorem:

$$(\forall s)(\forall s')(\exists \theta)\begin{bmatrix} \theta \text{ is a most-general, idempotent unifier of } s \text{ and } s' \;\; \mathit{and} \\ \theta \neq \mathit{nil} \\ \qquad \mathit{or} \\ s \text{ and } s' \text{ are not unifiable} \;\; \mathit{and} \\ \theta = \mathit{nil} \end{bmatrix}$$

In other words,

$$(\forall s)(\forall s')(\exists \theta)\begin{bmatrix} s \blacktriangleleft \theta = s' \blacktriangleleft \theta \;\; \mathit{and} & (\theta \text{ is a unifier }) \\ \quad (\forall \lambda)[\mathit{if} \;\; s \blacktriangleleft \lambda = s' \blacktriangleleft \lambda \;\; \mathit{then} \;\; \theta \succeq_{gen} \lambda] \;\; \mathit{and} & (\theta \text{ is most-general }) \\ \theta = \theta \diamond \theta \;\; \mathit{and} & (\theta \text{ is idempotent }) \\ \theta \neq \mathit{nil} & \\ \qquad \mathit{or} & \\ (\forall \lambda)[s \blacktriangleleft \lambda \neq s' \blacktriangleleft \lambda] \;\; \mathit{and} & (s \text{ and } s' \text{ are not unifiable }) \\ \theta = \mathit{nil} & \end{bmatrix}$$

This theorem will also establish that if two ℓ-expressions are unifiable, they have a most-general, idempotent unifier.

Before we give the proof of the theorem, let us look ahead at the program we will ultimately extract from the proof. For clarity, we present the program as a set of properties of the *unify* function; actually the deductive approach will produce the corresponding LISP-like applicative program. Of course, these properties will not be available to us during the proof.

- **general**

> For all ℓ-expressions s and s':
>
> $unify(s, s') = \{\}$ if $s = s'$ (*same*)
>
> $unify(s, s') = nil$ (*distinct*)
> if s and s' are nonvariables in distinct syntactic categories.

- **constant**

> For all constants c and c':
>
> $unify(c, c') = nil$ if $c \neq c'$ (*distinct*)

- **variable**

> For all variables x, expressions e, and lists of ℓ-expressions m:
>
> $unify(x, e) = \{x \leftarrow e\}$ if $x \not\preceq e$ (*not-in, left*)
>
> $unify(e, x) = \{x \leftarrow e\}$ (*not-in, right*)
> if $x \not\preceq e$ and e is not a variable
>
> $unify(x, e) = unify(e, x) = nil$ if $x \prec e$ (*in*)
>
> $unify(x, m) = unify(m, x) = nil$ (*list*)

- **function**

> For all function symbols f and f' (of arities n and n', respectively) and all lists of expressions ℓ and ℓ' (of lengths n and n', respectively):
>
> $unify(f \bullet \ell,\ f' \bullet \ell') = unify(\ell, \ell')$ if $f = f'$ (*same*)
>
> $unify(f \bullet \ell,\ f' \bullet \ell') = nil$ if $f \neq f'$ (*distinct*)

- **list**

For all ℓ-expressions t and t' and all lists of ℓ-expressions m and m':

$$\begin{array}{ll}
\text{let } \theta_{hd} = unify(t, t') & \\
\text{if } \theta_{hd} = nil & \\
\text{then } unify(t \circ m,\ t' \circ m') = nil & (no) \\
\text{else let } \theta_{tl} = unify(m \blacktriangleleft \theta_{hd},\ m' \blacktriangleleft \theta_{hd}) & \\
\quad \text{if } \theta_{tl} = nil & \\
\quad \text{then } unify(t \circ m,\ t' \circ m') = nil & (yes\text{-}no) \\
\quad \text{else } unify(t \circ m,\ t' \circ m') = \theta_{hd} \diamond \theta_{tl} & (yes\text{-}yes)
\end{array}$$

In expressing the list properties, we have used the notation

$$\begin{array}{l} \text{let } x = a \\ P(x) \end{array}$$

as an abbreviation for

$$P(a).$$

The virtue of this notation is that if $P(x)$ has many instances of x and if a is a lengthy expression, we would be required to rewrite a many times in writing $P(a)$. Thus, without this abbreviation, the final equality above would read

$$unify(t \circ m, t' \circ m') \quad = \quad unify(t,\ t') \diamond unify(m \blacktriangleleft unify(t, t'),\ m' \blacktriangleleft unify(t, t')).$$

In the *list* properties, we have given separate names for the three equalities, for easy reference. The *no* property corresponds to the case that $\theta_{hd} = nil$, the *yes-no* property to the case that $\theta_{hd} \neq nil$ and $\theta_{tl} = nil$, and the *yes-yes* property to the case that $\theta_{hd} \neq nil$ and $\theta_{tl} \neq nil$.

Now let us examine a proof of the theorem, to see how the above program can be constructed.

The proof is by well-founded induction over an ordering $\prec_{un}$ between pairs $\langle s,\ s' \rangle$ of ℓ-expressions. Rather than choose this ordering in advance, we will proceed with the proof, under the assumption that a satisfactory well-founded ordering can be defined. Afterwards, the proof will motivate the definition of an appropriate ordering $\prec_{un}$.

For two arbitrary ℓ-expressions s and s', we want to find an output θ that will satisfy the specification for s and s'. In other words, we want to prove the conclusion

$$P(s,s'):\quad (\exists\theta)\begin{bmatrix} s \blacktriangleleft \theta = s' \blacktriangleleft \theta \ \ \textit{and} \\ (\forall\lambda)[\textit{if}\ \ s \blacktriangleleft \lambda = s' \blacktriangleleft \lambda \ \ \textit{then}\ \ \theta \succeq_{gen} \lambda] \ \ \textit{and} \\ \theta = \theta \diamond \theta \ \ \textit{and} \\ \theta \neq \textit{nil} \\ \qquad \textit{or} \\ (\forall\lambda)[s \blacktriangleleft \lambda \neq s' \blacktriangleleft \lambda] \ \ \textit{and} \\ \theta = \textit{nil} \end{bmatrix}$$

We will be happy if θ satisfies either of the two disjuncts in this desired conclusion.

As our induction hypothesis, we assume that the program $unify(r, r')$ we are trying to construct will satisfy its specification for all inputs r and r' such that the pair $\langle r, r'\rangle$ is strictly less than $\langle s, s'\rangle$ in the selected ordering $\prec_{un}$. In other words, we assume

$$\textit{if}\ \ \langle r, r'\rangle \prec_{un} \langle s, s'\rangle$$
$$\textit{then}\ \begin{bmatrix} r \blacktriangleleft unify(r,r') = r' \blacktriangleleft unify(r,r') \ \ \textit{and} \\ (\forall\lambda)[\textit{if}\ \ r \blacktriangleleft \lambda = r' \blacktriangleleft \lambda \ \ \textit{then}\ \ unify(r,r') \succeq_{gen} \lambda] \ \ \textit{and} \\ unify(r,r') = unify(r,r') \diamond unify(r,r') \ \ \textit{and} \\ unify(r,r') \neq \textit{nil} \\ \qquad \textit{or} \\ (\forall\lambda)[r \blacktriangleleft \lambda \neq r' \blacktriangleleft \lambda] \ \ \textit{and} \\ unify(r,r') = \textit{nil} \end{bmatrix}$$

for all ℓ-expressions r and r'.

The proof distinguishes between several cases, corresponding to the properties of the apply function $\blacktriangleleft$. At the end of each case we will give the property of the *unify* function provided by the proof. Together, these properties constitute the final program.

GENERAL CASES

Case (same): $s = s'$.

Then any substitution θ is a unifier, *i.e.*,

$$s \blacktriangleleft \theta = s' \blacktriangleleft \theta.$$

- To find a unifier that is most-general, *i.e.*, such that

$$(\forall\lambda)[\mathit{if}\ s \blacktriangleleft \lambda = s' \blacktriangleleft \lambda\ \mathit{then}\ \theta \succeq_{gen} \lambda],$$

we note that (in this case) any substitution λ satisfies the antecedent $s \blacktriangleleft \lambda = s' \blacktriangleleft \lambda$. Therefore, θ must have the property that

$$\theta \succeq_{gen} \lambda$$

for any substitution λ. The *empty* property of the most-general relation $\succeq_{gen}$, *i.e.*,

$$\{\} \succeq_{gen} \lambda$$

for any substitution λ, suggests that, in this case, to satisfy the first disjunct of the desired conclusion $P(s,\ s')$, θ can be taken to be the empty substitution $\{\}$.

- The empty substitution $\{\}$ is idempotent, *i.e.*,

$$\{\} = \{\} \diamond \{\},$$

by either of the *empty* properties of composition.

- We ignore the final requirement, that $\{\} \neq nil$, because any substitution is distinct from *nil*.

Therefore, the empty substitution $\{\}$ is a most-general, idempotent unifier of s and s', and satisfies the first disjunct of the desired conclusion $P(s, s')$ in this case.

$\mathit{unify}(s, s') = \{\}$ if $s = s'$

Case (distinct): s and s' are nonvariables of distinct syntactic categories.

Then we can show that s and s' are not unifiable, *i.e.*

$$s \blacktriangleleft \lambda \neq s' \blacktriangleleft \lambda,$$

for any substitution λ. For, let λ be an arbitrary substitution. Recall that, because s and s' are not variables, we have (by the Syntactic Categories Proposition) that

$s \blacktriangleleft \lambda$ is in the same syntactic category as s,

$s' \blacktriangleleft \lambda$ is in the same syntactic category as s',

and therefore

$s \blacktriangleleft \lambda$ and $s' \blacktriangleleft \lambda$ are in distinct syntactic categories.

Hence,

$$s \blacktriangleleft \lambda \neq s' \blacktriangleleft \lambda;$$

i.e., s and s' are not unifiable.

It follows that, in this case, we can satisfy the second disjunct of the desired conclusion $P(s, s')$ by taking θ to be *nil*.

unify$(s, s') =$ *nil* if s and s' are nonvariables of distinct syntactic categories

CONSTANT CASE

Case (distinct): s and s' are distinct constants c and c' respectively.

Then we can show that c and c' are not unifiable, *i.e.*,

$$c \blacktriangleleft \lambda \neq c' \blacktriangleleft \lambda$$

for any substitution λ. For, let λ be an arbitrary substitution. Then (by the *constant* property of the apply function),

$$c \blacktriangleleft \lambda = c$$

and

$$c' \blacktriangleleft \lambda = c'.$$

But, since $c \neq c'$,

$$c \blacktriangleleft \lambda \neq c' \blacktriangleleft \lambda;$$

i.e., c and c' are not unifiable.

It follows that, in this case, we can satisfy the second disjunct of the desired conclusion $P(c, c')$ by taking θ to be *nil*.

$unify(c, c') = nil \qquad \text{if } c \neq c'$

VARIABLE CASES

Case (not-in, left): s is a variable x, s' is an expression e, and $x \not\preceq e$.

- We want to find a most-general unifier of x and e. However, by the Variable Unifier Proposition, $\{x \leftarrow e\}$ is a most-general unifier of x and e. This suggests taking

$$\theta \text{ to be } \{x \leftarrow e\}.$$

- We also want $\theta = \{x \leftarrow e\}$ to be idempotent, *i.e.*, that

$$\{x \leftarrow e\} \diamond \{x \leftarrow e\} = \{x \leftarrow e\}.$$

We show the equivalent condition (by the Idempotence Proposition) that

$$dom(\{x \leftarrow e\}) \cap range(\{x \leftarrow e\}) = \{\}.$$

We have (from the definition of the domain and range),

$$dom(\{x \leftarrow e\}) = \{x\} \text{ and}$$

$$range(\{x \leftarrow e\}) = vars(e).$$

Because $x \not\preceq e$, *i.e.* (by the Variables Proposition) $x \not\in vars(e)$, we have

$$dom(\{x \leftarrow e\}) \cap range(\{x \leftarrow e\}) = \{x\} \cap vars(e) = \{\},$$

as we wanted to show.

We have succeeded in showing that $\{x \leftarrow e\}$ is a most-general, idempotent unifier in this case.

$unify(x, e) = \{x \leftarrow e\} \qquad \text{if } x \not\preceq e$

Case (not-in, right): s is an expression e, s' is a variable x, *not* $x \preceq e$, and e is not a variable.

As in the previous case, we find that $\{x \leftarrow e\}$ is a most-general, idempotent unifier of e and x. The condition that the first argument e is not a variable is not required in the proof; it is included because the possibility that s and s' are both variables is covered by the previous case.

$unify(e, x) = \{x \leftarrow e\}$ if $x \not\preceq e$ and e is not a variable

Case (in): s is a variable x, s' is an expression e, and $x \prec e$.

In this case, we can show that x and e are not unifiable. For, let λ be an arbitrary substitution. Because $x \prec e$, we have (by the Monotonicity Proposition)

$$x \blacktriangleleft \lambda \prec e \blacktriangleleft \lambda$$

and hence (by the *irreflexivity* of $\prec$)

$$x \blacktriangleleft \lambda \neq e \blacktriangleleft \lambda.$$

Therefore, x and e are not unifiable and we can satisfy the second disjunct of the desired conclusion $P(x, e)$ by taking θ to be *nil*.

The symmetric case, in which s is an expression e, s' is a variable x, and $x \prec e$, is treated similarly.

$unify(x, e) = unify(e, x) = nil$ if $x \prec e$

Case (list): s is a variable x and s' is a list of ℓ-expressions m.

In this case, we can show that x and m are not unifiable. For, let λ be an arbitrary substitution. Then (by the Syntactic Categories Proposition), $x \blacktriangleleft \lambda$ is an expression but $m \blacktriangleleft \lambda$ is a list of ℓ-expressions; hence (by the disjointness of the syntactic categories)

$$x \blacktriangleleft \lambda \neq m \blacktriangleleft \lambda;$$

i.e., x and m are not unifiable. Therefore, we can satisfy the second disjunct of the desired conclusion $P(x, m)$ by taking θ to be *nil*.

The symmetric case, in which s is a list of ℓ-expressions m and s' is a variable x, is treated similarly.

$unify(x, m) = unify(m, x) = nil$

FUNCTIONAL CASES

Case(same): s and s' are functional expressions $f \bullet \ell$ and $f' \bullet \ell'$, respectively, where $f = f'$.

Recall that (in this case) we are attempting to prove the conclusion

$$P(f \bullet \ell, f \bullet \ell');$$

i.e., we want to find an output θ such that

$$\begin{aligned}
&(f \bullet \ell) \blacktriangleleft \theta = (f \bullet \ell') \blacktriangleleft \theta \ \mathit{and} \\
&(\forall \lambda)[\mathit{if}\ (f \bullet \ell) \blacktriangleleft \lambda = (f \bullet \ell') \blacktriangleleft \lambda \ \mathit{then}\ \theta \succeq_{gen} \lambda] \ \mathit{and} \\
&\theta = \theta \diamond \theta \ \mathit{and} \\
&\theta \neq \mathit{nil} \\
&\qquad \mathit{or} \\
&(\forall \lambda)[(f \bullet \ell) \blacktriangleleft \lambda \neq (f \bullet \ell') \blacktriangleleft \lambda] \ \mathit{and} \\
&\theta = \mathit{nil}.
\end{aligned}$$

This reduces (by the Application Unifier Proposition), to finding an output θ such that

$$(*) \qquad \begin{aligned}
&\ell \blacktriangleleft \theta = \ell' \blacktriangleleft \theta \ \mathit{and} \\
&(\forall \lambda)[\mathit{if}\ \ell \blacktriangleleft \lambda = \ell' \blacktriangleleft \lambda \ \mathit{then}\ \theta \succeq_{gen} \lambda] \ \mathit{and} \\
&\theta = \theta \diamond \theta \ \mathit{and} \\
&\theta \neq \mathit{nil} \\
&\qquad \mathit{or} \\
&(\forall \lambda)[\ell \blacktriangleleft \lambda \neq \ell' \blacktriangleleft \lambda] \ \mathit{and} \\
&\theta = \mathit{nil}.
\end{aligned}$$

Recall that we have assumed as our induction hypothesis that (in this case)

$$P(r, r') : \ \mathit{if}\ \langle r, r' \rangle \prec_{un} \langle f \bullet \ell, f \bullet \ell' \rangle$$

$$\text{then} \begin{bmatrix} r \blacktriangleleft unify(r, r') = r' \blacktriangleleft unify(r, r') \ \ and \\ (\forall \lambda)[if \ \ r \blacktriangleleft \lambda = r' \blacktriangleleft \lambda \ \ then \ \ unify(r, r') \succeq_{gen} \lambda] \ \ and \\ unify(r, r') = unify(r, r') \diamond unify(r, r') \ \ and \\ unify(r, r') \neq nil \\ \quad or \\ (\forall \lambda)[r \blacktriangleleft \lambda \neq r' \blacktriangleleft \lambda] \ \ and \\ unify(r, r') = nil \end{bmatrix}$$

for all ℓ-expressions r and r'.

The required condition (*) and consequent of the induction hypothesis are identical if we take r to be ℓ, r' to be ℓ', and θ to be $unify(\ell, \ell')$. Therefore, we can satisfy the conclusion if we can establish the appropriate instance of the induction hypothesis's antecedent, *i.e.*,

$$\langle \ell, \ell' \rangle \prec_{un} \langle f \bullet \ell, \ f \bullet \ell \rangle. \qquad (application\ ordering)$$

The well-founded ordering $\prec_{un}$ will be chosen subsequently. Assuming it will satisfy this condition, we have found that the desired conclusion $P(f \bullet \ell, \ f \bullet \ell')$ in this case is satisfied if we take θ to be $unify(\ell, \ell')$.

$unify(f \bullet \ell, \ f' \bullet \ell') = unify(\ell, \ell') \qquad \text{if } f = f'$

Case (distinct): s and s' are functional expressions $f \bullet \ell$ and $f' \bullet \ell'$, respectively, where $f \neq f'$.

In this case, we can show that $f \bullet \ell$ and $f' \bullet \ell'$ are not unifiable. For any arbitrary substitution λ, we have (by the *application* property of the apply function)

$$(f \bullet \ell) \blacktriangleleft \lambda \ = \ f \bullet (\ell \blacktriangleleft \lambda)$$

and

$$(f' \bullet \ell') \blacktriangleleft \lambda \ = \ f' \bullet (\ell' \blacktriangleleft \lambda).$$

Because f and f' are distinct, we have (by the *uniqueness* properties of ℓ-expressions),

$$f \bullet (\ell \blacktriangleleft \lambda) \ \neq \ f' \bullet (\ell' \blacktriangleleft \lambda),$$

and hence

$$(f \bullet \ell) \blacktriangleleft \lambda \neq (f' \bullet \ell') \blacktriangleleft \lambda.$$

Thus, $f \bullet \ell$ and $f' \bullet \ell'$ are not unifiable.

Therefore, we can satisfy the second disjunct of the desired conclusion, in this case, if we take θ to be *nil*.

$unify(f \bullet \ell,\ f' \bullet \ell') = nil \qquad \text{if } f \neq f'$

LIST CASES

In all of the list cases, s and s' are nonempty lists $t \circ m$ and $t' \circ m'$, respectively, where t and t' are ℓ-expressions and m and m' are lists of ℓ-expressions.

Recall that (in the list cases) we are attempting to prove the conclusion

$$P(t \circ m,\ t' \circ m');$$

i.e., we want to find an output θ such that

$$\begin{aligned}
&(t \circ m) \blacktriangleleft \theta = (t' \circ m') \blacktriangleleft \theta \ \ and \\
&(\forall \lambda)[if\ (t \circ m) \blacktriangleleft \lambda = (t' \circ m') \blacktriangleleft \lambda\ \ then\ \ \theta \succeq_{gen} \lambda]\ \ and \\
&\theta = \theta \diamond \theta\ \ and \\
&\theta \neq nil \\
&\qquad or \\
&(\forall \lambda)[(t \circ m) \blacktriangleleft \lambda \neq (t' \circ m') \blacktriangleleft \lambda]\ \ and \\
&\theta = nil.
\end{aligned}$$

By the Insertion Unifier Proposition, this decomposes into finding θ such that

$$(*) \qquad \begin{aligned}
&(1)\ \ t \blacktriangleleft \theta = t' \blacktriangleleft \theta\ \ and && (head\ unifier) \\
&(2)\ \ m \blacktriangleleft \theta = m' \blacktriangleleft \theta\ \ and && (tail\ unifier) \\
&(3)\ \ (\forall \lambda)\begin{bmatrix} if\ (t \circ m) \blacktriangleleft \lambda = (t' \circ m') \blacktriangleleft \lambda \\ then\ \ \theta \succeq_{gen} \lambda] \end{bmatrix}\ \ and && (most\text{-}generality) \\
&(4)\ \ \theta = \theta \diamond \theta\ \ and && (idempotence)
\end{aligned}$$

(5) $\theta \neq nil$ *(nonnil)*

or

(6) $(\forall\lambda)[t \blacktriangleleft \lambda \neq t' \blacktriangleleft \lambda \text{ or } m \blacktriangleleft \lambda \neq m' \blacktriangleleft \lambda]$ *and* *(ununifiability)*

(7) $\theta = nil.$ *(nil)*

The separate conditions of (*) are numbered for future reference. We attempt to establish conditions (1) to (5) or, alternatively, conditions (6) and (7).

Recall that we have assumed as our induction hypothesis (in this case)

$$P(r,r'):\ \ if\ \langle r,r'\rangle \prec_{un} \langle t \circ m,\ t' \circ m'\rangle$$
$$then \begin{bmatrix} r \blacktriangleleft unify(r,r') = r' \blacktriangleleft unify(r,r')\ \ and \\ (\forall\lambda)[if\ \ r \blacktriangleleft \lambda = r' \blacktriangleleft \lambda\ \ then\ \ unify(r,r') \succeq_{gen} \lambda]\ \ and \\ unify(r,r') = unify(r,r') \diamond unify(r,r')\ \ and \\ unify(r,r') \neq nil \\ \qquad or \\ (\forall\lambda)[r \blacktriangleleft \lambda \neq r' \blacktriangleleft \lambda]\ \ and \\ unify(r,r') = nil \end{bmatrix}$$

for all ℓ-expressions r and r'.

Let us compare this induction hypothesis with our required conclusion (*). A natural approach would be to observe that one of the first two conditions, say the *tail unifier* condition (2),

$$m \blacktriangleleft \theta = m' \blacktriangleleft \theta,$$

is identical to the condition that

$$r \blacktriangleleft unify(r,r') = r' \blacktriangleleft unify(r,r'),$$

asserted in our induction hypothesis, if we take r to be m, r' to be m', and θ to be $unify(m,m')$. However, we still would have to show the *head unifier* condition (1),

$$t \blacktriangleleft \theta = t' \blacktriangleleft \theta,$$

i.e.,

$$t \blacktriangleleft unify(m,m') = t' \blacktriangleleft unify(m,m').$$

But this condition is not necessarily true: a unifier of m and m' need not be a unifier of t and t'. Therefore, we would fail to prove the condition.

An attempt to do the same for the *head unifier* condition (1),

$$t \blacktriangleleft \theta = t' \blacktriangleleft \theta,$$

would fail for the same reason; these two required conditions are symmetric.

A less straightforward approach is to observe that one of the first two conditions, say the *tail unifier* condition (2),

$$m \blacktriangleleft \theta = m' \blacktriangleleft \theta,$$

is also equivalent to the same condition of the induction hypothesis

$$r \blacktriangleleft \mathit{unify}(r,\ r') = r' \blacktriangleleft \mathit{unify}(r,\ r')$$

under a more complex substitution than we considered earlier.

proof of the tail unifier condition (2):

Recall the definition of the composition of substitutions

$$m^* \blacktriangleleft (\theta_{hd} \diamond \theta_{tl}) = (m^* \blacktriangleleft \theta_{hd}) \blacktriangleleft \theta_{tl}$$

for all ℓ-expressions m^* and substitutions θ_{hd} and θ_{tl}. (We have renamed the variables to reflect how we intend to use them.)

Applying this equality to the left-hand side of our required condition

$$m \blacktriangleleft \theta = m' \blacktriangleleft \theta$$

(taking m^* to be m and θ to be $\theta_{hd} \diamond \theta_{tl}$) we obtain the condition

$$(m \blacktriangleleft \theta_{hd}) \blacktriangleleft \theta_{tl} = m' \blacktriangleleft (\theta_{hd} \diamond \theta_{tl}).$$

Applying the same equality to the right-hand side (taking m^* to be m') yields the condition

$$(m \blacktriangleleft \theta_{hd}) \blacktriangleleft \theta_{tl} = (m' \blacktriangleleft \theta_{hd}) \blacktriangleleft \theta_{tl}.$$

This condition is identical to the condition in our induction hypothesis

$$r \blacktriangleleft \mathit{unify}(r,\ r') = r' \blacktriangleleft \mathit{unify}(r,\ r')$$

if we take r to be $m \blacktriangleleft \theta_{hd}$, r' to be $m' \blacktriangleleft \theta_{hd}$, and θ_{tl} to be $\mathit{unify}(m \blacktriangleleft \theta_{hd},\ m' \blacktriangleleft \theta_{hd})$.

Let us retain the abbreviation θ_{tl} for the term $unify(m \blacktriangleleft \theta_{hd},\ m' \blacktriangleleft \theta_{hd})$. Thus, the above match has suggested taking

$$\theta \quad \text{to be } \theta_{hd} \diamond \theta_{tl},$$

where

$$\theta_{tl} \quad \text{is } unify(m \blacktriangleleft \theta_{hd},\ m' \blacktriangleleft \theta_{hd})$$

and θ_{hd} is any substitution.

Let us rewrite the induction hypothesis for this case, making the substitutions suggested by the above match:

$$\begin{array}{ll} \textit{if} & \langle m \blacktriangleleft \theta_{hd},\ m' \blacktriangleleft \theta_{hd}\rangle \prec_{un} \langle t \circ m,\ t' \circ m'\rangle \\ \textit{then} & \left[\begin{array}{l} (m \blacktriangleleft \theta_{hd}) \blacktriangleleft \theta_{tl} = (m' \blacktriangleleft \theta_{hd}) \blacktriangleleft \theta_{tl} \ \textit{and} \\ (\forall \lambda)[\textit{if}\ (m \blacktriangleleft \theta_{hd}) \blacktriangleleft \lambda = (m' \blacktriangleleft \theta_{hd}) \blacktriangleleft \lambda \ \textit{then}\ \theta_{tl} \succeq_{gen} \lambda] \ \textit{and} \\ \theta_{tl} = \theta_{tl} \diamond \theta_{tl} \ \textit{and} \\ \theta_{tl} \neq nil \\ \qquad\qquad \textit{or} \\ (\forall \lambda)[(m \blacktriangleleft \theta_{hd}) \blacktriangleleft \lambda \neq (m' \blacktriangleleft \theta_{hd}) \blacktriangleleft \lambda] \ \textit{and} \\ \theta_{tl} = nil \end{array}\right] \end{array}$$

We will refer to this as the *tail induction hypothesis*.

To apply this tail induction hypothesis, we must establish

- the antecedent

  $$(8) \quad \langle m \blacktriangleleft \theta_{hd},\ m' \blacktriangleleft \theta_{hd}\rangle \prec_{un} \langle t \circ m,\ t' \circ m'\rangle, \qquad (\textit{tail ordering})$$

  to ensure that the consequent of the tail induction hypothesis will be true;

- the condition

  $$(9) \quad \theta_{tl} \neq nil, \qquad (\textit{tail nonnil})$$

  to ensure that the second disjunct of the consequent of the tail induction hypothesis will be false, so that the first disjunct must be true.

Regardless of the choice of θ_{hd}, this will establish the *tail unifier* condition (2) of the required conclusion (*). We must also establish the *head unifier* condition

(1), the *most-generality* condition (3), and the idempotence condition (4), where θ is taken to be $\theta_{hd} \diamond \theta_{tl}$. We know $\theta_{hd} \diamond \theta_{tl}$ will satisfy the *nonnil* condition (5), since any substitution is distinct from *nil*. In cases where we fail to establish one or more of these conditions, we can alternatively attempt to establish the *ununifiability* condition (6) and the *nil* condition (7) of the required conclusion (*), taking θ to be *nil*.

We consider these conditions one by one, but not in the given order.

proof of the head unifier condition (1):

We must find a substitution θ_{hd} such that $\theta_{hd} \diamond \theta_{tl}$ is a unifier of t and t'; *i.e.*, that

$$t \blacktriangleleft (\theta_{hd} \diamond \theta_{tl}) = t' \blacktriangleleft (\theta_{hd} \diamond \theta_{tl}),$$

or, equivalently (by the definition of composition), that

$$(t \blacktriangleleft \theta_{hd}) \blacktriangleleft \theta_{tl} = (t' \blacktriangleleft \theta_{hd}) \blacktriangleleft \theta_{tl}.$$

It thus suffices to find θ_{hd} such that

$$t \blacktriangleleft \theta_{hd} = t' \blacktriangleleft \theta_{hd}.$$

We observe that the above condition is identical to the condition that

$$r \blacktriangleleft unify(r,\ r') = r' \blacktriangleleft unify(r,\ r'),$$

asserted in our induction hypothesis, if we take r to be t, r' to be t' and θ_{hd} to be $unify(t,\ t')$. We retain the abbreviation

$$\theta_{hd} \text{ is } unify(t,\ t').$$

Let us rewrite the induction hypothesis for this case, making the substitutions suggested by the above match:

$$\begin{array}{l} \mathit{if}\ \langle t,\ t'\rangle \prec_{un} \langle t \circ m,\ t' \circ m'\rangle \\ \mathit{then} \left[\begin{array}{l} t \blacktriangleleft \theta_{hd} = t' \blacktriangleleft \theta_{hd}\ \mathit{and} \\ (\forall \lambda)[\mathit{if}\ t \blacktriangleleft \lambda = t' \blacktriangleleft \lambda\ \mathit{then}\ \theta_{hd} \succeq_{gen} \lambda]\ \mathit{and} \\ \theta_{hd} = \theta_{hd} \diamond \theta_{hd}\ \mathit{and} \\ \theta_{hd} \neq nil \\ \qquad\quad \mathit{or} \\ (\forall \lambda)[t \blacktriangleleft \lambda \neq t' \blacktriangleleft \lambda]\ \mathit{and} \\ \theta_{hd} = nil \end{array}\right] \end{array}$$

We will refer to this as the *head induction hypothesis.*

To apply this head induction hypothesis, we must establish

- the antecedent

  $$\langle t,\ t' \rangle \prec_{un} \langle t \circ m,\ t' \circ m' \rangle, \qquad \textit{(head ordering)}$$

  to ensure that the consequent of the head induction hypothesis will be true.

- the condition

  $$\theta_{hd} \neq nil \qquad \textit{(head nonnil)}$$

  to ensure that the second disjunct of the consequent of the head induction hypothesis will be false, so that the first disjunct must be true.

As usual, we defer discussion of the *head ordering* condition, that

$$\langle t,\ t' \rangle \prec_{un} \langle t \circ m,\ t' \circ m' \rangle,$$

until we have accumulated all such conditions, so that we can define an ordering $\prec_{un}$ to satisfy them all at once.

The *head nonnil* condition $\theta_{hd} \neq nil$ is not necessarily true: t and t' need not be unifiable. Let us now consider the alternate possibility.

Case(no): $\theta_{hd} = nil$.

Then, by our head induction hypothesis, t and t' are not unifiable; *i.e.*,

$$t \blacktriangleleft \lambda \neq t' \blacktriangleleft \lambda$$

for all substitutions λ. Therefore, we can satisfy the *ununifiability* condition (6) and the *nil* condition (7) of the required conclusion (*), in this case, by taking θ to be *nil*.

$unify(t \circ m,\ t' \circ m') = nil \qquad \text{if } \theta_{hd} = nil$

Case: $\theta_{hd} \neq nil$.

That is, $unify(t, t') \neq nil$. In this case, our head induction hypothesis establishes that θ_{hd} is indeed a most-general, idempotent unifier of t and t', and therefore the *head unifier* condition (1) is satisfied. It remains to show the *tail ordering* condition (8) and the *tail nonnil* condition (9); these conditions ensure that we can apply the tail induction hypothesis to establish the *tail unifier* condition (2) of the required conclusion (*). It also remains to show the original *most-generality* condition (3) and *idempotence* condition (4). As usual, we assume that we can establish the *tail ordering* condition (8), and defer its proof.

proof of the tail nonnil condition (9):

The condition that $\theta_{tl} \neq nil$, *i.e.*, $unify(m \blacktriangleleft \theta_{hd}, m' \blacktriangleleft \theta_{hd}) \neq nil$, is not necessarily true: $m \blacktriangleleft \theta_{hd}$ and $m' \blacktriangleleft \theta_{hd}$ need not be unifiable. Let us consider the alternate possibility.

Subcase (yes-no): $\theta_{tl} = nil$.

That is, $unify(m \blacktriangleleft \theta_{hd}, m' \blacktriangleleft \theta_{hd}) = nil$. Then, by our tail induction hypothesis (where r and r' were taken to be $m \blacktriangleleft \theta_{hd}$ and $m' \blacktriangleleft \theta_{hd}$, respectively), $m \blacktriangleleft \theta_{hd}$ and $m' \blacktriangleleft \theta_{hd}$ are not unifiable, *i.e.*,

$$(m \blacktriangleleft \theta_{hd}) \blacktriangleleft \lambda \neq (m' \blacktriangleleft \theta_{hd}) \blacktriangleleft \lambda,$$

for all substitutions λ. We establish, in this case, the *ununifiability* condition (6), that $(\forall \lambda)[t \blacktriangleleft \lambda \neq t' \blacktriangleleft \lambda \;\; or \;\; m \blacktriangleleft \lambda \neq m' \blacktriangleleft \lambda]$. For suppose, to the contrary, that, for some substitution λ,

$$t \blacktriangleleft \lambda = t' \blacktriangleleft \lambda,$$

i.e., λ is a unifier of t and t', and

$$m \blacktriangleleft \lambda \;=\; m' \blacktriangleleft \lambda,$$

i.e., λ is a unifier of m and m'.

Because λ is a unifier of t and t', and because θ_{hd} is a most-general, idempotent unifier of t and t', we have (by the Most-General, Idempotent Unifier Proposition)

$$\lambda \;=\; \theta_{hd} \diamond \lambda.$$

Hence, because λ is a unifier of m and m', *i.e.*, $m \blacktriangleleft \lambda = m' \blacktriangleleft \lambda$, we have

$$m \blacktriangleleft (\theta_{hd} \diamond \lambda) \;=\; m' \blacktriangleleft (\theta_{hd} \diamond \lambda),$$

or equivalently (by the definition of composition)

$$(m \blacktriangleleft \theta_{hd}) \blacktriangleleft \lambda \;=\; (m' \blacktriangleleft \theta_{hd}) \blacktriangleleft \lambda.$$

i.e., λ is a unifier of $m \blacktriangleleft \theta_{hd}$ and $m' \blacktriangleleft \theta_{hd}$. But this contradicts our earlier finding, that $m \blacktriangleleft \theta_{hd}$ and $m' \blacktriangleleft \theta_{hd}$ are not unifiable. Hence, the *ununifiability* condition (6) is established and, in this case, we can establish the required conclusion (*) by taking θ to be *nil*.

$unify(t \circ m,\ t' \circ m') = nil$ if $\theta_{hd} \neq nil$ and $\theta_{tl} = nil$

Subcase (yes-yes): $\theta_{tl} \neq nil$.

That is, the *tail nonnil* condition (9), that $unify(m \blacktriangleleft \theta_{hd},\ m' \blacktriangleleft \theta_{hd}) \neq nil$, is true. Let us again retrace our steps, to see what we have established.

Since we have assumed the *tail ordering* condition (8), we know that the consequent of the tail induction hypothesis is true (where r and r' were taken to be $m \blacktriangleleft \theta_{hd}$ and $m' \blacktriangleleft \theta_{hd}$, respectively.) Because (in this case), the *tail nonnil* condition (9) is true, we know that the second disjunct of the consequent is false, and therefore that the first disjunct must be true. This implies that θ_{tl} is a most-general, idempotent unifier of $m \blacktriangleleft \theta_{hd}$ and $m' \blacktriangleleft \theta_{hd}$, and hence (by applying the definition of composition twice) that $\theta_{hd} \diamond \theta_{tl}$ is a unifier of m and m'. This establishes the *tail unifier* condition (2), which was our reason for applying the tail induction hypothesis in the first place.

proof of the most-generality and idempotence conditions (3 *and* 4):

It still remains to show the *most-generality* condition (3),

$$(\forall\lambda)[\mathit{if}\ \ (t \circ m) \blacktriangleleft \lambda = (t' \circ m') \blacktriangleleft \lambda\ \ \mathit{then}\ \ (\theta_{hd} \diamond \theta_{tl}) \succeq_{gen} \lambda]$$

and the *idempotence* condition (4), that

$$\theta_{hd} \diamond \theta_{tl} = (\theta_{hd} \diamond \theta_{tl}) \diamond (\theta_{hd} \diamond \theta_{tl}).$$

For this purpose, it suffices (by the Most-General, Idempotent Unifier Proposition) to show the single condition that

$$(\forall\lambda)\begin{bmatrix} \mathit{if}\ \ (t \circ m) \blacktriangleleft \lambda = (t' \circ m') \blacktriangleleft \lambda \\ \mathit{then}\ \ \lambda = (\theta_{hd} \diamond \theta_{tl}) \diamond \lambda \end{bmatrix}$$

i.e. (by the Insertion Unifier Proposition),

$$(\forall\lambda)\begin{bmatrix} \textit{if } t \blacktriangleleft \lambda = t' \blacktriangleleft \lambda \textit{ and } m \blacktriangleleft \lambda = m' \blacktriangleleft \lambda \\ \textit{then } \lambda = (\theta_{hd} \diamond \theta_{tl}) \diamond \lambda \end{bmatrix}$$

Suppose that λ is an arbitrary substitution such that

$$t \blacktriangleleft \lambda = t' \blacktriangleleft \lambda$$

and

$$m \blacktriangleleft \lambda = m' \blacktriangleleft \lambda.$$

We would like to show that then

$$\lambda = (\theta_{hd} \diamond \theta_{tl}) \diamond \lambda.$$

Because λ is a unifier of t and t', and because θ_{hd} is a most-general, idempotent unifier of t and t', we have (by the Most-General, Idempotent Unifier Proposition, again) that

$$\lambda = \theta_{hd} \diamond \lambda.$$

Therefore, because $m \blacktriangleleft \lambda = m' \blacktriangleleft \lambda$ we have

$$m \blacktriangleleft (\theta_{hd} \diamond \lambda) = m' \blacktriangleleft (\theta_{hd} \diamond \lambda),$$

i.e. (by the definition of composition),

$$(m \blacktriangleleft \theta_{hd}) \blacktriangleleft \lambda = (m' \blacktriangleleft \theta_{hd}) \blacktriangleleft \lambda.$$

In other words, λ is a unifier of $m \blacktriangleleft \theta_{hd}$ and $m' \blacktriangleleft \theta_{hd}$.

But then, because θ_{tl} is a most-general, idempotent unifier of $m \blacktriangleleft \theta_{hd}$ and $m' \blacktriangleleft \theta_{hd}$, we have (by the Most-General, Idempotent Unifier Proposition, yet again) that

$$\lambda = \theta_{tl} \diamond \lambda.$$

Therefore,

$$\lambda = \theta_{hd} \diamond \lambda = \theta_{hd} \diamond (\theta_{tl} \diamond \lambda) = (\theta_{hd} \diamond \theta_{tl}) \diamond \lambda.$$

In short, we obtain the condition,

$$\lambda = (\theta_{hd} \diamond \theta_{tl}) \diamond \lambda,$$

that we wanted to show.

We conclude that $\theta_{hd} \diamond \theta_{tl}$ satisfies the *most-generality* condition (3) and the *idempotence* condition (4) of the required conclusion (*), and thus in this case we are justified in taking θ to be $\theta_{hd} \diamond \theta_{tl}$.

$unify(t \circ m,\ t' \circ m') = \theta_{hd} \diamond \theta_{tl}$ if $\theta_{hd} \not\doteq nil$ and $\theta_{tl} \not\doteq nil$

This concludes the final case.

THE ORDERING

We have deferred the choice of an ordering $\prec_{un}$ to satisfy the ordering conditions we have accumulated during the proof. The choice of this ordering is not so well-motivated formally as the other steps of this derivation. The ordering conditions to be satisfied by $\prec_{un}$ are as follows:

the *application ordering* condition

$$\langle \ell,\ \ell' \rangle \prec_{un} \langle f \bullet \ell,\ f \bullet \ell' \rangle,$$

the *head ordering* condition

$$\langle t,\ t' \rangle \prec_{un} \langle t \circ m,\ t' \circ m' \rangle,$$

and the *tail ordering* condition

$$\langle m \blacktriangleleft \theta_{hd},\ m' \blacktriangleleft \theta_{hd} \rangle \prec_{un} \langle t \circ m,\ t' \circ m' \rangle$$

for all function symbols f, lists of expressions ℓ and ℓ', ℓ-expressions t and t', and lists of ℓ-expressions m and m', where $\theta_{hd} \not\doteq nil$, *i.e.*, $unify(t,\ t') \not\doteq nil$.

It would be natural to attempt to use as the definition of $\prec_{un}$ the subexpression ordering on one of the two arguments. However, if we take $\prec_{un}$ to be, say, the ordering $\prec_1$ on the first argument, defined by

$$\langle r,\ r' \rangle \prec_1 \langle s,\ s' \rangle$$

if and only if

$$r \prec s,$$

it will satisfy the first two of these conditions, and will satisfy the third condition in the case that $m \blacktriangleleft \theta_{hd} = m$. However, this ordering may fail to satisfy the third condition if $m \blacktriangleleft \theta_{hd} \neq m$, because $m \blacktriangleleft \theta_{hd}$ may no longer be a subexpression of $t \circ m$, and may in fact be much larger. For example, if

$$t \text{ is } x$$

$$t' \text{ is } g(a, y, b)$$

and

$$m \text{ is } [f(x, x, x)],$$

$$m' \text{ is } [z]$$

then

$$t \circ m \text{ is } [x,\ f(x, x, x)]$$

$$t' \circ m' \text{ is } [g(a, y, b),\ z].$$

In this case

$$\theta_{hd} \text{ is } \{x \leftarrow g(a, y, b)\}$$

and

$$m \blacktriangleleft \theta_{hd} \text{ is } [f(g(a, y, b),\ g(a, y, b),\ g(a, y, b))]$$

$$m' \blacktriangleleft \theta_{hd} \text{ is } [z].$$

Thus, $m \blacktriangleleft \theta_{hd}$ is not a subexpression of $t \circ m$.

In the case that $m \blacktriangleleft \theta_{hd} \neq m$, however, it can be shown that the variables of $m \blacktriangleleft \theta_{hd}$ and $m' \blacktriangleleft \theta_{hd}$ are a proper subset of the variables of $t \circ m$ and $t' \circ m'$; *i.e.*,

$$vars(m \blacktriangleleft \theta_{hd}) \cup vars(m' \blacktriangleleft \theta_{hd}) \quad \subset \quad vars(t \circ m) \cup vars(t' \circ m').$$

In other words, θ_{hd} removes variables from m and m' without introducing any that are not in $t \circ m$ or $t' \circ m'$. Thus, the ordering $\prec_{vars}$, defined by

$$\langle r,\ r' \rangle \prec_{vars} \langle s,\ s' \rangle$$

if and only if

$$vars(r) \cup vars(r') \quad \subset \quad vars(s) \cup vars(s'),$$

will satisfy the *tail ordering* condition in this case. Thus, in the above example,

$$vars(m \blacktriangleleft \theta_{hd}) \cup vars(m' \blacktriangleleft \theta_{hd}) \quad = \quad \{y, z\}$$

and

$$vars(t \circ m) \cup vars(t' \circ m') \quad = \quad \{x, y, z\},$$

and hence

$$\langle m \blacktriangleleft \theta_{hd},\ m' \blacktriangleleft \theta_{hd} \rangle \quad \prec_{vars} \quad \langle t \circ m,\ t' \circ m' \rangle.$$

However, this ordering will fail to satisfy the first two conditions of $\prec_{un}$, and will also fail to satisfy the third condition in the case that $m \blacktriangleleft \theta_{hd} = m$ and $m' \blacktriangleleft \theta_{hd} = m'$. For example (by the *application* property of *vars*),

$$vars(\ell) = vars(f \bullet \ell)$$

$$vars(\ell') = vars(f \bullet \ell')$$

and hence

$$vars(\ell) \cup vars(\ell') \quad = \quad vars(f \bullet \ell) \cup vars(f \bullet \ell'),$$

i.e.,

$$not\ [\langle \ell, \ell' \rangle \prec_{vars} \langle f \bullet \ell,\ f \bullet \ell' \rangle].$$

In other words, the first condition, $\langle \ell,\ \ell' \rangle \prec_{un} \langle f \bullet \ell,\ f \bullet \ell' \rangle$, is never satisfied under the $\prec_{vars}$ ordering.

The successful ordering $\prec_{un}$ is a lexicographic combination of these two orderings $\prec_{vars}$ and $\prec_1$, defined by the property

$$\langle r,\ r' \rangle \prec_{un} \langle s,\ s' \rangle$$

if and only if

$$vars(r) \cup vars(r') \quad \subset \quad vars(s) \cup vars(s')$$

or

$$vars(r) \cup vars(r') \quad = \quad vars(s) \cup vars(s') \quad and \quad r \prec_1 s.$$

- To see that, under this definition, the *application ordering* condition,

$$\langle \ell,\ \ell' \rangle \prec_{un} \langle f \bullet \ell,\ f \bullet \ell' \rangle,$$

is satisfied, note that (as we mentioned above)

$$vars(\ell) \cup vars(\ell') \;=\; vars(f \bullet \ell) \cup vars(f \bullet \ell')$$

and (by a *component* property of $\prec$)

$$\ell \prec f \bullet \ell.$$

Hence, by the definition of the ordering $\prec_{un}$, the *application ordering* condition is indeed satisfied.

- To see that, under this definition, the *head ordering* condition,

$$\langle t,\ t' \rangle \prec_{un} \langle t \circ m,\ t' \circ m' \rangle,$$

is satisfied, note that (by the *insertion* property of *vars*)

$$vars(t) \;\subseteq\; vars(t) \cup vars(m) \;=\; vars(t \circ m)$$

$$vars(t') \;\subseteq\; vars(t') \cup vars(m') \;=\; vars(t' \circ m'),$$

and hence

$$vars(t) \cup vars(t') \;\subseteq\; vars(t \circ m) \cup vars(t' \circ m').$$

In case the inclusion is proper, *i.e.*,

$$vars(t) \cup vars(t') \;\subset\; vars(t \circ m) \cup vars(t' \circ m'),$$

we have

$$\langle t,\ t' \rangle \prec_{un} \langle t \circ m,\ t' \circ m' \rangle$$

immediately. On the other hand, if the two sets are equal, *i.e.*,

$$vars(t) \cup vars(t') \;=\; vars(t \circ m) \cup vars(t' \circ m'),$$

we note that (by a *component* property of $\prec$)

$$t \prec t \circ m.$$

Hence, by the definition of the ordering $\prec_{un}$, the *head ordering* condition is also satisfied in this case.

- Finally, we must show that the *tail ordering* condition,

$$\langle m \blacktriangleleft \theta_{hd},\ m' \blacktriangleleft \theta_{hd}\rangle \prec_{un} \langle t \circ m,\ t' \circ m'\rangle,$$

is satisfied, where $\theta_{hd} = unify(t,\ t') \neq nil$. First, we have

$$\begin{aligned}
&vars(m \blacktriangleleft \theta_{hd}) \cup vars(m' \blacktriangleleft \theta_{hd})\\
&\quad\subseteq\ vars(m) \cup range(\theta_{hd}) \cup vars(m') \cup range(\theta_{hd})\\
&\qquad\qquad\text{by the Variable Introduction Proposition}\\
&\quad=\ vars(m) \cup vars(m') \cup range(\theta_{hd})\\
&\quad\subseteq\ vars(m) \cup vars(m') \cup vars(t) \cup vars(t')\\
&\qquad\qquad\text{by the Domain and Range Proposition,}\\
&\qquad\qquad\text{because } \theta_{hd} \text{ is a most-general, idempotent unifier for } t \text{ and } t'\\
&\quad=\ vars(t \circ m) \cup vars(t' \circ m')\\
&\qquad\qquad\text{by the } insertion \text{ property of } vars
\end{aligned}$$

In short,

$$vars(m \blacktriangleleft \theta_{hd}) \cup vars(m' \blacktriangleleft \theta_{hd}) \ \subseteq\ vars(t \circ m) \cup vars(t' \circ m').$$

By the definition of the ordering $\prec_{un}$, we must either show that this inclusion is proper, *i.e.*,

(*) $\quad vars(m \blacktriangleleft \theta_{hd}) \cup vars(m' \blacktriangleleft \theta_{hd}) \ \subset\ vars(t \circ m) \cup vars(t' \circ m'),$

or show that

(**) $\quad m \blacktriangleleft \theta_{hd} \prec t \circ m.$

For this purpose, we distinguish between two subcases.

Subcase: $m \blacktriangleleft \theta_{hd} = m$.

Then (by a *component* property of $\prec$)

$$m \blacktriangleleft \theta_{hd} \ =\ m \ \prec\ t \circ m.$$

Subcase: $m \blacktriangleleft \theta_{hd} \neq m$.

In this case, we will show (∗), that the inclusion is proper, *i.e.*,

$$vars(m \blacktriangleleft \theta_{hd}) \cup vars(m' \blacktriangleleft \theta_{hd}) \quad \subset \quad vars(t \circ m) \cup vars(t' \circ m').$$

We have already shown the $\subseteq$ inclusion; therefore, it suffices to show the existence of a variable z such that

(‡) $$z \in vars(t \circ m) \cup vars(t' \circ m')$$

but

(††) $$z \notin vars(m \blacktriangleleft \theta_{hd}) \cup vars(m' \blacktriangleleft \theta_{hd}).$$

First, because $m \blacktriangleleft \theta_{hd} \neq m$, we know (by the Invariance Corollary) that

$$vars(m) \cap dom(\theta_{hd}) \neq \{\},$$

i.e., there is a variable z such that

$$z \in vars(m)$$

and

$$z \in dom(\theta_{hd}).$$

We know (by the *insertion* property of *vars*) that

$$vars(m) \subseteq vars(t) \cup vars(m) = vars(t \circ m).$$

Then, because $z \in vars(m)$, we have the desired property (†),

$$z \in vars(t \circ m) \cup vars(t' \circ m').$$

Next, because θ_{hd} is idempotent, we have (by Idempotence Proposition)

$$dom(\theta_{hd}) \cap range(\theta_{hd}) = \{\},$$

and, thus, because

$$z \in dom(\theta_{hd}),$$

we have

$$z \not\in range(\theta_{hd}).$$

It follows (by the Variable Elimination Proposition) that

$$z \not\in vars(m \blacktriangleleft \theta_{hd})$$

$$z \not\in vars(m' \blacktriangleleft \theta_{hd}),$$

and hence

$$z \not\in vars(m \blacktriangleleft \theta_{hd}) \cup vars(m' \blacktriangleleft \theta_{hd}).$$

This is the desired property (††). We have thus established the proper inclusion

$$vars(m \blacktriangleleft \theta_{hd}) \cup vars(m' \blacktriangleleft \theta_{hd}) \quad \subset \quad vars(t \circ m) \cup vars(t' \circ m').$$

In both subcases, we can conclude that

$$\langle m \blacktriangleleft \theta_{hd},\ m' \blacktriangleleft \theta_{hd} \rangle \prec_{un} \langle t \circ m,\ t' \circ m' \rangle.$$

Thus, the *tail ordering* condition for $\prec_{un}$ is satisfied.

This concludes the entire derivation proof. ∎

ALTERNATE DERIVATIONS

We have followed only one proof of the specification of the desired theorem. Had we followed other proofs, different programs would have resulted. For example, in the *list* cases of the above derivation, we first matched the *tail unifier* condition $m \blacktriangleleft \theta = m' \blacktriangleleft \theta$ against the induction hypothesis (after applying the definition of composition); had we instead matched the symmetric *head unifier* condition $t \blacktriangleleft \theta = t' \blacktriangleleft \theta$, another proof would have been obtained, and the *list* cases of the resulting program would have been as follows:

let $\theta^*_{tl} = unify(m,\ m')$
if $\theta^*_{tl} = nil$
then $unify(t \circ m,\ t' \circ m') = nil$
else let $\theta^*_{hd} = unify(t \blacktriangleleft \theta^*_{tl},\ t' \blacktriangleleft \theta^*_{tl})$

if $\theta^*_{hd} = nil$
then $unify(t \circ m,\ t' \circ m') = nil$
else $unify(t \circ m,\ t' \circ m') = \theta^*_{tl} \diamond \theta^*_{hd}$.

This program will also satisfy the same specification as the original program but, because it examines the list from right to left rather than left to right, it may produce a different most-general, idempotent unifier.

In general, by exploring different branches of the proof tree, we may obtain families of different unification algorithms analogous to the families of different sort programs obtained from a single specification by Clark and Darlington [1980]. The particular derivation we obtained did not take the efficiency of the final program into account. Other branches of the derivation tree lead to more efficient unification algorithms.

AUTOMATION OF THE PROOF

Our primary objective in examining the above derivation in such detail is to consider the computational prerequisites for discovering the proof automatically. Let us review the proof from this point of view.

The first requirement of a theorem-proving system for program synthesis is that it be able to prove theorems that contain existential quantifiers and that require mathematical induction. Existential quantifiers are necessary to transform the specification into a theorem, and induction is necessary to introduce repetitive constructs into the target program. Although resolution theorem provers, say, can prove theorems with existential quantifiers, and several theorem provers (*e.g.*, Boyer and Moore [1975] Huet and Hullot [1980]) can do proofs by induction, it is rare to see these abilities combined.

The amount of knowledge about ℓ-expressions and substitutions necessary to produce the above proof is formidable. If such knowledge were built into the system, the system would then be specially tailored to this subject domain, and would loose generality. On the other hand, if the knowledge were provided to the system as a set of axioms, the system would also need to know how to use the knowledge efficiently.

Much of the derivation proof is fairly mechanical. At each stage, one must decide which property to apply next, from a finite collection of legal next steps. However, certain steps are not straightforward, and are motivated only by their ultimate success. For example, in the list cases, the straightforward use of the induction hypothesis failed, but the application of the definition of composition allowed us to use the induction hypothesis in a more general way, and resulted in the introduction of the composition $\theta_{hd} \diamond \theta_{tl}$ in the final program.

In finding the well-founded ordering $\prec_{un}$, the use of the sets $vars(s)$ and $vars(s')$ of variables in the ℓ-expressions s and s' was not suggested by the specification, which makes no reference to this notion.

The idempotence condition $\theta = \theta \diamond \theta$ was included in the initial specification. This condition played a vital part in the proof; however, the unification algorithm would be equally useful without this property. Had the idempotence condition not been required initially, it or an equivalent condition would have had to be invented and added to the specification in the middle of the proof.

Even with the idempotence condition provided, the proof seems somewhat more difficult than current theorem-provers can produce. Our hope is that studying hand derivations of this sort will enable us to improve the power of automatic systems.

INTERACTIVE SYNTHESIS

Although the above proof may be beyond the power of current automatic systems, a partially interactive system could be used to produce it with known techniques. This approach requires more human effort, but it still would convey many of the benefits of automatic synthesis:

- The person would provide those steps that require cleverness but the system would take care of the routine details.
- Whatever mistakes the person might make, the system would not permit him to produce a program that did not meet its specification.
- The program would be accompanied by a full proof of its correctness.
- The derivation could be retained, so that if the program needed modification, the appropriate portions of the program could be updated without endangering its correctness.
- The assumptions on which the correct operation of the program depends would be made explicit.

Of course, for an interactive system to be successful, it would have to communicate in terms the person would be able to understand.

ACKNOWLEDGMENTS

We would like to thank Yoni Malachai, Pierre Wolper, and Frank Yellin, for their careful reading of the manuscript and their many suggestions; and Evelyn Eldridge-Diaz for her TEXing of the paper.

REFERENCES

Bledsoe, W. W. [Aug. 1977], "Non-resolution theorem proving," Artificial Intelligence Journal, Vol. 9, No. 1, pp. 1-35.

Boyer, R. S. and J S. Moore [Jan. 1975], "Proving theorems about LISP functions," JACM, Vol. 22, No. 1, pp. 129-144.

Clark, K. and J. Darlington [Feb. 1980], "Algorithm classification through synthesis," Computer Journal, Vol. 23, No. 1, pp. 61-65.

Colmerauer, A., H. Kanoui and M. van Caneghem [1979], "Etude et realisation d'un systeme Prolog," Internal Report, Groupe d'Intelligence Artificielle, U.E.R. de Luminy, Université d' Aix-Marseille II.

Green, C. C. [May 1969], "Application of theorem proving to problem solving," Proceedings of the International Joint Conference on Artificial Intelligence, Washington DC, pp. 219-239.

Herbrand, J. [1930], "Researches in the theory of demonstration," in *From Frege to Gödel: a source book in mathematical logic*, 1879-1931 (J. Van Heijenoort, ed.), Harvard University Press, Cambridge, MA, 1967, pp. 525-581.

Hewitt, C. [Apr. 1971], "Description and theoretical analysis (using schemata) of PLANNER: A language for proving theorems and manipulating models in a robot," Ph.D. thesis, MIT, Cambridge, MA.

Huet, G. P. [June 1975], "A unification algorithm for typed λ-calculus," Theoretical Computer Science, Vol. 1, No. 1, pp. 27-57.

Huet, G. P. and J.-M. Hullot [Oct. 1980], "Proofs by induction in equational theories with constructors," Proceedings of Symposium on Foundations of Computer Science, Syracuse, NY, pp. 96-107.

Livesay, M., J. Siekmann, P. Szabó and E. Unvericht [Feb. 1979], "Unification problems for combinations of associativity, commutativity, distributivity and idempotence axioms," Proceedings of the Fourth Workshop on Automated Deduction, Austin, Texas, pp. 175-184.

Manna, Z. and R. Waldinger [Summer 1975], "Knowledge and reasoning in program synthesis," Artificial Intelligence Journal, Vol. 6, No. 2, pp. 175-208.

Manna, Z. and R. Waldinger [Jan. 1980], "A deductive approach to program synthesis," ACM Transactions on Programming Languages and Systems, Vol. 2, No. 1, pp. 92-121.

Manna, Z. and R. Waldinger [1982], *Deductive Basis for Computer Programming*, forthcoming.

Martelli, A. and U. Montanari [July 1976], "Unification in linear time and space: a structured presentation," Internal Report, IEI, Pisa.

Paterson, M. S. and M. N. Wegman [April 1978], "Linear unification," Journal of Computer and System Sciences, Vol. 16, No. 2, pp. 158-167.

Prawitz, D. [1960], "An improved proof procedure," Theoria, Vol. 26, pp. 102-139.

Robinson, J. A. [Jan. 1965], "A machine-oriented logic based on the resolution principle", JACM, Vol. 12, No. 1, pp. 23-41.

Stickel, M. E. [Sept. 1975], "A complete unification algorithm for associative-commutative functions," Proceedings of the Fourth International Joint Conference on Artificial Intelligence, Tbilisi, Georgia, USSR, pp. 71-76.

Von Henke, F. W. and D. C. Luckham [April 1974], "A methodology for verifying programs," Proceedings of the International Conference on Reliable Software, Los Angeles, CA, pp. 156-164.

Waldinger, R. J. and R. C. T. Lee [May 1969], "PROW: a step toward automatic program writing," Proceedings of the International Joint Conference on Artificial Intelligence, Washington, DC, pp. 241-252.

Waldinger, R. J. and K. N. Levitt [1974], "Reasoning about programs," *Artificial Intelligence*, Vol. 5, pp. 235-316.

Warren, D., L. M. Pereira and F. Pereira [Aug 1977], "PROLOG – the language and its implementation compared with LISP," Proceedings of Symposium on Artificial Intelligence and Programming Languages, SIGPLAN Notices (ACM), Vol. 12, No. 8, pp. 109-115.

The Synthesis of Implementations for Abstract Data Types,

A Program Transformation Tactic

John Darlington

Department of Computing
Imperial College London

Introduction

Program transformation offers a systematic way of developing correct and efficient programs from specifications. The formal system, outlined in Burstall and Darlington [1977], that has become known as the unfold/fold system, offers a very simple formal base which is nevertheless capable of achieving a wide class of transformations. However significant transformations expressed solely in terms of these rules became too complex to be communicated or achieved by a system. To overcome these problems we are developing a transformation meta-language that allows transformations to be explicated in a structured way without removing the formal base of these transformations. Constructs in this meta-language consist of higher level transformation processes or tactics such as combine loops, remove recursion or re-use storage. These are ultimately expressible in terms of the unfold/fold steps but provide a richer vocabulary for communicating transformation plans either to man or machine. One of the more useful of these tactics is one that enables implementations for abstract data types to be synthesised automatically. In this paper we consider this transformation process in some detail and use it to exemplify our ideas on structuring program transformations and the sort of systems that we envisage being developed.

Abstract data types provide a powerful means of structuring the program development process. The scenario proposed is

A. W. Biermann and G. Guiho (eds.), Computer Program Synthesis Methodologies, 309–334.

that one first considers the problem at the appropriate 'abstract' level, choosing and specifying appropriate data types and writing the program in terms of these data types. Only as a second stage is any consideration made of an efficient representation for the data types used. Here implementations are written for the basic operations of the abstract data type using structures available in the high level language chosen. If one wants additional security it is proposed by Hoare [1972] that one verifies that the implementations written do in fact correctly simulate the action of the abstract operations. This is done by specifying a function showing how the more concrete data type represents the more abstract and checking that, for every abstract operation implemented, the implementation performs its intended purpose under this 'representation relationship'. One method of specifying the behaviour of data types is the algebraic or equational, Burstall and Goguen [1977], Liskov and Zilles [1977], and Guttag [1977]. In this paper we show how program transformation techniques, Burstall and Darlington [1977], can be combined with the algebraic definition of data types to achieve the synthesis of implementations, achieving the same ends as verification with less effort from the user. Here if one simply specifies the abstract data type and gives the representation relationship, transformation techniques can be used to automatically produce implementations in the chosen representation that are guaranteed to be correct.

This technique was first demonstrated in Burstall and Darlington [1977], section 8, the novelty here is that the use of algebraic methods of specification allows the use of more sophisticated data types both at the 'abstract' and 'concrete' level and fits in smoothly with the programming language and transformation rules used.

In section 1 we introduce the reader to the techniques and terminology involved in the specification of abstract data types, a first order recursion equation language and the unfold/fold transformation system. Section 2 shows in detail how implementations for abstract data types can be synthesised and section 3 discusses the transformation meta-language concept as exemplified by the change data type tactic and describes the advantages this brings with regard to communicating transformations and developing systems to assist in achieving transformational developments.

1. Preliminaries

1.1 Formal Definition of Abstract Data Types

We follow Guttag [1977] and Guttag and Horning [1978] in defining abstract data types algebraically by specifying the types of the functions provided and giving the behaviour of these functions by a set of equations or axioms. Thus the type list of integers can be defined (ignoring error conditions),

```
Type list of integers
   types
        nil:->list
        cons:integerxlist->list
        head:list->integer
        tail:list->list
      member:integerxlist->boolean
        join:listxlist->list

   axioms
        head(cons(n,l))=n
        tail(cons(n,l))=l
        member(x,nil) =false
        member(x,cons(y,l))=if x=y then true
                                   else member(x,l)
        join(nil,l)=l
        join(cons(n,l1),l2)=cons(n,join(l1,l2))
```

For such a specification to be useful it must convey all the necessary information about the operations of the type. This concept is formalised in Guttag and Horning [1978] and given the name sufficient completeness. Here an abstract type is treated as an heterogeneous algebra with the carrier consisting of several phyla, one being the type of interest (lists in the above example) and the others the domains required in the definition (booleans and integers). Similarly the operations are partitioned into those with ranges in the type of interest, S, and those with ranges in the other phyla,O. S is further subdivided into the constructors C (nil and cons) and the extensions E (join). The former are the operations basic to the type and the latter are those that can be expressed in terms of these constructor functions.

A type definition is said to be sufficiently complete if for every formula of the form $f(x_1,\ldots,x_n)$ where $f \in O$ and the x_i are arbitrary terms in the appropriate domains there is a theorem derivable from the axioms of the type of the form $f(x_1,\ldots,x_n)=u$ where u is an element of the apropriate domain. That is it must be possible, using the axioms, to completely evaluate any term with range not the type of interest. In the above example this means that any term of the form member(x,t) for any t must be reducible to either true or false. Guttag and Horning [1978] gives a method for constructing a set of sufficiently complete axioms. We will not give it here. All the type specification given in this paper are sufficiently complete and the transformation methods used depend on the specifications being sufficiently complete.

1.2 Recursion Equation Languages

In this section we briefly introduce the recursion equation language, NPL, developed at Edinburgh with R. M. Burstall. Further details can be found in Burstall [1977]. NPL has been developed into a more advanced language, HOPE, Burstall et al [1980]. All the transformations shown here on NPL programs can equally well be performed on HOPE programs.

Programs in a recursion equation language consist of sequences of equations defining reduction rules for the functions being defined. In NPL separate equations can be written for separate cases of a parameter. For example factorial can be written

```
fact(n)<=1 if n=0
         n*fact(n-1)otherwise
```

or

```
fact(O)<=1
fact(n+1)<=(n+1)*fact(n)
```

In NPL rules must be exclusive (no overlapping cases) as the order of the rules is unimportant. Running an NPL program involves reducing an expression according to the rules until no further reductions are possible. Data stuctures in NPL are simply terms in the constructors of the structure. For example lists are just terms in the constructors nil and cons and a function length, to count the number of elements in a list, would be

```
length(nil)<=0
length(cons(n,l))<=1+length(l)
```

NPL is a strongly typed language and full type information has to be given for all he functions used, thus the fact and length functions above have type

```
fact:integer->integer
length:list of integers->integer
```

Thus programs in NPL look very much like equational definitions of abstract types. The difference is that NPL equations are meant to be used in a left to right manner as reduction rules therefore it is the programmers job to ensure that chains of such reductions always terminate (that is the rules are Noetherian) while in the abstract type definitions the equations need not be thus constrained. Another difference is that in NPL recursively defined functions can only apear as the outermost function on the left hand side of a rule while equational specifications of abstract data types involve no such constraint. For example in defining sets with constructors ∅ and consset the following rule is needed.

```
consset(nl consset(n2,S)) =
    consset(n2,consset(nl,S))
```

Nevertheless we may usefully incorporate abstract type definitions into NPL programs in many ways. Firstly many abstract type definitions obey the above restrictions and may thus be incorporated directly into NPL programs.
As we shall see below this has the important advantage that such type definitions can be run directly in NPL the axioms forming an 'abstract' implementation. Thus programs using the abstract type can be developed and tested before any concrete implementation is considered. Here we ae viewing programs in NPL as specifications or initial versions of programs that will be transformed into more efficient versions and, perhaps, finally translated into a more conventional procedural language such as PASCAL. Thus we use the terms abstract and concrete to mean types closer or further from the problem domain respectively as well as the more conventional usage to indicate types that are or are not provided by the programming language chosen.

Even if the abstract data type specification does not conform to the rules necessary for it to be run directly it can still be usefully incorporated into an NPL program. It is possible to envisage 'clever' interpreters that would allow programs employing such definitions to be exercised, if not run to conclusion, checking out their behaviour before proceeding to an implementation.

As we shall see later in either case we can use transformation methods to produce implementations for these abstract programs that act efficiently.

1.3 A Transformation Methodology

In this section we introduce the unfold/fold transformation system described in more detail in Burstall and Darlington [1977]. The aim of this methodology is to improve the efficiency of programs written in a functional language such as NPL without changing their meaning. We will introduce it via a simple example, say that we have lists as above and define functions length and join as before thus,

```
length(nil)=0                          (1)
length(cons(n,l))=1+length(l)          (2)
```

join(nil,11)=11 (3)
join(cons(n,11),I2)=cons(n,join(11,12)). (4)

Say that we wanted to write a program to join two lists together and calculate the length of the resulting list. Given the above equations we could, naively, write this as

lengthof2(11,12) = length(join(11,12)) (5)

This is a perfectly adequate program but it contains some avoidable inefficiency. Let us see if we can improve it. All our manipulations will take equations and produce further equations that do not change the meaning of the program.

Firstly we can <u>instantiate</u> equation (5) by letting 11 be nil getting

lengthof2(nil,12) = length(join(nil,12))

Equation (3) allows us to rewrite this as

lengthof2(nil,12) = length(12) (6)

Returning to (5) we now instantiate 11 to cons(n,11) getting

lengthof2(cons(n,11),12) = length(join(cons(n,11),12))

Equation (4) allows us to rewrite this as

lengthof2(cons(n,11),12) = length(cons(n,join(11,12)))

Equation (2) allows us to rewrite this as

lengthof2(cons(n,11),12) = 1 + length(join(11,12))

Finally equation (5) allows us to replace the subexpression length(join(11,12)) on the right hand side by lengthof2(11,12) getting

lengthof2(cons(n,11),12) = 1 + lengthof2(11,12) (7)

Thus we have produced two new equations, (6) and (7), which are true statements about lengthof2

lengthof2(nil,12) = length(12)

lengthof2(cons(n,l1),l2) = 1 + lengthof2(l1,l2)

what is more these two equations constitute a complete program for lengthof2 which is more efficient than the one given by equation (5), so we can replace (5) by (6) and (7).

More formally we have the following transformation rules which act on equations and produce further equations

(i) Definition Introduce a new recursion equation whose left hand expression is not an instance of the left hand expression of any previous equation.

(ii) Instantiation Introduce a substitution instance of an existing equation.

(iii) Unfolding If E <= E' and F <= F' are equations and there is some occurrence in F' of an instance of E, replace it by the corresponding instance of E' obtaining F", then add the equation F <= F".

(iv) Folding If E <= E' and F <= F' are equations and there is some occurrence in F' of an instance of E', replace it by the corresponding instance of E obtaining F", then add the equation F <= F".

(v) Abstraction We may introduce a where clause, by deriving from a previous equation E <= E' a new equation

$$E <= E'[u_1/F_1,\dots,u_n/F_n]^{+} \quad \underline{\text{where}}\ \langle u_1,\dots,u_n\rangle = \langle F_1,\dots,F_n\rangle$$

(vi) Laws We may transform an equation by using on its right hand expression any laws we have about the primitives (associativity ,comutativity etc).

+E[E1/E2] means E with all occurrences of subexpression E2 replaced by E1.

Further details and examples can be found in Burstall and Darlington [1977].

As mentioned earlier these rules preserve correctness. Unfortunately they only preserve partial correctness, that is the new equations may not terminate in some cases when the original equations did, so that we have to check that the new program does indeed terminate, a task that is usually straightforward given the simple forms of our programs. Theoretical studies of this system can be found in Kott [1980].

Despite their simple nature these rules provide a basis for a wide variety of transformations as we shall see in the following section.

As mentioned above one way we see transformation being applied aims at eventually producing an efficient program in a conventional high level language that can be compiled and run in the normal way. However as the nature of these languages preclude any significant transformations being performed after the translation from an applicative language it is important that as much work as possible is done within the applicative language. Thus the final translation to an imperative language is straightforward and no improvement remains to be done beyond that which can be achieved by existing optimising compilers. It is important to stress that the improvements that are needed to go from a specification to a runnable program go way beyond those achievable by present day optimising compilers, involving complete changes in the algorithm used. Paradoxically it seems to be the case that one can reach a more efficient final program if one starts with a very inefficient one that is understandable enough to allow the necessary improvements to be systematically introduced rather than starting with a partially hand optimised program that is too convoluted to allow any further significant progress to be achieved.

Progress in machine architecture may soon do away with the need for the final translation to an imperative language.

Functional languages are very well suited to parallel evaluation and the declining cost of hardware has made multi-processor implementations practical. Thus for example the work on data flow machines, Gurd et al [1978], and reduction machines, Darlington and Reeve [1981], promises to make direct evaluation of functional programs feasible. This would not remove the need for transformation but would drastically alter what would be considered an efficient program. The transformation methodology outlined above can be used to increase the benefit gained from parallel evaluation just as easily as it can interweave separate computations for efficient evaluation on a single processor.

2. Synthesis of Implementations for Abstract Data Types

2.1 Implementation of a Queue by a Circular Buffer

This example is taken from Guttag et al [1978]. The data type is a queue of integers. This is built up by the constructors emptyq and addq. The selectors frontq and removeq select and remove the integer that has been in the queue the longest.

```
type queue

    types

        emptyq:->queue
        addq: queue x integer->queue
        removeq: queue->queue
        frontq: queue->integer
        isemptyq: queue->boolean

    axioms

        isemptyq(emptyq)=true
        isemptyq(addq(q,i))=false
        frontq(addq(q,i)) = if isemptyq(q)
                               then i
                               else front(q)


        removeq(add(q,i)) = if isemptyq(q)
                               then emptyq
                               else addq(removeq(q),i)
```

The implementation we will look at for this data type is the circular list or buffer. This is a conventional list in that insertion and selection are made at the front of the list by the operations insert, value and removecirc but it has an additional operation, right, that performs a right circular shift moving the front element to the rear of the list and shifting all the other elements up one. Guttag, Horowitz and Musser specify it thus.

```
type circular list

    types

        create:->circular list
        insert: circular list x integer->circular list
        value: circular list->integer
        removecirc: circular list->circular list
        right: circular list->circular list
        isemptycirc: circular list->boolean

    axioms

        isemptycirc(create) = true
        isemptycirc(insert(c,i)) = false
        value(insert(c,i) = i
        removecirc(insert(c,i)) = c
        right(insert(create,i)) = insert(create,i)
        right(insert(insert(c,i1),i2))
           = insert(right(insert(c,i2)),i1)
```

We can use a circular list to represent a queue efficiently by keeping the element that has been on the queue the longest at the front of the circular list. Addition to the back of the queue can then be made by inserting the element at the front of the circular list and right shifting the resulting list once. Can we synthesise such an implementation for addq in terms of our circular list primitives? The first step is to make more precise the representation relationship described above. We do this by defining a function rep mapping circular lists to queues thus

```
rep(create) = emptyq

rep(insert(circ,i)) = joinq(addq(emptyq,i)
                           ,rep(circ))
```

where joinq is the function that joins two queues together (the queue of the first argument being put at the front) defined thus

```
joinq(q,emptyq) = q

joinq(q,addq(r,i)) = addq(joinq(q,r),i)
```

We want to synthesise a function, addcirc say, that simulates the behaviour of addq under this representation. That is we require an addcirc that satisfies

addcirc: circular list x integer->circular list

rep(addcirc(c,i)) = addq(rep(c),i) *

Our technique is to take the last equation defining addcirc implicitly and use transformation techniques to produce an explicit, runnable, definition of addcirc.

We first instantiate * by setting c to create getting

rep(addcirc(create,i)) = addq(rep(create),i)

We now use our existing equations to develop the right hand side of the above equation. Thus we successively develop

rep(addcirc(create,i)) = addq(rep(create),i).

= addq(emptyq,i)

unfolding using the equation for rep

= joinq(addq(emptyq,i),emptyq)

folding using the equation for joinq

= joinq(addq(emptyq,i),rep(create))

folding using the equation for rep

= rep(insert(create,i))

folding using the equation for rep.

Thus we have the equation

rep(addcirc(create,i)) = rep(insert(create,i))

As rep is normally a many-one function it would not normally be legitimate to strip of the rep's from both sides of the above equation i.e. use the inference rule rep(E1) = rep(E2) => E1 = E2. However in this case different concrete objects that represent the same abstract object can be considered equal i.e. rep forms equivalence classes over the concrete objects and members of the same equivalence class under rep are indistinguishable as far as the implementation is concerned. Thus in this case we can remove the rep's from both sides of the above equation and get

addcirc(create,i) = insert(create,i)

However we must be careful to use this derived equality only in the context of implementations for queues as its unrestricted use can lead to inconsistencies.

Returning to our defining equation for addcirc we next instantiate c to insert(c,i1) getting

rep(addcirc(insert(c,i1),i2))

= addq(rep(insert(c,i1)),i2)

= addq(joinq(addq(emptyq,i1),rep(c)),i2)

unfolding using the equation for rep

= joinq(addq(emptyq,i1),addq(rep(c),i2))

folding using the equation for joinq

= joinq(addq(emptyq,i1),rep(addcirc(c,i2)))

folding using *

= rep(insert(addcirc(c,i2),i1))

folding using the equation for rep.

Thus we have

```
rep(addcirc(insert(c,il),i2))
 = rep(insert(addcirc(c,i2),il))
```

and we can again strip off the rep's getting

```
addcirc(insert(c,il),i2) = insert(addcirc(c,i2)
                                  ,il)
```

We have derived two equations for addcirc

```
addcirc(create,i) = insert(create,i)

addcirc(insert(c,il),i2) = insert(addcirc(c,i2),il)
```

These equations constitute an implementation for addq in terms of the list primitives. The adcirc we have synthesised is equivalent to an addcirc defined thus

```
addcirc(c,i) = right(insert(c,i))
```

3. Systems for Transformation

For reasons of convenience and accuracy it seems imperative that transformational developments are supported by some system. The simplest system that could be provided would be one where the user indicates at each step which transformation is to be applied and the system checks that the transformation is indeed applicable and if it is produces the new version of the program. Such systems are perfectly feasible but still require a lot of detailed intervention from the user.

The ultimate goal is of course a fully automatic system that would take the program specification and without any user intervention produce an efficient version of the program. Such systems are not at present capable of tackling realistically sized programs. We have developed a prototype of such a system which is described in Darlington [1977] and is capable of semi-automatically improving small NPL programs. Below we show how it is capable of achieving the implementation synthesis described in the previous section. This will illustrate the complexity of attempting to achieve transformations solely at the unfold/fold level and lead us on to a discussion of the idea of a transformation meta-language.

3.1 Abstract Data Types in NPL

The definition of queues given earlier translates to the NPL program below. Lines starting with +++ indicate type definitions while --- lines correspond to the axioms.

```
DEF

+++ emptyq = queue

+++ addq(queue,integer) = queue

+++ removeq(queue) = queue

+++ frontq(queue) = integer

+++ isemptyq(queue) = truval

var q:queue,i: integer


--- isemptyq(emptyq) = true

--- isemptyq(addq(q,i)) = false

--- frontq(addq(q,i)) = i if isemptyq(q)

                        frontq(q) otherwise

--- removeq(addq(q,i)) = emptyq if isemptyq(q)

                         addq(removeq(q),i) otherwise


END
```

Given the above equations we can evaluate queue expressions by enclosing them in a VAL ... END block in the normal NPL manner.

```
VAL

    frontq(removeq(addq(addq(addq(emptyq,1),2),3)))

END

:2    (System's response)
```

```
VAL

     removeq(addq(addq(emptyq,1),2))

END

:addq(emptyq,2)
```

One can define and run derived functions such as joinq in exactly the same way

```
DEF

+++ joinq(queue,queue) = queue

var q1,q2:queue

--- joinq(q1,addq(q2,i)) = addq(joinq(q1,q2),i)

END

VAL

joinq(addq(emptyq,1),addq(addq(emptyq,2),3))

END

: addq(addq(addq(emptyq,1),2),3)
```

Having developed and tested one's high level function the next stage in developing an efficient implementation is to define the lower level data type. For circular lists the NPL program is

```
DEF

+++ isemptycirc(circularlist) = truval

+++ create = circularlist

+++ insert(circularlist,integer) = circularlist

+++ value(circularlist) = integer

+++ removecirc(circularlist) = circularlist

+++ right(circularlist) = circularlist


var c: circularlist i,i2,i2:integer
```

```
--- isemptycirc(create) = true

--- isemptycirc(insert(c,i)) = false

--- value(insert(c,i)) = i

--- removecirc(insert(c,i)) = c

--- right(create) = create

--- right(insert(create,i)) = insert(create,i)

--- right(insert(insert(c,i1),i2))

                  = insert(right(insert(c,i2)),i1)

END
```

To define the representation one gives the representation relationship directly as a NPL function, thus

```
DEF

+++ rep(circularlist) = queue

--- rep(create) = emptyq

--- rep(insert(c,i))

            = joinq(addq(emptyq,i),rep(c))

END
```

We define addcirc by giving the NPL equation

```
DEF

rep(addcirc(c,i)) = addq(rep(c),i)

END
```

This last equation is the only one of all the NPL equations given that will not be accepted by the NPL interpreter. However we can use a transformation system to produce acceptable equations as in section 2.

3.2 Semi-Automatic Transformation

The dialogue reproduced below shows an interaction with the semi-automatic transformation system described fully in Darlington [1977]. This system has several in built heuristics to guide it to developing a sequence of unfold/fold steps to improve the given program. It is an interactive system and the responsibility of the user is to firstly indicate what equations he wants developed and what instantiations of these equations are to be considered. The system takes these suitably instantiated equations and develops them. When it thinks it has made a significant improvement it asks the user's opinion by printing out the current version of the program. The user has three options, he can either accept this as a suitable final version, accept it as being on the right path and ask the system to continue this line of development or reject it and ask the system to try another approach. This is of course unsatisfactory in many ways, often it requires knowledge of the final program being aimed for to interact successfully with the system. However if one does produce an acceptable final program it is guaranteed correct (or more exactly partially correct) as all the steps the system carries out preserve partial equivalence. One way to regard this system is as a transformation checker in another guise. It would be possible to get the system to behave more purposefully by adding further heuristics and checks but it was not our intention to develop this line of approach beyond the development of a simple prototype system. For any realistically sized example the combinatorial explosion would always defeat such a system at present.

The dialogue reproduced below is transcribed directly from a terminal with the system running on the Edinburgh DEC-10 with a KI processor. The previous NPL program had been compiled.

:

```
: START;                                              user enters
                                                      system.
TYPE IN INSTANTIATED L.H.S.BASE CASES FIRST.          system.
: END                                                 user-no
                                                      "base cases"

NOW TYPE IN RECURSIVE BRANCHES.

REP(ADDCIRC(CREATE,I)),
REP(ADDCIRC(INSERT(CIRC,I1),I2)) END
                                                      two cases to
                                                      consider
HOW DOES

REP(ADDCIRC(CREATE, I)) <=
   REP(INSERT(CREATE, I))                             solution for
                                                      firstcase
                                                      accepted

LOOK TO YOU
:OK

HOW DOES

REP(ADDCIRC(INSERT(CIRC, I1), I2))
   <= ADDQ(REP(INSERT(VALUE(INSERT(CIRC, CIRC)),
         VALUE(INSERT(VALUE(INSERT(CIRC,CIRC)),I1))))
        ,VALUE(INSERT(VALUE(INSERT(CIRC, CIRC)),I2)))
                                                      on a
                                                      wrong
                                                      path

LOOK TO YOU
: NOGOOD

HOW DOES

REP(ADDCIRC(INSERT(CIRC,I1),I2))
  <= ADDQ(REP(INSERT(JOINQ(CIRC,EMPTYQ),
           JOINQ(I1,EMPTYQ))),JOINQ(I2,EMPTYQ))

LOOK TO YOU
: NOGOOD
```

```
          .
          .
          .

     Five more proposed solutions rejected

          .
          .
          .

HOW DOES

REP(ADDCIRC(INSERT(CIRC,I1), I2))
 <= ADDQ(JOINQ(ADDQ(VALUE(INSERT(CIRC,EMPTYQ)),
     VALUE(INSERT(CIRC,I1))), REP(VALUE(INSERT
     (CIRC,CIRC)))),VALUE(INSERT(VALUE(INSERT(CIRC,
     CIRC)),I2)))

LOOK TO YOU
: NOGOOD

HOW DOES

REP(ADDCIRC(INSERT(CIRC,I1),I2))
  <=    JOINQ(ADDQ(EMPTYQ,I1),ADDQ(REP(CIRC,I2))
                                              on the
                                              right path
LOOK TO YOU
:CONTINUE

HOW DOES

REP(ADDCIRC(INSERT(CIRC,I1),I2))
  <= JOINQ(ADDQ(EMPTYQ,I1), REP(ADDCIRC(CIRC,I2)))

LOOK TO YOU
:CONTINUE

HOW DOES

REP(ADDCIRC(INSERT(CIRC,I1),I2))
  <= REP(INSERT(ADDCIRC(CIRC,I2),I1))

LOOK TO YOU
: OK

FINISHED.CPU TIME TAKEN 0 MINS 15 SECS
```

3.3 Use of Meta-language Programs

The increasing complexity of transformations expressed solely in terms of unfold/fold steps points to the need for a structured way of describing them.

This can be achieved by the development of a meta-language to explicate transformations. This idea was first developed, in the program proving context, by the LCF project, Gordon et al [1977]. The meta-language can be viewed as a control language for a transformation system, programs written in this meta-language describe how transformations should be performed. Just as programs in a conventional language can be structured so can meta-language programs. The basic constructs are the unfold/fold steps, out of these can be built higher level transformation steps, or tactics in the LCF terminology. These higher level tactics form a richer vocabulary for describing transformations. The analogy is with mathematics where proofs are rarely communicated in terms of the basic axioms but their correctness rests on the ability to ultimately express the higher level concepts in terms of basic axioms. In our transformation work a wide range of tactics have been developed these include various varieties of loop combination, the removal of recursion, conversion from an implicit specifications to explicit recursions and conversion of data type. We choose the latter to give an example of a meta-language program.

All syntheses of implementations follow a similar pattern. The parameters of this tactic are the abstract and concrete type specifications, the particular representation relationship to be used, the particular abstract function or functions for which an implementation is required, and the name of the concrete function to be synthesised. Given this information to achieve an implementation one has to follow a similar pattern of setting up the implicit equation defining the concrete function, suitably instantiating it and then investigating sequences of unfold, folds until an equation of the correct form is developed. 'Correct form' is characterised by amongst other things the equation having rep as the outermost function on the left hand and right hand side and otherwise only mentioning concrete functions. Given such a description of the form being aimed for the amount of searching necessary to achieve it is drastically reduced.

Below we give the outline of a meta-language program embodying the above tactic. We will not explain all the features of the meta-language which is still under development but hope that the reader will be able to understand the principlese

```
tactic changedatatype(abtype, conctype, rep,abfn, concfn)

let impeqn be

        'rep(concfn(x))=abfn(rep(x))'

    in

            {develop(instantiate(impeqn,inst),datapred)
                | inst∈instancesof(typeof(concfn))}

endtactic;

tactic   develop(eqn,pred)

    / applies unfold, rewrite and fold rules to the right
hand side of eqn until it is of a form that satisfies pred
/
endtactic

tactic   datapred(term)

    / checks whether term satisfies the conditions
outlined above /

endtactic
```

Typeof returns the type of the function it is given as argument and this information is used by instancesof to decide what cases to try (in this case create and insert(c,il))

Thus a call of changedatatype either fails, i.e. does not achieve a successful transformation, or returns a set of new equations to be incorporated into the original program. For the example given in section 2 the call would be changedatatype(queue,circularlist,rep,addq,addcirc) where queue, circularlist, addq and rep are previously defined. This would return the set of equations

```
{rep(addcirc(create,i))=rep(insert(create,i)),
 rep(addcirc(insert(c,il),i2))=rep(insert(addcirc(c,i2)
                                                  ,il))}
```

The meta-language idea has been applied to transformation by Martin Feather.His ZAP system, Feather [1979], employs a simple one level meta-language but is nevertheless capable of achieving the transformation of moderately sized programs

As well as acting as guidance to a system the higher levels of such a meta-language programs constitute a succinct way of communicating transformational developments. Meta-language programs can be written so as to isolate the main ideas behind a particular algorithm so that it can be understood or communicated easily without having to comprehend complex, highly optimised code. This is discussed further in Darlington [1981].

Conclusion

We hope that we have introduced the reader to some of the power and scope of transformation and applicative languages. We believe that transformations explicated in a structured way probvide a formal but understandable way of communicating algorithm development and give the opportunity of developing useable program development systems. The tactic of synthesising implementations for abstract data types is a fundamental tool for software development and the use of functional languages allowing abstract programs to be run prior to any implementation offers the opportunity of raising the level of program development and testing.

Acknowledgements

Related work on the synthesis of implementations is reported in Gaudel and Terine [1978] and Wirsing et al [1979]. We have benefited from conversations with them and John Guttag. The meta-language idea obviously owes much to the LCF project,its application to transformation is being investigated jointly with Ian Moor. Our work on transformation rests on co-operation with Rod Burstall and we are indebted to him and David MacQueen for the continuing development of NPL.The author gratefully acknowledges the continuing support of the Science Research Council of Great Britain.

References

Backus J. [1978]. Can programming be liberated from the von Neumann style? A functional style and its algebra of programs.Turing Lecture CACM 21,8 pp 613-641.

Burstall, R.M. [1977]. Design considerations for a functional programming language. Proc. of Infotech State of the Art Conference, Copenhagen pp 45-57.

Burstall, R.M. and Darlington, J. [1977]. A transformation System for developing recursive programs. JACM Vol 24 No. 1 pp 44-67.

Burstall, R.M. and Goguen J.A. [1977]. Putting theories together to make specifications. Proc. 5th IJCAI Boston pp 1045-1058.

Burstall R.M., MacQueen D.B. and Sanella D.T. [1980]. HOPE an experimental applicative language. Internal Report, Department of Computer Science University of Edinburgh.

Darlington, J. [1977]. Program transformation and synthesis. Present capabilities. Report 77/43 Dept. of Computing and Control Imperial College, London. To appear in The Artificial Intelligence Journal.

Darlington J. and Reeve M. [1981]. ALICE a multi-processor reduction machine for the parallel evaluation of applicative languages. Internal Report, Department of Computing, Imperial College.

Feather, M.S. [1979]. 'ZAP' program transformation system, primer and users manual. Report No. 54 Dept. of Artificial Intelligence, University of Edinburgh.

Gordon M.J., Milner R.A. and Wadsworth C. [1977]. Edinburgh LCF. Report CSR-11-77. (Part 1). Department of Computer Science, University of Edinburgh.

Gurd J.R., Watson I. and Glauert J.R.W. [1980]. A multilayered data flow computer architecture. Internal Report, Department of Computer Science, University of Manchester.

Guttag J.V. [1977]. Abstract data types and the development of data structures CACM Vol 20 pp 297-304.

Guttag J.V. and Horning J.J. [1978]. The algebraic specification of abstract data types. Acta Informatica Vol 10 No 1.

Guttag J.V., Horowitz E. and Musser D.R. [1978]. The design of data type specifications. In Current Trends in Programming Methodology Vol 1. Prentice Hall.

Hoare C.A.R. [1972]. Proof of correctness of data representations. Acta Informatica Vol 4 pp 271-281.

Liskov B. and Zilles S. [1977]. An introduction to formal specification of data abstractions. In Current Trends in Programming Methodology Vol 1. Prentice Hall. pp 1-32.

Wirsing M. Pepper P. Partsch H. Dosch W. and Broy M. [1979]. On heirarchies of abstract data types. Internal Report Project CIP. Technical University of Munich.

NATURAL LANGUAGE PROGRAMMING[1]

Alan W. Biermann

Duke University, Durham, N.C. 27706, USA

A procedural semantics system is described for English imperative sentences in natural language programming. Issues related to the handling of dialog focus, noun group resolution, quantifier processing, and imperative verb execution are discussed. Sequences of imperative sentences may be assembled to build natural language programs and techniques are given for processing such programs. The final sections include a discussion of related research and a brief overview of the field.

1. INTRODUCTION

Computer programming has traditionally been an activity for trained specialists who work with pencil and paper in the careful construction of sections of code. Each program is necessarily the result of a conscious design procedure and the individual lines of code may be revised many times before the programming task is complete. Such programs must be semantically and syntactically perfect (or nearly so) before they can be expected to be useful so the programmer must strive to achieve an unusually high standard of excellence in comparison with most other human endeavors.

However, progress in natural language processing systems and in voice recognition and synthesis systems will begin to change our concept of computer programming. Systems are beginning to come into experimental use which allow nonspecialists to obtain useful work from machines using natural language input. At first these systems are only able to respond to simple one

A. W. Biermann and G. Guiho (eds.), Computer Program Synthesis Methodologies, 335–368.

sentence requests but as the technologies improve sequences of sentences will be processed. Then a facility will be added to allow one to store a set of sentences and call them as a subroutine. Finally, hierarchies of such subroutines will be constructable and natural language programming will have come into existence.

Natural language programming will be very different from traditional programming. It will be used for obtaining quick and convenient service from machines often by nonspecialists in computing and where the efficiency of execution is not an issue. The programs will be created in a much more casual manner and will tend to be less precise than traditional programs both syntactically and semantically. Still, users will expect a high degree of reliability in the machine behaviors they obtain.

This paper will present a theory of natural language processing which is especially designed for the natural language programming task. Specifically, we will describe techniques for processing imperative sentences in English and programs which have been constructed from them. We will also briefly survey the literature in this field.

2. THE LANGUAGE TO BE PROCESSED

It is important at the beginning to have a philosophical understanding of the designer's goals for a natural language processing system. Linguists agree that the only reasonable definition of a natural language is that it is the set of all sentences that the speakers of the language agree are grammatically correct. But as pragmatic computer scientists, we must accept the fact that this set of sentences cannot be discovered and cannot be implemented. So we must put limitations on what we are going to attempt to implement; yet any such constraints impose a degree of artificiality on the language. It seems that by the definition of natural language we are doomed to fail in an implementation as long as we lack a complete simulator for the human speaker. That is, it appears that the only true natural language is a complete natural language and we do not accept this as a realistic goal for an implementation.

Thus we should acknowledge at the outset that we cannot pretend to handle, for example, all of English and that any subset of that language will be necessarily artificial to some extent. However, we can attempt to build a facility sufficiently broad in capabilities so that it will properly handle most user inputs for particular constrained situations. Thus we will call any language that can be actually implemented a formal-natural or fortual language and we will attempt to design such a language to have the following properties.

(1) It should be a subset of a natural language.
(2) It should be approximately learnable from a small amount of instruction, a few rules and/or examples.

We say "approximately learnable" because we accept the fact that even a user who conscientiously tries to follow the rules and examples will occasionally input sentences that will not be processed. Still such languages should have most of the advantages of true natural language including the facts that they can be spoken as well as written, and are fast, efficient, and convenient for users.

Two examples of programs written in fortual language appear below. This paper will describe how programs of this type can be processed.

Program 1.

"Here is a way to compute the prime numbers from 2 to N;

(1) Fill a 1 by N-1 array with the integers from 2 to N.

(2) Delete each entry which contains a multiple of i for each i in the array."

Program. 2.

"When I ask for a status report on a doctoral student, give me his or her year in grad school, source and amount of financial support, and which core exams have been passed. If the student has begun a thesis, give me the advisor and thesis topic."

3. SENTENCE PROCESSING

This section will describe the world model for the theory to be presented. Then it will discuss syntax, the resolution of noun groups, and the processing of imperative sentences.

3.1. The World Model

We will want to process programs that manipulate objects in a world. This world will have a finite or infinite <u>universe</u> of objects which is partitioned into a finite set of <u>types</u>, subsets on the universe. Relations will be defined on objects in the universe, and operations will be defined which can change the universe by creating, deleting, or changing objects or by altering relations.

For the sake of making the discussions more concrete, one world will be used for a series of examples in this section, and that world will be the text of this paper. The universe in this case will be the union of the following sets (types). Each object in a set or type will be considered as an abstract entity. However, in an actual implementation, each object would include the address to the physical locations it represents.

Set or type	Type name
P	paper
T	title
A	author
S_1, S_2, S_3, ---	section
b_1, b_2, b_3, ---	subsection
h_1, h_2, h_3, ---	section heading
p_1, p_2, p_3, ---	paragraph
s_1, s_2, s_3, ---	sentence
w_1, w_2, w_3, ---	word
c_1, c_2, c_3, ---	character
S_r	references

Intuitively, c_1, c_2, c_3, --- are the representations including the addresses of all of the characters in the paper, w_1, w_2, w_3, --- are the representations including the addresses of all the words of the paper, and so forth. Of course, these representations may not be stored explicitly in the machine but they will be generated as needed in the execution of a sentence.

Thus the universe needs a representation for each concrete entity in the world which might be referred to. If, for example, one were to mention "the first word in the paper", the noun group processor would be expected to produce the object w_1. If one said

"Delete the first word in the paper."

the noun group processor would pass w_1 to the "delete" routine.

But w_1 would include the addresses of certain characters in the paper so that the delete routine could access them. If one said

"Delete the first character of the second word."

the noun group might resolve to, say, c_9 and that address would be passed to the "delete" routine.

The most important type of relation in this world will be the referential relations which enable us to verbally address items in the universe. The "in" relation, for example, enables us to address individual characters in the paper as in

"the second character in the first word in the third sentence in the ninth paragraph in the last section in the paper."

That is, moving from the end of the sentence backward, the "in" relationship enables us to reference smaller and smaller entities until a character is designated.

As a part of the implementation of the domain dependent semantics, it will be necessary to create a routine typegen (x,r,y) where

x is the name of a type,

r is the name of a referential relation, and

y is the name of an object.

The value of typegen (x,r,y) is the set of all objects of type x which have relation r to object y. Thus, since the first word w_1 of this paper is the word in the title "NATURAL", the value of typegen (character, in, w_1) is the set $\{c_1, c_2, c_3, c_4, c_5, c_6, c_7\}$ where c_1 addresses the first character N, and so forth. That is, typgen (character, in, w_1) gives the characters in w_1.

The next sections will discuss the syntax and semantics of noun groups which resolve to yield sets of objects in the universe.

3.2 Syntax

This paper will assume that the problem of parsing incoming sentences is solved (Woods (30), Biermann and Ballard (5)) and will concentrate on some of the difficult issues of semantic processing. Typical sentences in this domain will include an imperative verb and its operands, and the task of the parser is to find all constituents and mark them properly.

As an illustration, the sentence

"Delete the first long word that contains the letter A in the sentence."

has an imperative verb and one operand, the noun group which is analyzed as follows:

the	article (definite)
first	ordinal
long	adjective
word	head noun
that contains the letter A	nonreferential qualifying phrase
in the sentence	referential qualifying phrase

The primary construct in this noun group is the head noun "word" and it has three prenominal modifiers and two postnominal qualifying phrases. Of course, each qualifying phrase may contain noun groups that must be similarly parsed and the noun groups can be nested to considerable depth. Most of the noun groups discussed here will have this form, and the difficult problems are concerned with finding their meaning as will be discussed next.

3.3 Noun Group Processing

In traditional computer programs, objects are always referred to by proper name. In fact, at the introduction to most programs one will find dozens or hundreds of declared names to be used in that program. In natural language programming, there may be few if any proper names, and objects are referenced by an interaction between the focus mechanism and the noun group processor. Intuitively, the focus mechanism yields a list of items that are currently being discussed in the dialog. The noun groups in the dialog then select, through a variety of mechanisms, individual items or sets using the focus list as a point of reference.

Thus in a traditional programming language, the eighteenth character in the paper might be specified as C(18) whereas in natural language one might set the focus by saying

"Consider the third word."

and then use the noun group "the first character" to reference what is actually the eighteenth character in the paper. The designated object will be the first character within the entity in focus.

The details of designing a good focus mechanism are immensely complex and beyond the scope of this paper but are described in Ballard (1), Grosz (13), and Walker (30). However, for the sake of our examples, we will use a naive mechanism and simply assume the focus list includes no more than one object in each type from the universe,namely the most recently mentioned object of that type (if any). Thus if the first sentence in a dialog is

"Underline the second sentence in the first paragraph."

the focus list will be $\{S_2, p_1\}$.

We will also need a function that refers to the focus list, containerof (X) where X is the name of a type. This function will yield the smallest object on the focus list which can contain objects of type X. Thus, with the above focus list,

containerof (word) = S_2

containerof (sentence) = p_1

and such evaluations will be needed in noun group resolution. Thus if one follows the above sentence with

"Delete the first word."

then containerof (word) will be used to find the set of words in focus. In this case, it would yield S_2 and the first word in S_2 would be deleted.

A noun group processor will now be described largely by examining its action on a series of noun groups. Throughout this paper we will discuss only semantic processing which begins with a parsed sentence and all constituents marked.

The first noun group to be discussed is

"the first word"

which has a singular head noun, an ordinal, and a definite article. Resolution of this type of noun group begins with the head noun whose meaning is found by the function

typegen(word, in, containerof(word))

The function containerof invokes the focus mechanism to find the smallest entity in focus capable of containing its argument, a word. Then typegen will find all the words in that containing entity. Thus if one input

"Consider the title of the paper."

"Print the first word."

then focus from the first sentence would include T for title. So execution of containerof would yield T and execution of typegen would give $\{w_1, w_2, w_3\}$.

Next processing moves to the ordinal "first" and the function apply calls a procedure to realize the meaning of "first". Specifically apply(ordinal, first, $\{w_1, w_2, w_3\}$) would be invoked yielding the value $\{w_1\}$. Finally, the article is processed with the function apply(artdef, the, $\{w_1\}$) which checks that the obtained set $\{w_1\}$ is singular since the head noun was singular and then yields the same result $\{w_1\}$. The meaning of the whole noun group "the first word" will be written

```
for X ε apply(artdef, the,
          apply(ordinal, first,
            typegen(word, in,
              containerof(word))))
     collect(X)
```

Thus the set of all X found in the set obtained by executing apply is collected together and formatted by the collect routine.

This meaning representation for the noun group "the first word"

may seem to be unnecessarily complex yet we will see it is a special case of a very general mechanism. Consider, for example, the noun group "a word."

Here the processing would be similar except that the definition of the indefinite article "a" is quite different from a definite article.

```
for X ε apply (artindef, a,
            typegen(word, in, containerof (word))
   collect (X)
```

Indefinite noun groups have three possible meanings in our system, (1) randomly select an object, (2) determine whether there exists such an object, or (3) create an object. If a person says

"Consider the title."

"Delete a word."

they probably mean that a word is to be randomly selected from the title (meaning (1)) and deleted. If one says

"Delete the sentence containing a word with 10 characters."
then semantics must search each sentence for a word with the desired characteristics (meaning (2)). If one says

"Add a page to the paper."

the page being referred to is not to be found in the paper. In fact, processing "a page" must result in a new entity being created and added to the universe (meaning (3)).

Assuming one has mentioned "a word" in the first context given ("Delete a word."), resolution of the noun group would proceed as follows:

containerof(word) would yield T

typgen(word, in, T) would yield $\{w_1, w_2, w_3\}$

apply(artindef, a, $\{w_1, w_2, w_3\}$) might randomly select w_2 and thus yield $\{w_2\}$

collect ($\{w_2\}$) would format the set as a noun group

Since the indefinite article has different meanings in different contexts, apply clearly has arguments related to context that are not given explicity in our representation.

If the user had said

"Delete two words."

this would again be an indefinite noun group with apply making two random choices rather than one.

Inclusion of a superlative prenominal modifier results in similar processing. Thus

"the longest word"

has the following meaning.

```
for X ε apply(artdef, the
          apply(superlative, longest,
            typegen(word, in,
              containerof(word) )))
   collect(X)
```

Here, typegen creates a set of words, apply-longest removes all but the longest from the set, and apply-the will check that the correct number of items has been found (one since "word" is singular).

Furthermore, long strings of prenominal modifiers are processed in the same way. For example, if we have written procedural definitions of the words "important" and "long," that have the effect of deleting from a set the items that are not "important" or "long", the meaning of

"the first two important long words"

is

```
for X ε apply(artdef, the,
          apply(ordnum, first-two,
            apply(adjective, important,
              apply(adjective, long,
                typegen(word, in,
                  containerof(word) )))))
   collect(X)
```

We next examine postnominal modifiers that use referential relations such as "in". Consider

"the words in the sentences."

Syntax processing will mark "in the sentences" as a modifier of "words" and our strategy will be to process postnominal modifiers which use referential relationships before processing head nouns and other constructions. The meaning of "the sentences" is, therefore,

```
for X ε apply(artdef, the,
          typegen(sentence, in,
            containerof(sentence) ))
```

Then the main portion of the noun group is processed with its meaning nested in the already constructed for-loop,

```
for X ε apply(artdef, the,
          typegen(sentence, in
            containerof(sentence) ))
for Y ε apply(artdef, the,
          typegen(word, in, X))
  collect(Y)
```

These two nested for-loops operate as in traditional programming languages. Variable X would sequentially point to the individual sentences in focus. For each value of X, Y would sequentially point to the words in X and collect would form the set of all such Y encountered.

Now one can see how deeply nested noun groups can be handled. For example, in the following noun group, each successive qualifier modifies the immediately preceding noun. So processing will start within the deepest nested parentheses and move outward.

```
"(the important words

    (in a sentence

      (in the first two paragraphs

        (in each section) )))"
```

The meaning representation thus will be:

```
for W ε apply(qualifier, each,
          typegen(section, in,
           containerof(section)))
 for X ε apply(artdef, the,
          apply(ordnum, first-two,
            typegen(paragraph, in, W)
  for Y ε apply(artindef, a,
            typegen(sentence, in, X)
   for Z ε apply(artdef, the,
```

```
        apply(adjective, important,
          typgen(word, in, Y)
    collect(Z)
```

Some postnominal modifiers, however, act more like adjectives than referential relations. These are called <u>subsetting relations</u> and are processed after the head noun. An example appears in

"the words in the sentence longer than the first word"

where "words" has two qualifying phrases, "in the sentence" and "longer than the first word". The order of processing is

(1) referential relations,

(2) head noun,

(3) subsetting relations, and

(4) prenominal modifiers.

The subsetting relation is processed like a prenominal modifier and is represented in a separate structure.

```
  for X ε apply(artdef, the,
           typegen(sentence, in
            containerof(sentence)))
    for Y ε apply(artdef, the
             apply(relation, R1,
              typegen(word, in, X)))
       collect (Y)

R1:  (longer than NG)

NG:  for X ε apply(artdef, the,
              apply(ord, first,
               typegen(word, in,
                 containerof(word))))
        collect (X)
```

3.4 Dealing With Large or Infinite Sets

The functions typegen and apply described above are conceptually simple but must include some features that in an implementation can be complex. Specifically, a problem arises if typegen is ever required to generate a very large or infinite set. In the noun group

"the first character,"

typegen will be called upon to create a representation of all of the characters in focus. If a word is in focus, typegen can produce explicitly the complete set of characters. But if a complete text is in focus, typegen must yield a set that could include millions of entities. To add to the embarrassment, we find that the next operation (apply-first) will remove all items in the set except the first one.

The obvious solution is that sets of objects must be representable both explicitly and procedurally throughout the system. Typegen must be able to create both representations, and apply and all higher level processing must be able to deal with both. Thus if the characters of the text are at addresses 100 to 40000 and the user says

"Capitalize the first character."

then typegen would create a procedural representation of character

(CHAR (100,+1,40000))

and apply-first would convert this to an explicit set {CHAR (100)}. However, if the user said

"Capitalize all the characters."

the procedural representation would pass out of the noun group without modification to the imperative verb processor.

While efficiency considerations for large sets demand a movement to procedural representations, infinite sets require such a feature in order to be represented at all. Consider

"the integers between 10 and 20"

where typegen will be called upon upon to create the set of all possible integers. Not until the qualifying phrase is processed can the set be explicitly enumerated. In some cases, limitations on the infinite set will come from indirect sources as in

"the integers i such that the ith word begins with a."

Here i spans the range from 1 to the number of words in focus and processing of the qualifier will produce a subset from that range.

The above sections outline a theory of noun group processing that is satisfactory for a variety of imperative sentences. The next section will discuss the execution of complete sentences.

3.5 Imperative Verb Processing

The noun group processor creates the nested loop programs described in the previous section and these programs can be executed to obtain lists of domain objects to be operated on. The imperative verb processor then executes the appropriate actions on these operands. Such computations are typically straightforward but, as will be shown below, adverbial modifiers may also appear in the sentence with some associated complications.

An example of straightforward imperative processing occurs for the sentence

"Capitalize the first letter of each word."

The imperative processor will receive a list of objects, specifically, the addresses of the characters to be affected, and a program which implements the operation "capitalize" will modify those objects appropriately. The only complexity comes from the fact that the verb may receive different types of objects and must do the correct operation in each case. Thus one might say

"Capitalize the word."

"Capitalize the paragraph."

"Capitalize the paper."

Of course, many imperatives can take more than one operand, and the roles of the individual operands are usually indicated by attached prepositions.

"Move ----- to -----."

"Move ----- from ----- to -----."

"Exchange ----- with -----."

"Add ----- to -----."

And each of these sentence templates may have many forms such as

"Move ----- into -----."

"Move into ----- -----."

"Into ----- move -----."

The imperative verb may have modifiers of significant complexity. Here are some examples:

"Capitalize each word using boldface type."

"Capitalize each word if the sentence is important."

"Capitalize each word for all sentences that are underlined."

The first example requires that a parameter be set for the capitalize routine, use boldface type. The second and third examples are actually flow-of-control statements similar to those of ALGOL-like languages. In these cases, the adverbial phrases translate to ALGOL-like code as follows:

if (declarative sentence) then (imperative verb) (operands)

for variable = (set of objects) do (imperative verb) (operands)

Adverbial modifiers have the peculiarity that they can appear in many different locations in a sentence. Thus the word "quickly" can appear in four slots in the sentence

"Capitalize each word in the sentence."

as follows:

"Quickly capitalize each word in the sentence."

"Capitalize quickly each word in the sentence."

"Capitalize each word quickly in the sentence."

"Capitalize each word in the sentence quickly."

Adverbial phrases can appear in any or all of these slots in ordinary English.

"For each paragraph, capitalize if the paragraph is important each word using boldface type in the first sentence if it is longer than five words."

In a practical implementation, it appears to be unwise to allow a user this much freedom because the ambiguity in the flow-of-control structure quickly gets out of hand. Fink (9) and Sigmon (26) have chosen the strategy of allowing flow-of-control adverbials to precede the imperative sentence or to follow it. Then the pre-imperative adverbials are processed from left to right and the post-imperative adverbials are processed from right to left unless special conditions within the phrases indicate otherwise. Following this strategy, the sentence

"For each paragraph, if it is important, capitalize the first word if it is a verb unless the sentence contains conjunctions."

is processed as

```
for each paragraph
 if it is important
  unless the sentence contains conjunctions
   if it is a verb
    capitalize the first word
```

Of course, it is easy to construct sentences where this simple strategy breaks down and then the designer must decide what to do. Either special code must be inserted to look for cases where the rule is not correct and to invoke appropriate action, or one must admit that this is a place where the realized fortual language will deviate from natural language.

Another point made clear by the above example is that the focusing mechanisms can become extremely complicated. What is meant by noun groups such as "it" and "the sentence" in such nested constructions? In some cases, focus comes from the higher level construction in a flow-of-control nesting, and in other cases it comes from earlier phrases in a left to right reading of the sentence.

4. SPECIAL PROBLEMS IN NATURAL LANGUAGE PROCESSING

Three topics of concern and extensive research in computational linguistics will be mentioned here, pronominal processing, conjunctions, and ambiguity.

4.1 Pronouns

As indicated above, noun groups select objects using the focus mechanism. A specific noun group should include enough information to designate the desired objects but in some cases the focus is so clear that mention of "it" or "them" is an adequate specification. Pronominal processing can thus be thought of as an extension to ordinary noun group processing except that greater reliance is placed on focus mechanisms.

Pronoun resolution does include one additional consideration, the roles of the various possible resolvents in related sentences. Thus in the sequence

"Move the title to page 4."

"Print it."

the pronoun probably refers to "the title" rather than to "page 4" even though the latter is the most recently mentioned object. However, "the title" is in some sense the subject of the first sentence and is more likely to be the resolvent of a pronoun. Extensive discussions of problems associated with pronoun appear in Ballard (1), Grosz (13), Hobbs (16), and others.

4.2 Conjunctions

Conjunctions have never received comprehensive treatment by computational linguists but some facilities are included in most systems. Conjunctions are complicated syntactically because almost any group of words in a sentence template can be replaced by a conjunction of such groups. For example, in the following sentence, the strings of words bracketed by 1's, 2's, and 3's can be replaced by conjoined strings.

```
 Capitalize the first word in the paragraph.
1          1  2          3 2             3
```

1. Capitalize and print the first word in the paragraph.

2. Capitalize the first word in and second word following the paragraph.

3. Capitalize the first word in the paragraph and beginning with an A.

While many such constructions are unusual, most are possible in one situation or another and thus must be considered for implementation. Woods et al. (31) and Biermann and Ballard (5) give implementation strategies for conjunctive structures.

The semantics of conjunction is also complicated. For example, the word "and" can mean either sequence, union, or intersection as is demonstrated, respectively, by sentences 1, 2, and 3 above.

4.3 Ambiguity

Ambiguity is quite common throughout natural language because of the vagueness of individual word definitions, because multiple parses of sentences are common, because of the unpredictability of human focusing behaviors, and many other causes. The designer of a language processor can follow one of the following three strategies in dealing with ambiguity:

(1) Attempt to discover each ambiguity and query the user about which meaning is desired.

(2) Attempt to guess from context which meaning is desired and select that meaning without comment to the user.

(3) Implement only one meaning and require that the user learn what that meaning is.

The first option is the one traditionally preferred by computer scientists. This would require that a system in, for example, a numerical domain to respond to

"Add X to Y."

with this type of message:

WHICH MEANING DO YOU DESIRE?

(1) PUT X+Y INTO Y LEAVING X UNCHANGED.

(2) COMPUTE X+Y AND WAIT TO BE TOLD WHAT TO DO WITH THE RESULT. LEAVE BOTH X AND Y UNCHANGED.

(3) FORM A PAIR CONTAINING THE VALUES OF BOTH X AND Y.

(4) NONE OF THE ABOVE.

However, in a natural language environment, this type of behavior is difficult to implement because there are so many conceivable meanings to a sentence and because it is so expensive computationally to find all of them. Furthermore, it is not even desirable because a system user will not want to receive a list of every possible meaning for each input sentence. Users typically want immediate and correct response to a command and no back-talk whatsoever.

The second alternative given above is the one humans use almost exclusively. However, this is difficult to implement so the third alternative is the one which seems to be most feasible in the near future. If this choice is made, one must admit that the implemented (fortual) language does deviate from true natural language in a significant way. It corresponds in a way to the implementation of a specialized jargon where many of the ordinary meanings for words and expressions are replaced by well defined meaning. This alternative, however, does leave most of the advantages of natural language communication intact and increases the probability that a successful processor can actually be built.

5. SEQUENTIAL SENTENCES

Once a technique has been developed for processing individual sentences, one can assemble several of them into a sequence and create a program. In order to clarify the points made in earlier sections and demonstrate the execution of a natural language program, we will examine the interpretation of Program 1 given in Section 2.

We begin the example by assuming that Program 1 has been entered at some earlier time and the user has just input

"Compute the prime numbers from 2 to 5 and print them."

This sentence conjoins two commands and most of our considerations will be on the first. Employing the strategy of this paper, Program 1 would be invoked and translated into the following:

```
        Fill NG1 with NG2.

        For i ε NG3, delete NG4,

     NG1:  for X ε apply(artindef, a,
                     apply(adjective, 1-by-4,
                      typegen(array, in,
                       containerof(array))))
                collect(X)

     NG2:  for X ε apply(artdef, the,
                     apply(relation, R1,
                       typegen(integer, in,
                        containerof(integer))))
                collect(X)

     R1:  (from 2 to 5)

     NG3:  for X ε apply(artdef, the
                      typegen(array, in
                        containerof(array)))

           for i ε apply(quantifier, each,
                      typegen(integer, in, X))

                collect(i)

     NG4:  for X ε apply(quantifier, each,
                      apply(relation, R2,
                        typegen(entry, in,
                          containerof(entry))))

                collect(X)

     R2:  (which contains NG5)
```

```
NG5:  for X ε apply(artindef, a,
                  noungroupresult(multiple, i))
```

We now trace the execution of the above code that results from the user's command. First the imperative "fill" evaluates its arguments and executes its own associated operation.

EXECUTE NG1:

typegen - array finds the set of existing arrays.
apply-1-by-4 removes those which are not 1 by 4 in dimension,
apply-a selects randomly from the remaining set of arrays,
If none exists, apply-a creates one.

RESULT: array 0 0 0 0

EXECUTE NG2:

typegen-integer creates a procedural representation of all integers,
apply-R1 reduces this to the set {2,3,4,5},
apply-the returns the set {2,3,4,5} since a plural entity was being sought.

RESULT: { 2,3,4,5}

EXECUTE Fill NG1 with NG2:

RESULT: array 2 3 4 5

Then execution passes to the second sentence:

For i ε NG3, delete NG4.

EXECUTE NG3:

typegen-array finds the set of all arrays,
apply-the will expect a singleton set. (If it finds more than one array, it uses the focus list to select the correct array.)
typegen-integer will collect the set of integers in the selected array {2,3,4,5}
apply-each essentially acts as a no-op here.

RESULT: {2,3,4,5}

Thus for each value i in NG3, the delete NG4 operation will be performed. Following the first and only nontrivial pass through the loop (i=2), we will evaluate NG 4 and then the deletion operation.

EXECUTE NG4:

typegen-entry finds the set of entries in the array in focus: {(E 1), (E 2), (E 3), (E 4)}
apply-R2 calls for evaluation of NG5

EXECUTE NG5:

noungroupresult is a special function that evaluates noun groups that produce results such as "sum of", "square of" and "absolute value of". In this case, it generates a procedural representation of the multiples of i.

RESULT: (INT (4,+2,infinity))

apply-R2 removes from consideration the set of entries not containing a value in NG5: {(E 3)}

apply-each passes the result as the noun group meaning.

RESULT: {(E 3)} (the third entry)

EXECUTE delete NG4:

RESULT: array 2 3 5

Subsequent turns through the loop produce no further modifications to the array and so the first of the user's requests is complete.

However, the second command "print them" could yield the embarassing result that the system does not know where the prime numbers are that have been computed. It turns out that the result of a computation is usually easy to find and can be referenced without explicit aid. If one describes how to square a number, invert a matrix, compute a salary, etc., the answer desired is usually the contents of a data structure that has been loaded and manipulated in an overt way. If the result of the computation were not clear, the user might be forced to conclude Program 1 with a sentence like

"The prime numbers from 2 to N are the numbers in the array."

In the current example, "them" references the numbers in the array and these numbers would be printed.

6. NONSEQUENTIAL AND NONDETERMINISTIC SENTENCES

One of the noticeable characteristics of human dialog is the seeming fragmentation in the interactions. Overview statements are given, minor details are mentioned, earlier points are clarified, other information is omitted, and so forth. We cannot hope to process such unpredictable behavior but we can attempt to relax some of the constraints that might be implied by the earlier discussions.

One variation from the format of sequential statements considered thus far is the inclusion of demon commands which continuously look for a condition to occur and then self activate to do a task. Such commands are usually marked with an adverbial such as

"whenever (declarative)"

or "on (declarative)".

In fact, even some conventional languages such as PL/I include this type of feature. An example of a program using this feature is the following:

"Here is a way to sort a linear array into ascending order. If every entry except the last entry is less than or equal to its sucessor, halt. Whenever there is an integer i such that the ith entry is larger than the $(i+1)^{th}$ entry, exchange those entries."

We will not attempt to analyze the execution of this type of program but will only comment that it appears that straightforward modifications to our existing mechanism should be enough to handle this type of construction. We introduce it here because of some interesting ways that it interacts with another feature of interest, nondeterministic execution.

Search and backtrack strategies have been used for decades to increase the apparent intelligence of systems (Nilsson (22), Samuel (25), Winograd (29)). It is therefore worth investigating what advantages might accrue from coding all of the mechanisms described in Sections 3 through 5 in a nondeterministic language such that guesses can be made as to which path to follow and failures with backup can occur whenever an unacceptable path is chosen.

The value of this approach can be seen by tracing the method of this paper in handling the following sentence and then observing the usefulness of a fail and backtrack feature.

"Put 1,2,3, and 4 into an array."

Following the strategy described above, processing would first evaluate the first noun group to 1,2,3,4 and then process "an array". Let us suppose several arrays are in existence but each is too small to hold the four numbers. Unfortunately, the indefinite article routine would select randomly from the available arrays and pass control to the imperative verb. But the "put" operation would fail because the selected array is too small, and the user would receive an error message for even this innocuous command.

However, with a fail and backtrack procedure, the "put" failure could cause a return to the indefinite article processor yielding a new selection and another attempt by the verb routine. If all the available arrays were tried and found to be too small, the indefinite article handler could then begin creating arrays of size 2,3, and so forth until "put" eventually succeeded. In this case, the user would achieve a satisfying response to the command, a new array created of correct size and with all entries properly loaded. (Of course, one would prefer that the backtrack mechanism be more informed than the one we describe here and carry back the cause of failure, array size must not be less than four. Biermann et al. (7), Truscott (27) and others give examples of such failure memories.)

A program which uses this type of nondeterminism is the following solution to the N-queens problem.

> "Create an N by N matrix.
>
> Put a 1 into the first or second or third or ---- or N^{th} entry of column i for i=1 to N.
>
> Never put a 1 into a row, a column, or a diagonal which contains a 1."

Here the nondeterminism is explicity programmed by the inclusion of the or-conjunction in the second sentence. The third sentence is a nonsequential demon command that will cause occasional failures with backup in the execution of the second command.

The nondeterminism implemented in the indefinite article processor allows a slightly improved version of the above program. The second command could be changed to

> "Put a 1 into an entry of column i for i=1 to N."

Or nondeterminism could be allowed to occur inside the verb processor. In the following version, the "put" routine nondeterministically selects which location in column i to place the 1 and changes its selection on backup until a satisfactory one is found.

"Put a 1 into column i for i=1 to N."

A more difficult to implement but also more attractive version would be the following:

"Put N ones into the matrix."

7. BACKGROUND

The techniques described here for processing natural language programs are evolving and this paper describes the current status of the approach. The concepts of typegen, the apply function, and verb and adverbial processing were developed and tested in the Natural Language Computer (NLC) system (Ballard (1), Biermann and Ballard (5), Biermann, Ballard and Holler (6)). This system allows a user to display one or more matrices on a computer terminal and to execute simple operations with typed fortual-English commands. The system responds in about 2 to 4 seconds displaying the result of each input to allow the user to confirm that correct action has occurred. If an undesired behavior is observed, the user can type "back up" and retype any command in clarified language. This system was tested under controlled conditions with subjects being asked to solve problems typical of a first course in computer science. In one experiment, twenty-three subjects typed a total of 1581 commands and 81 percent of these were processed immediately and correctly by the system.

However, the NLC system does not represent noun group meaning with nested looping structures of the type described here and does not handle indefinite noun groups and quantifiers as gracefully or as completely as the current method. So some of the mechanisms given here are improvements to a basic design which will be incorporated into a later implementation.

The method of handling quantifiers is somewhat controversial and at variance with accepted methods in computational linguistics. Hendrix, for example, in Walker (28) describes quantifier processing in terms of a search for correct scope. The method of the current paper requires the quantifier in the sentence

"Underline each word."

to be processed as a part of the noun group. Hendrix would process this sentence as

"For all i, underline word i."

claiming that the scope of the quantifier includes the imperative. However, this approach has associated dangers because movement of the quantifier can produce unforeseen side effects. For example, the method of this paper easily handles

"Underline each word and the page number."

whereas the movement of the quantifier yields

"For all i, underline word i and the page number."

meaning that the page number will be underlined many times. Problems of this kind can be fixed but it is desirable if the mechanism handles them naturally.

Objections can be raised against the nested loop noun group semantics given here and its handling of quantifiers and qualifiers. Consider the sentence

"Congratulate the second person to meet every president."

The relation "to meet" would be classified by our method as nonreferential and thus would be processed after the head noun "person". Our sentence would construct the set of all people, remove those who had not met "every president", select the second one (presumably using time as the ordering relation), and congratulate him (or her). This is probably correct processing.

However, if the sentence is changed to

"Congratulate the second person to meet each president."

most humans would expect each president to be considered separately. For each president, the set of persons who met him or her would be constructed and the second selected. Here the Hendrix method would be preferred because it allows a different scoping rule for each quantifier and could process both sentences correctly. The method of this paper could, of course, be revised to deal with this contingency, but again, one prefers that the same mechanism handle all problems uniformly.

The NLC system has a procedure declaration facility that acts like a user's word definition feature. For example, if the user wishes to define a new operation on matrices, a new imperative verb such as xcompute, can be created. The user might type

"Display a 5 by 5 matrix."

"Call it M."

"Define a way to xcompute M."

(procedure body)

"End the definition."

The new verb could then be used like any other imperative verb.

While it is desirable to have a way to define new words, it is also quite common to describe procedures in natural language without giving them an explicit name. This was done with the example programs 1 and 2 in this paper. It would be desirable that if such programs were input to a system, one could say things like

"Find the 100th prime number."

or "Print status reports on each student in CS225."

and have the needed "how to" programs properly utilized to give the desired results. In other words, the naming conventions that computer scientists are so accustomed to using for addressing data and procedures are not used very much in natural language either for data or for procedures.

8. OTHER WORK

The approach to natural language programming described in this paper concentrates on the development of a uniform and domain independent theory which can be used in the construction of practical systems. An attempt has been made to keep the theory simple enough so that an implementation can be as robust as any traditional compiler and yet deliver a nearly natural language capability. This approach, however, does make the requirement on the user that the algorithm for the desired computation must be essentially given by the input sentences.

A more ambitious goal would be to allow the user to engage in a more fragmented and thus more naturalistic dialog concerning the desired program. Such a dialog would not necessarily indicate the method of computation but would concentrate primarily on specifying the behavior. In this case, the program synthesis system would have to construct an internal model of the specifications from the dialog, and query the user to fill in any perceived omission. Then the object program would be synthesized from this model, and it might not have any resemblance in structure to the original user inputs. It could conceivably be an efficient piece of object code, a goal not attempted by the method of this paper. This approach to the program synthesis problem has been studied by Balzer (2), Green (11), Heidorn (14), Martin et al. (18), and Mikelsons (20,21).

The earliest project along these lines was by Heidorn (14) who built a system capable of acquiring a model for a queueing problem and then translating the model into code for a GPSS simulation. The model was stored in terms of a semantic network which was built up sequentially from user sentences. The system had knowledge of what information was necessary for a GPSS simulation and made requests of the user if the information was not complete. The system had abilities to translate the internal net back into English for verification by the user as well as to do a translation from the net into the GPSS language. The following sentences give a sample of the system performance.

1. when a vehicle arrives at a station, it leaves there immediately if the length of the line at a pump in the station is not less than 2.

2. 75 percent of the vehicles are cars and a fourth are trucks.

3. there is just one pump.

4. a simulation run time of 8 hours is desired.

5. ask questions for further information.

6. HOW OFTEN DO THE VEHICLES ARRIVE AT THE STATION?

7. the arrivals of vehicles are normally distributed with a mean of 8 minutes.

Etc.

The first four sentences were typed and translated into the network representation. The sixth sentence was a system response generated because of certain input requirements for the GPSS simulator. The dialog continued for a total of twenty-nine sentences before the model was completed and ready for translation to the GPSS language.

Because of the use of the semantic net as an intermediate data structure, it was not necessary for the user to input information in any special order. The code generator could produce the correct sequence from the network regardless of the order of its construction. Notice also that the sentences are all declaratives stating properties of the problem and not the method of solution.

This very high level of performance was made possible partially because of the characteristics of queueing problems and the simulation functions provided by the GPSS system. If a theory for general problem representation and algorithm synthesis could be devised as satisfactory as Heidorn's theory for queueing problems, comparable behavior would be possible in the more general case.

Such a more general theory is being attempted by Green and his associates (Green (11), Green et al. (12), Barstow (4), Kant (17), McCune (19), Ginsparg (10)). They have built a system which interacts in English dialog with the user and which also accepts examples or traces of the desired behavior. The system attempts to gather all available types of information together from the user and from an internal domain knowledge base to assemble a high level solution algorithm. Then a coding phase is designed to reduce the algorithm into a standard programming language taking into account efficiency considerations.

The system has been demonstrated in the synthesis of an artificial intelligence "concept formation" program. It attempts to handle completely free format English as follows (with the machine outputs in capitals):

18. PLEASE DESCRIBE THE TOP LEVEL OF TF.

19. It inputs a scene, tests whether it fits the concept, verifies the result of this test with the user, and updates the concept. Then it repeats the process.

20. I TAKE IT YOU MEAN THAT IT REPEATS UNTIL THE USER PROVIDES AN EXIT SIGNAL.

21. Yes, stop when the user types "quit" instead of an input scene.

22. OK. NOW WE - - - -

Etc.

User inputs are translated into a semantic network form and additions and disabmiguations are made using a domain knowledge base. A "discourse expert" module monitors input information and provides prompts to encourage adequate flow of information from the user. The core of the system is the "model-building expert" that constructs high level data structures and control structures to implement the target process. Finally, translation of the algorithm into an object module in a traditional programming language completes the process.

This project has attempted to explore the complete process undertaken by a human programmer in interacting with a client and generating executable code. As such it has led to many insights concerning the nature of the various kinds of knowledge needed in programming and their interactions in the synthesis of code. Many of these ideas will undoubtedly be incorporated into practical synthesis systems as the technology advances.

Related to this work was a project at ISI that emphasized the processing of informal program specifications in a domain independent system (Balzer (2), Balzer *et al.* (3)) and a project at MIT that concentrated on natural language processing and associated cognitive structures (Martin *et al.* (18)). Also, there was a project at IBM aimed at customizing software packages using natural language dialog (Mikelsons (20,21)). All of this work is reviewed by Heidorn (15).

9. CONCLUSIONS

Natural language programming has been a controversial subject for many years (Balzer (2), Dijkstra (8), Green (11), Heidorn (14,15), Petrick (23), Sammet (24)) with proponents arguing that natural language is an extremely powerful and convenient communication technique which everyone knows and which is easily spoken as well as written. Because humans communicate with each other so easily and so successfully with natural language, they should communicate if possible with machines in the same way.

Opponents argue, on the other hand, that machines are different from people and require a more precise and exacting communication method. They point to the ubiquitous ambiguity and vagueness in natural language constructions and object that machines cannot be expected to give reliable behavior from such instructions. They also point to the likelihood that true natural language probably will not be implemented and that any approximation to a natural language might be impossibly difficult to learn. For example, the user might not remember the vocabulary of one thousand words or in what ways words or phrases may be allowed to be assembled. Also traditional programmers tend to be very fond of their concise notations and beautifully structured code and have little attraction to what they conceive of as stilted, loose, and verbose natural language. Even worse, the natural language programs might also be inefficient and unpredictable. Finally, opponents have objected to the unesthetic aspects of having a machine appear to be human by attempting to understand and respond to natural language when, in fact, it is actually a passive automaton.

Unfortunately, the above argumentation has taken place without people knowing what natural language programs may eventually look like, how they may be processed, or how they might be used. There has been a lack of working systems about which one can make judgments and a lack of experience with machine executable programs. However, on the basis of the last decade of work, one can make some guesses concerning the validity of the various arguments.

Concerning the pro arguments, it is true that natural language is very powerful but it is difficult to implement all the facilities that make it powerful. One can implement a noun group processor making it possible to easily access objects, adjectives enabling one to insert a conditional test with only a single added token, and quantifiers which add a loop to the code again at the cost of only one word in the program. But many words and constructions have immensely complicated meanings and will not reliably be available on machines. For example, one might say

"Do the same thing to all other ---- except ----."

Any system built in the near future is likely to have difficulties successfully handling this type of command.

Concerning the argument that everyone knows natural language and would thus be able to use machines, we have concluded above that only approximations to true natural language will be achieved and that some training will always be needed. However, we have presented experimental evidence (Biermann, Ballard, and Holler (6)) that the amount of training need not be large and that a system can be implemented that realistically can handle most of the vocabulary and constructions needed by users if the domain is limited.

The fact that natural language can be spoken as well as written may become an important factor in the next few years because commercial voice recognition equipment is improving at a surprising rate. Isolated word recognition equipment is now available that can handle grammar branching factors in the range of a few dozen to one hundred at reasonable reliabilities. Such systems allow one to speak at rates of one or two words per second but require that a space be left after each word of approximately 200 milliseconds. Limited connected speech is also becoming available.

The most common argument against natural language programming concerns the vagueness and ambiguity in the language. However, in the narrow domains for which real systems are built, this has not tended to be a problem. For example, in the experiment reported in Biermann, Ballard, and Holler (6), subjects typed a total of 1581 sentences and only two were ambiguous with the system selecting the wrong meaning.

The complaints that natural language programs will make extremely inefficient use of computer resources are probably well founded for most systems built in the near future. The reliability of such programs remains an unknown but one might expect that it could be very high for short programs of a few sentences. But there is no significant data on this issue.

Concerning esthetic issues, one would suspect that if humans can do their work more easily and efficiently using natural language in interactions with machines, they will be glad to do so. The abilities to output sentences such as

"Please begin your command with an imperative verb."

or to properly execute commands of the kind given in this paper will not likely offend people if the machines are doing a job effectively.

REFERENCES

(1) Ballard, B. W.: 1979, Semantic Processing for a Natural Language Programming System (Ph.D. Dissertation), Report CS-1979-5, Duke University, Durham, N.C.

(2) Balzer, R.M.: 1973, "A Global View of Automatic Programming," Proc. Third Joint Conf. on Artificial Intelligence, pp. 494-499.

(3) Balzer, R. M., Goldman, N., and Wile, D.: 1978, "Informality in Program Specifications," IEEE Trans. on Software Engineering, Vol. SE-4, No. 2, pp. 94-103.

(4) Barstow, D. R.: 1977, Automatic Construction of Algorithms and Data Structures Using a Knowledge Base of Programming Rules (Ph.D. Dissertation), Stanford Artificial Intelligence Memo AIM-308, Stanford, Calif.

(5) Biermann, A. W. and Ballard, B. W.: 1980, "Towards Natural Language Programming," American Journal of Computational Linguistics, Vol. 6, No. 2, pp. 71-86.

(6) Biermann, A. W., Ballard, B. W., and Holler, A.: 1979, "An Experimental Study of Natural Language Programming," Report CS-1979-9, Duke University, Durham, N. C.

(7) Biermann, A. W., Baum, R. I., and Petry, F. E.: 1975, "Speeding Up the Synthesis of Programs from Traces," IEEE Trans. on Computers, Vol. C-24, No. 2.

(8) Dijkstra, E. W.: 1978, "On the Foolishness of 'Natural Language Programming,'" Unpublished Report.

(9) Fink, P. K.: 1981, Thesis in preparation, Duke University, Durham, N. C.

(10) Ginsparg, J. M.: 1978, Natural Language Processing in an Automatic Programming Domain, Stanford Artificial Intelligence Laboratory Memo AIM-316, Stanford, Calif.

(11) Green, C. C.: 1967, "The Design of the PSI Program Synthesis System," Proc. of the Second Inter. Conference on Software Engineering, San Francisco, Calif.

(12) Green, C. C., Waldinger, R. J., Barstow, D. R., Elschlager, R., Lenat, D. B., McCune, B. P., Shaw, P. E., and Steinberg, L. I.,: 1974, "Progress Report on Program Understanding Systems," Memo AIM-240, Stanford Artificial Intelligence, Stanford, Calif.

(13) Grosz, B. J.: 1978, "Focus Spaces: A Representation of the Focus of Attention of a Dialog," in Understanding Natural Language (Ed. D. E. Walker), North Holland, pp. 269-284.

(14) Heidorn, G. E.: 1972, Natural Language Inputs to a Simulation Programming System, Technical Report, Naval Postgraduate School, Monterey, Calif.

(15) Heidorn, G. E.: 1976, "Automatic Programming through Natural Language Dialogue: A Survey," IBM Journal of Research and Development, pp. 302-313.

(16) Hobbs, J. R.: 1976, "Pronoun Resolution," Report 76-1, Dept. of Computer Sciences, City University of New York, New York.

(17) Kant, E.: 1979, Efficiency Considerations in Program Synthesis: A Knowledge Based Approach, Memo AIM-331, Stanford Artificial Intelligence Laboratory, Stanford, Calif.

(18) Martin, W. A., Ginzberg, M. J., Krumland, R., Mark, B., Morgenstern, M., Niamir, B., and Sunguroff, A.: 1974, Internal Memos, Automatic Programming Group, Massachusetts Institute of Technology, Cambridge, Mass.

(19) McCune, B.: 1978, Building Program Models Incrementally from Informal Descriptions, Technical Report, Stanford Artificial Intelligence Laboratory, Stanford, Calif.

(20) Mikelsons, M.: 1973, Computer Assisted Application Definition, Unpublished Memo, IBM Thomas J. Watson Research Center, YorktownHeights, New York.

(21) Mikelsons, M.: 1975, "Computer Assisted Application Definition," Proc. of Second ACM Symposium on Principles of Programming Languages, Palo Alto, Calif.

(22) Nilsson, N. J.: 1980, Principles of Artificial Intelligence, Tioga Publishing Co., Palo Alto, Calif.

(23) Petrick, S. R.: 1976, "On Natural Language Based Computer Systems," IBM Journal of Research and Development, Vol. 20, No. 4, pp. 314-325.

(24) Sammet, J. E.: 1966, "The Use of English as a Programming Language," Communications of ACM, Vol. 9, pp. 228.

(25) Samuel, A.: 1967, " Some Studies in Machine Learning Using the Game of Checkers II, IBM Journal of Research and Development, Vol. 11, pp.601-617.

(26) Sigmon, A. H.: 1981, Thesis in preparation, Duke University, Durham, N. C.

(27) Truscott, T.: 1981, Thesis in prepartion, Duke University, Durham, N. C.

(28) Walker, D. E.: 1978, Understanding Spoken Language, North Holland, New York.

(29) Winograd, T.: 1972, Understanding Natural Language, Academic Press, New York.

(30) Woods, W. A.: 1970, "Transition Network Grammers for Natural Language Analysis," Communications of ACM, Vol. 13, pp. 591-606.

(31) Woods, W. A., Kaplan, R. M. and Nash-Weber, B.: 1972, The Lunar Sciences Natural Language Information System: Final Report, Report Number 2378, Bolt, Beranek, and Newman, Inc., Cambridge, Mass.

[1] This material is based upon work partially supported by the National Science Foundation under Grant Number MCS-7904120 at Duke University and partially supported by the Naval Postgraduate School, Monterey, California while the author was visiting in the Spring of 1981.

AUTHOR INDEX

SUBJECT INDEX

GPSR Compliance

The European Union's (EU) General Product Safety Regulation (GPSR) is a set of rules that requires consumer products to be safe and our obligations to ensure this.

If you have any concerns about our products, you can contact us on ProductSafety@springernature.com

In case Publisher is established outside the EU, the EU authorized representative is:

Springer Nature Customer Service Center GmbH
Europaplatz 3
69115 Heidelberg, Germany

Batch number: 10816188

Printed by Printforce, the Netherlands